Biotherapeutics

Recent Developments using Chemical and Molecular Biology

RSC Drug Discovery Series

Titles in the Series:

1: Metabolism, Pharmacokinetics and Toxicity of Functional Groups
2: Emerging Drugs and Targets for Alzheimer's Disease; Volume 1
3: Emerging Drugs and Targets for Alzheimer's Disease; Volume 2
4: Accounts in Drug Discovery
5: New Frontiers in Chemical Biology
6: Animal Models for Neurodegenerative Disease
7: Neurodegeneration
8: G Protein-Coupled Receptors
9: Pharmaceutical Process Development
10: Extracellular and Intracellular Signaling
11: New Synthetic Technologies in Medicinal Chemistry
12: New Horizons in Predictive Toxicology
13: Drug Design Strategies: Quantitative Approaches
14: Neglected Diseases and Drug Discovery
15: Biomedical Imaging
16: Pharmaceutical Salts and Cocrystals
17: Polyamine Drug Discovery
18: Proteinases as Drug Targets
19: Kinase Drug Discovery
20: Drug Design Strategies: Computational Techniques and Applications
21: Designing Multi-Target Drugs
22: Nanostructured Biomaterials for Overcoming Biological Barriers
23: Physico-Chemical and Computational Approaches to Drug Discovery
24: Biomarkers for Traumatic Brain Injury
25: Drug Discovery from Natural Products
26: Anti-Inflammatory Drug Discovery
27: New Therapeutic Strategies for Type 2 Diabetes: Small Molecules
28: Drug Discovery for Psychiatric Disorders
29: Organic Chemistry of Drug Degradation
30: Computational Approaches to Nuclear Receptors
31: Traditional Chinese Medicine
32: Successful Strategies for the Discovery of Antiviral Drugs
33: Comprehensive Biomarker Discovery and Validation for Clinical Application
34: Emerging Drugs and Targets for Parkinson's Disease
35: Pain Therapeutics; Current and Future Treatment Paradigms
36: Biotherapeutics: Recent Developments using Chemical and Molecular Biology

How to obtain future titles on publication:
A standing order plan is available for this series. A standing order will bring delivery of each new volume immediately on publication.

For further information please contact:
Book Sales Department, Royal Society of Chemistry, Thomas Graham House, Science Park, Milton Road, Cambridge, CB4 0WF, UK
Telephone: +44 (0)1223 420066, Fax: +44 (0)1223 420247,
Email: booksales@rsc.org
Visit our website at www.rsc.org/books

Biotherapeutics

Recent Developments using Chemical and Molecular Biology

Edited by

Lyn H. Jones
Pfizer R&D
WorldWide Medicinal Chemistry
Cambridge, MA
USA
Email: lyn.jones@pfizer.com

and

Andrew J. McKnight
AnaptysBio, Inc.
San Diego, CA
USA
Email: amcknight@anaptysbio.com

RSC Publishing

RSC Drug Discovery Series No. 36

ISBN: 978-1-84973-601-5
ISSN: 2041-3203

A catalogue record for this book is available from the British Library

Published by The Royal Society of Chemistry,
Thomas Graham House, Science Park, Milton Road,
Cambridge CB4 0WF, UK

Registered Charity Number 207890

For further information see our web site at www.rsc.org

Preface

Biotherapeutic modalities are often considered beyond the reach of the medicinal chemist. The language and terminology can differ significantly between small and large molecule drug discovery, and yet the aims are essentially the same. We believe chemistry has an essential role in the future success of this exciting area, and this book was conceived as an attempt to illustrate the successful partnership of chemical and molecular biology to enable and advance biotherapeutics (so-called 'chemologics').

The design–synthesis–screen–design cycle, an engine for successful small molecule drug discovery, is not usually a component of biotherapeutic discovery, yet our challenge to this community is that that need not be the case – a deeper molecular understanding should be brought to bear in the biotherapeutics field, such that the empiricism that currently persists can be addressed. The challenge will be to generate knowledge and apply those learnings prospectively to avoid making the same mistakes and accelerate our decision making – the delineation of structure–function or structure–toxicology relationships will find increasing value in the biotherapeutic space. Small molecule drug discovery has already evolved past the period of 'make lots of stuff, screen lots of stuff, and see what pops out!' It is neither inspiring nor cost effective – design strategies are now far more sophisticated, augmented significantly by advances in biophysical techniques, computational sciences and the accuracy of predictive *in silico* tools. Our belief is that these methods will be harnessed to a greater extent in the advancement of biotherapeutic discovery and optimization approaches in the future.

This book approaches the huge area of biotherapeutics from the perspective of improved *molecular design*, which draws from the synergies between chemical biology, medicinal chemistry and molecular biology in particular. Recent developments in these disciplines that have delivered drugs, clinical candidates or significantly advanced biotherapeutic discovery and design will

RSC Drug Discovery Series No. 36
Biotherapeutics: Recent Developments using Chemical and Molecular Biology
Edited by Lyn H. Jones and Andrew J. McKnight

Published by the Royal Society of Chemistry, www.rsc.org

be described. A broad range of modalities are highlighted that will appeal to those working in a number of biomolecular areas (oligonucleotides, sugars, proteins and peptides). The chapters, written by an impressive list of world experts in their respective fields, detail a number of diverse therapeutic opportunities, including immunopharmacotherapy, optimized fully human or humanized antibodies, bicyclic peptide phage libraries, synthetic proteins and vaccines, micro-RNA, bacterial toxins, stabilized cyclotides, antibody–drug conjugates, peptide epitope mimicry and synthetic immunology.

Additionally, we believe this book will serve as inspiration for the medicinal chemistry community, particularly when presented with examples of how their expertise can make considerable impact in the biotherapeutics arena. Much has been made of the need to chose the 'best target' in drug discovery, but as much emphasis should then be placed on choosing the 'best modality', really, our approach should be 'modality agnostic'.

Our vision is that all biopharmaceutical chemists, whether in industry or academia, are equipped with capabilities both in small and large molecule drug discovery (and as a minimum can speak the language of, and engage in, 'biotherapies') and we hope this book will help towards that goal. In some ways, this book is a call to the traditional small molecule medicinal chemistry community to ask broader questions of their projects and therapeutic programmes. At the earliest stage of interest in a biological target we should be asking 'what therapeutic modalities shall we apply?' and both chemists and biologists are fundamental to the success of those strategic discussions, as well as the successful prosecution of the programme.

We are extremely grateful to the authors of the chapters in this book. They have not only described their areas of interest and expertise with great skill, but they have also shared compelling insights into the future opportunities for biotherapeutics. We also thank Rosalind Searle and Cara Sutton, RSC Publishing, for their editorial support and encouragement.

Lyn H. Jones and Andrew J. McKnight

Contents

RSC Drug Discovery Series No. 36
Biotherapeutics: Recent Developments using Chemical and Molecular Biology
Edited by Lyn H. Jones and Andrew J. McKnight

Published by the Royal Society of Chemistry, www.rsc.org

CHAPTER 1

Synthetic Immunology

THIHAN R. PADUKKAVIDANA,[a,†]
PATRICK J. McENANEY[a,†] AND DAVID A. SPIEGEL*[a,b]

[a] Department of Chemistry, Yale University, 225 Prospect Street, PO Box 208107, New Haven, CT 06520-8107, USA; [b] Department of Pharmacology, Yale University School of Medicine, 333 Cedar Street, SHM B350B, New Haven, CT 06520, USA
*Email: david.spiegel@yale.edu

1.1 Introduction and Scope

The field of immunology has become increasingly lucid at the level of atoms and molecules. Owing also to advances in synthetic chemistry, the rational design and construction of synthetic systems that perform complex immunological functions – an area termed synthetic immunology – has come within reach. Here we will highlight one facet of synthetic immunology concerned with the development of low-molecular weight ("small"), synthetic molecules that are capable of functionally mimicking biological molecules. It is important to note at the outset that this article does not aim to be comprehensive in scope. At the expense of being all-inclusive, we focus on several specific contributions, which highlight how advances in immunology and chemistry have proven mutually complementary. We have divided this chapter into four subsections: (1) synthetic ligands for pattern recognition receptors, including toll-like receptors (TLRs), NOD-like receptors (NLRs), nuclear family receptors and C-type lectins, (2) synthetic molecules that modulate the complement system,

[†]These authors contributed equally to this work.

RSC Drug Discovery Series No. 36
Biotherapeutics: Recent Developments using Chemical and Molecular Biology
Edited by Lyn H. Jones and Andrew J. McKnight

Published by the Royal Society of Chemistry, www.rsc.org

(3) synthetic systems for controlling cell–cell communication, including chemokine/cytokine mimetics, and (4) synthetic ligands for modulating adaptive immune processes, including T-cell and B-cell functions.

We regret not being able to cover all of the exciting developments that might be classified into the area of Synthetic Immunology. Such areas include: synthetic vaccine development, as this topic is covered elsewhere in this book;[1] synthetic modulators of cellular signaling processes, as the functions of such molecules extend beyond the immune system;[2–5] protein-based and cellular immunotherapies, including therapeutic monoclonal antibodies;[6–9] immunomodulator strategies involving nanoparticles or virus-like particles;[10–13] DNA- and RNA-based therapeutics;[14–16] and strategies for controlling cellular differentiation.[17–20] Lead references to each of these areas are provided for interested readers.

It is our hope that this chapter will serve as a broad-based introduction for biomedical scientists, including chemists interested in extending their activities into the immunological realm, as well as immunologists looking to learn about how modern synthetic chemistry can enhance fundamental biological understanding. Ultimately, we believe that the intellectual perspective residing at the interface between synthetic chemistry and immunology will enable scientific advances that were never before thought possible, thus proving critical to the furtherance of basic biomedical research and patient care.

1.2 Synthetic Ligands for Pattern Recognition Receptors (PRRs)

Pattern recognition receptors (PRRs) are a diverse class of proteins that function canonically as part of the innate immune response. These receptors recognize pathogen-associated molecular patterns (PAMPs), which are widely conserved, repeating motifs found within pathogens and not within hosts, as well as damage-associated molecular patterns (DAMPs), which are host-derived molecules that arise from tissue damage.[21,22] Improper activation of these receptors has been shown to cause hyper-inflammatory disease states, and extensive research efforts have focused on identifying PRR antagonists. Such developments have been reviewed elsewhere.[23–33] In this section we discuss a few select examples of molecules that functionally mimic the natural ligands of PRRs.

1.2.1 Synthetic Mimics of TLR Ligands

Toll-like receptors (TLRs) were originally discovered as important receptors for *Drosophila melanogaster* embryonic development,[34] and were later found to play a critical role in innate immunity in humans.[35] Extensive research during the past two decades has revealed the TLR superfamily to contain more than ten different family members (TLRs 1–13).[22]

TLR2 is capable of binding various bacterial lipopeptides,[22] as well as self-derived ligands such as high mobility group box 1 (HMGB1) and products of lipid oxidation.[36,37] Interestingly, TLR2 preferentially forms heterodimers with other TLR members (e.g., TLRs 1 and 6), thus expanding its ligand binding profile.[38] Indeed, synthetic lipopeptide collections, designed to mimic the acylated amino terminus of bacterial lipoproteins,[39,40] have proven critical to understanding the different adaptor molecules involved in TLR2 function. These studies have led to the understanding that the TLR2 heterodimer formed in response to a given ligand depends both on the arrangement of fatty acyl groups and the sequence of peptide backbones in that ligand.[38,41] For example, synthetic tri-acylated lipopetpides, such as Pam_3CSK_4 (Figure 1.1, Panel A, **1**), selectively engage the TLR2–TLR1 heterodimer, while di-acylated analogs, such as Pam_2CSK_4 (**2**), bind both TLR2–TLR1 and TLR2–TLR6 heterodimers.[40] Modification of the peptide backbone in Pam_2CSK_4 with the decapeptide GDPKHSPKSF provides fibroblast-stimulating lipopeptide-1 (FSL-1, **3**),[42] which is selective for the TLR2–TLR6 complex.[38] A crystal structure of the TLR1–TLR2–lipopeptide provides some insight into the molecular origin of this ligand-dependence.[43] Furthermore, the specific TLR2 heterodimer formed in response to ligand stimulation can have a significant effect on downstream biological outputs. For example, synthetic ligands that induce TLR2–TLR1 heterodimer formation (e.g., Pam_3CSK_4) have been

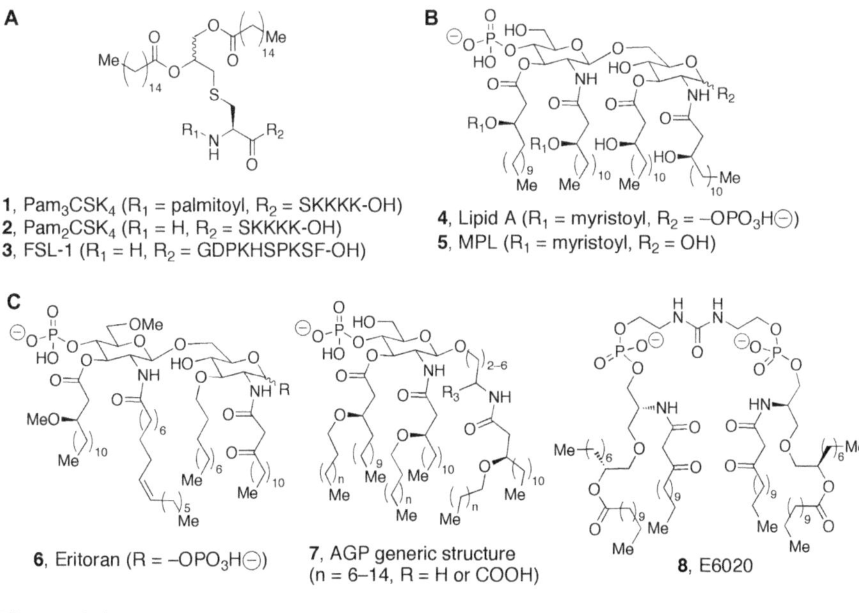

Figure 1.1 TLR 2 and TLR4 ligands. (A) Synthetic ligands capable of agonizing TLR2-containing heterodimers. (B) Lipid A-derived ligands for TLR4 including natural *E. coli* lipid A (4) and monophosphoryl lipid A (5). (C) Lipid A derivatives eritoran (6), aminoalkyl glucosaminide 4-phosphates (AGPs, 7), and non-carbohydrate ligand E6020 (8).

shown to enhance regulatory T-cell (T_{reg}) proliferation to a larger extent than those leading to other heterodimeric TLR2 complexes.[44] Conversely, a small molecule inhibitor of the TLR2–TLR1 complex was recently discovered through a cell-based high-throughput small molecule screen.[45] Efforts to expand the repertoire of TLR2-targeted compounds offer hope of providing chemical insights into immunomodulatory functions, while also delivering useful lead molecules for studying pathogen recognition, cancer progression, and other disease-relevant biological processes.[21,46]

TLR4 recognizes lipopolysaccharide (LPS), a component of the Gram-negative bacterial cell wall. Stimulation of TLR4 leads to rapid induction of various pro-inflammatory processes including cytokine release and immune cell differentiation.[22] TLR4 ligands can serve as useful adjuvants for vaccine development,[47] however activation of this receptor is also associated with autoimmune diseases, neuropathies, and septic shock.[48] Therefore there has been significant interest in identifying both synthetic agonists and antagonists of TLR4.[49,50] Efforts to this end have primarily started from lipid A (Figure 1.1, Panel B, **4**) – the hydrophobic portion of LPS – and have led to the development of numerous synthetic immunomodulators. For example, early studies demonstrated that removal of the anomeric phosphate group in **4** can ameliorate some of the molecule's toxic effects in humans. The resulting compound, monophosphoryl lipid A (MPL, **5**), retains its TLR4 agonist activity and has been approved by the FDA, for use as a vaccine adjuvant.[51] Further investigations into synthetic lipid A analogs have led to a general model relating ligand structure with TLR4-modulating activity: agonists are believed to form "conical" structures in solution – wherein the cross-section of the hydrophobic portion of the molecule is greater than that of the hydrophilic one – whereas antagonists are believed to form "cylindrical" shapes such that hydrophobic and hydrophilic cross-sections are nearly identical.[47] Indeed, this model is supported by extensive structure-activity relationships along with recent crystallographic studies.[52–54] For example, compounds lacking hydrophobic acyl substituents – such as eritoran (Figure 1.1, Panel C, **6**) – serve as receptor antagonists.[55,56] On the other hand, agonist structures have been obtained from replacement of labile phosphate moieties with negatively charged bioisosteres (e.g., sulfate), substitution of ester-branched acyl groups with corresponding alkyl ethers, and removal of the lipid A reducing sugar to afford aminoalkyl glucosaminide 4-phosphates (AGPs, **7**).[50,57] Complete replacement of the lipid A disaccharide with a simple phosphatidylethanolamine-based construct provides agonist E6020 (**8**), which is believed to retain the "conical" conformation and is capable of eliciting TNF-α production from whole blood, as well as IL-6 production and NF-κB activation in cell lines expressing TLR4.[47,58–60] Additional TLR4 modulators have been obtained using "glycocluster" strategies,[61] attaching LPS to magnetic nanoparticles,[62] and other approaches.[63–65] Furthermore, the discovery of TLR4-active compounds structurally unrelated to lipid A, such as paclitaxel, heme derivatives, and opioids, will undoubtedly lead to new opportunities for developing synthetic immunomodulators.

TLR7/8 are expressed in a variety of immune cells of the myeloid lineage and are intimately involved in anti-viral immunity by triggering the release of type 1 interferons.[66] The first synthetic ligands for TLR7/8 were discovered by 3M scientists in cellular screens of nucleoside analogs for antiviral applications.[67] Interestingly, although these efforts led to the development of the class of imidazoquinoline derivatives – including Imiquimod (Figure 1.2, Panel A, **9**), which was first marketed in 1997 for treating complications of human papilloma virus (HPV) infection[68] – it was not until much later that the mechanism of these compounds as TLR7/8 agonists was confirmed.[69] Over the past decade, derivatives of imidazoquinolines, including resiquimod **(10)** and S-27609 **(11)**,[70] have proven to be effective TLR7-stimulating adjuvants resulting in anti-tumor activity[71] and polarization of immune cells towards T_H1 responses.[70,72] Like single-stranded RNA (ssRNA), which is the natural ligand for TLR7/8, small molecule agonists such as **9**–**11** activate macrophages to produce pro-inflammatory cytokines (IFN, TNF-α and others) via the TLR7 MyD88-dependent signaling pathway.[69] Furthermore, homology modeling studies suggest that both ssRNA and small molecule **10** have overlapping binding sites on TLR8, suggesting that imidazoquinolines are both structural and functional mimics of natural ligands.[68,73,74] Further progress in illuminating the structural details of TLR7/8 interactions with their ligands will undoubtedly lead to improvements in synthetic ligand mimics.

Like TLR7/8, **TLR9** is localized to endolysosomal compartments, recognizes foreign nucleic acids, and induces a type-1 IFN response upon activation.[75]

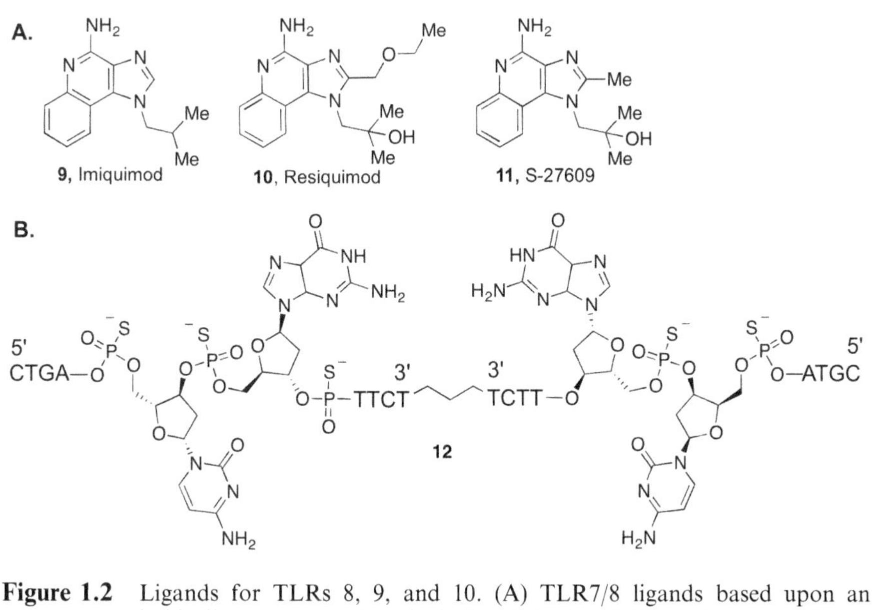

Figure 1.2 Ligands for TLRs 8, 9, and 10. (A) TLR7/8 ligands based upon an imidazilone core structure, including Imiquimod (9), Resiquimod (10) and S-27609 (11). (B) One example of the "immunomer" class of TLR9 ligands.

Unlike TLR7/8, however, the natural ligand for TLR9 is unmethylated CpG oligodeoxynucleotides (CpG DNA), which is present in both bacterial genomes and DNA viruses. Synthetic agonists for TLR9 that take advantage of structural similarities with natural nucleic acid ligands have been developed. For example, having determined that an accessible 5'-end of CpG-containing oligodeoxynucleotides is necessary for immunomodulatory activity of synthetic TLR9 ligands,[76] researchers at Hybridon have developed a class of "immunomers" (**12**) containing short CpG DNA sequences chemically connected through synthetic 3'–3' linkers.[77] Interestingly, both the chemical composition and the length of "immunomer" derivatives were found to impact their immunomodulatory effects.[78–80] Indeed, immunomers containing only 5 or 6 nt in each segment possess immunostimulatory properties in both mice and humans.[81] Also, replacement of phosphate backbone with phosphorothioate motifs,[82,83] or incorporation of 5'-dinucleotides containing non-natural bases,[78] can confer enhanced stability and alter patterns of cytokine secretion in tissue culture systems. Evaluation of synthetic TLR9 agonists in human clinical trials as vaccine adjuvants, and immunotherapies for allergy, cancer, and infectious disease are currently underway.[84]

1.2.2 Synthetic NOD Agonists

Nucleotide-binding oligomerization domain (NOD)-like receptors (NLRs) are a family of soluble cytosolic proteins consisting of over twenty members. NLRs are believed to detect cytosolic PAMPs derived from foreign agents as well as danger-associated molecular patterns (DAMPs) from injured host tissue. NLRs promote NF-κB and mitogen-activated protein kinase (MAPK) signaling and complex with the inflammasomes to upregulate pro-inflammatory IL-18, IL-1β, IL-6, TNF, and type I IFN secretion.[85]

NOD1 has been shown to recognize the dipeptide, D-glutamyl-meso-diaminopimelic acid (iE-DAP),[86] present as a component of bacterial peptidoglycan (**13**), which activates the NF-κB pathway, enhancing the production and secretion of several cytokines.[87] Hasegawa and coworkers have chemically synthesized iE-DAP derivatives to identify minimal motifs responsible for NOD1 activation.[88] It was found that the iE-DAP structure serves as a core motif for NOD-1 activation; functionalization of this group with substituted alkyl chains (**14**), benzamides (**15, 16**), or short peptides (**17**) enhances NOD-1 activation. The authors speculate that this enhancement may result from increased hydrophobicity of the molecules compared to the parent compound (**18**), which enhances their interactions with the cell membrane. Notably, compound **14** was found to be several hundred fold more active than the original iE-DAP.[89] Moreover, it was found that iE-meso-DAP stereoisomer showed a 10 to 10,000 fold greater ability to stimulate NOD-1-mediated NF-κB activation as compared to other stereoisomers.[88]

NOD2, a close relative of NOD1, is expressed in several phagocytic cells, including monocytes, granulocytes, and dendritic cells.[90] Its natural ligand is believed to be the MurNAc-L-Ala-D-isoGln muramyl dipeptide (MDP,

Figure 1.3, Panel A, red box),[91] which is found in both Gram-positive and Gram-negative bacteria. In a similar manner to NOD1, NOD2 ligation leads to the secretion of IL-12, IL-8, IL-6, and TNF-α. Almost four decades of research on MDPs have led to the development of numerous analogs,[92] including mifamurtide (Figure 1.3, Panel C, **19**), approved for combination chemotherapy treatments.[93] One of the primary reasons for the development of these synthetic molecules was the potent pyrogenicity and somnogenicity induced by the natural ligand that made it relatively unattractive as an adjuvant. These derivatives of MDP have been shown to have greater biological activity and lack the

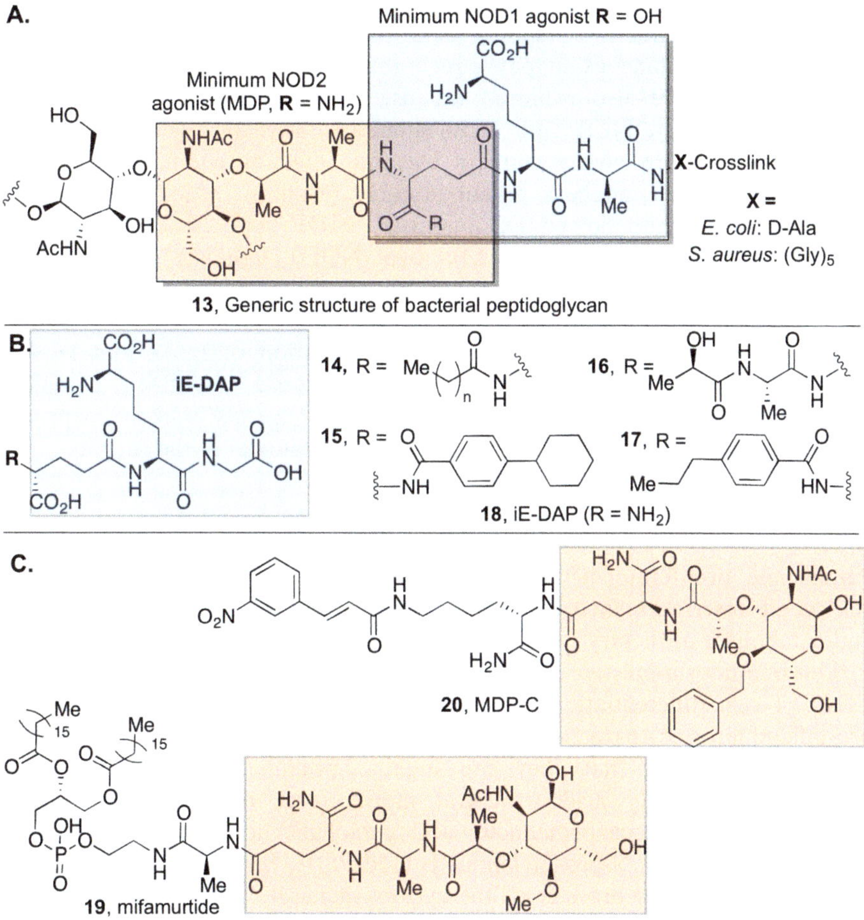

Figure 1.3 Natural and synthetic ligands for NOD 1 and NOD2. (A) Generic structure of bacterial peptidoglycan (13) with boxes indicating the components recognized by NOD1 (blue) and NOD2 (red). (B) Synthetic NOD1 agonists with various substitution on the N-terminus of iE-DAP (blue box). (C) Two NOD2 agonists mifamurtide (19) and MDP-C (20). Red boxes indicate the regions of natural peptidoglycan that are being mimicked.

undesirable characteristics of MDP, and have been extensively discussed elsewhere.[92,94] Other efforts to optimize the biological effects of MDP have led to the identification of MDP-C (**20**), which is both apyrogenic and nonallergenic, yet potently immunostimulatory. Here, Yang *et al.* utilized a novel mesh-bagged gathered-bunch combinatorial chemistry strategy (abbreviated MBGB) to synthesize MDP analogs on solid supports.[95] They produced over 2300 MDP derivatives and tested their ability to stimulate murine macrophages to eliminate tumor cells. This work has resulted in the discovery of the novel compound MDP-C (**20**), which significantly enhances dendritic cell (DC)-mediated, IL-2, IL-12 release and the induction of cytotoxic activity by cytotoxic T-cells (CTLs) *in vitro*. Finally, MDP-C has also been shown to have Hepatitis B Surface Antigen specific antibody response *in vivo*, with minimal pyrogenicity (in rabbits), passive cutaneous anaphylaxis (in rats), and low toxicity (in mice).[96]

Researchers have also explored the use of cocktails of synthetic PRR agonists to stimulate NLRs and TLRs simultaneously. Such a combinatorial treatment may more closely mimic a bacterial infection compared to single agents, and in turn induce a potent immune reaction. Indeed, Tada *et al.* combined two synthetic NOD2 agonists (MDP and N-acetylmuramyl-L-aranyl-L-isoglutamine (MDP-LL)), two NOD1 agonists (FK565 and FK156),[97] and various TLR agonists (TLR4- lipid A, TLR2- Pam_3CSSNA, TLR3- poly (I:C), and TLR9- CpG DNA) to stimulate DC cultures. Combination treatments led to the synergistic upregulation of IL-12 and IFN-γ production by T-cells, indicative of a T_H1-response.[98] Utilizing complex cocktails to mimic the natural repertoire of PRR ligands is likely to prove increasingly important to synthetic immunology strategies in the future.

1.2.3 Peroxisome Proliferator-activated Receptor Gamma

Peroxisome proliferator-activated receptor gamma (PPAR-γ) is a ligand-activated transcription factor that plays an important role in regulating storage and catabolism of dietary fats, and whose stimulation is also associated with various immunosuppressive effects. These include inhibition of DC maturation, $CD4^+$ T-cell differentiation into T_H17 cells,[99] and the suppression of pro-inflammatory cytokine release.[100] Endogenous PPAR-γ ligands are similar to one another in that they are highly lipophilic and contain long, polyunsaturated fatty acid structural motifs (e.g., docosahexanoic acid, arachidonic acid, eicosapentaenoic, and gamolenic acid).[101] Among these ligands, 15-deoxy-$\Delta^{12,14}$-prostaglandin J_2 (15d-PGJ_2, Figure 1.4, **21**), which is formed as a terminal product of the cyclooxygenase-2 pathway, has proven especially useful in studying PPAR-γ.[102] Perhaps the most widely studied synthetic PPAR-γ ligands were not originally developed for their immunomodulatory effects; the class of thiazolidinediones (TZDs) or "glitazones" – which include pioglitazone (**22**), troglitazone (**23**) and others – were discovered from screening efforts in rodent models for insulin resistance and have been used for over 28 years as therapeutics for diabetes.[101] Their mechanisms of action have only recently been shown to involve PPARs.[103] The critical features

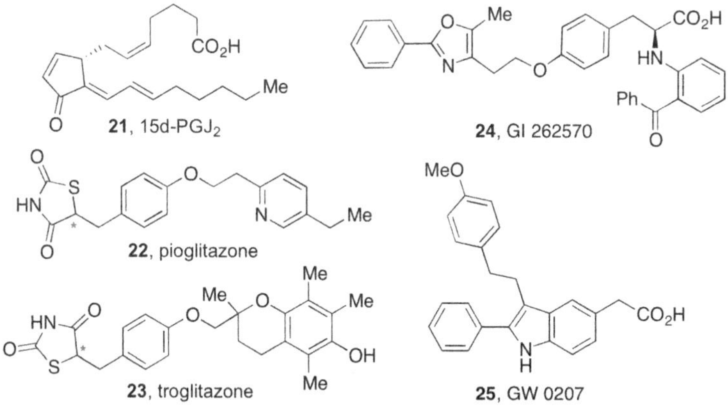

Figure 1.4 Natural and synthetic ligands for PPAR-γ. Red stars in structures 22 and 23 indicates the C-5 position of the thiazolidinedione core, which is prone to epimerization.

of these synthetic ligands that allow binding to PPAR-γ are believed to include the 2,4-thiazolidinedione structural motif, which mimics the head-group of a lipid molecule, and the aromatic, unsaturated, tail, which mediates hydrophobic contacts with the receptor. This structural model is supported by apo- and ligand-bound crystal structures of PPAR-γ.[104,105] Optimization of TZD compounds using direct PPAR-γ binding assays have revealed that only compounds in the (*S*)-enantiomeric series at C-5 (indicated with red stars in Figure 1.4) bind the receptor.[106] Because the thiazolidinedione nucleus is prone to epimerization at this position, next-generation PPAR-γ modulators have incorporated head groups of increasing kinetic stability such as tyrosine derivatives (e.g., GI 262570, **24**),[107] and the achiral GW 0207 (**25**).[108] Explorations into the immunomodulatory activities of PPAR-γ ligands are still at an early stage; future studies that combine efforts of chemists and immunologists will surely provide insight into fundamental and biomedical applications for PPAR-γ ligands.[109]

1.2.4 C-Type Lectin Receptors (CLRs)

C-type lectin receptors (CLRs) comprise a group of membrane-bound Ca^{2+}-dependent carbohydrate-binding proteins found on "professional" antigen presenting cells (APCs) such as macrophages, DCs, and B-cells.[110] Although there are up to 14 different types of CLRs, we deal here with synthetic strategies for modulating only one of these family members – the DC-specific intercellular adhesion molecule-3-grabbing non-integrin (DC-SIGN).[111,112] Like other CLRs, DC-SIGN can recognize glycosylated peptides and proteins,[113] leading to internalization, antigen processing, presentation on MHC class II molecules, and cross-presentation on MHC class I molecules.[114]

Notably, DC-SIGN expression is restricted to DCs, it possesses a dual specificity for both high-mannose and Lewis-type carbohydrates,[115] and functions to maintain immune homeostasis through cell adhesion and intracellular signaling. Interest in identifying synthetic DC-SIGN modulators has arisen due to the protein's intriguing function as a receptor for pathogen entry; binding to DC-SIGN has been shown to promote infections by both viral and bacterial pathogens. Structural studies have provided insight into DC-SIGN–ligand binding, and numerous DC-SIGN modulators have developed. For example, researchers have explored the binding of various carbohydrates to DC-SIGN, most of which bind with relatively low affinity, and other groups have aimed to enhance interaction strength using multivalent ligands. Prost and colleagues have disclosed an approach wherein functionalization of a (–)-shikimic acid core structure with mannose derivatives, followed by bioconjugation with bovine serum albumin (BSA), afforded a ligand capable of stimulating DC-SIGN-mediated endocytosis as well as JNK signaling.[116] Kiessling and coworkers have identified non-carbohydrate quinoxalinone-based DC-SIGN ligands using a high-throughput screening assay, followed by focused medicinal chemistry efforts.[117,118]

In a separate study, Srinivas *et al.* explored whether a synthetic glyco-targeting approach could be utilized to induce antigen uptake and cross-presentation by DCs via the MR or DC-SIGN.[119] To achieve this, the authors developed a fully synthetic glycocluster conjugate containing a 25 amino acid peptide conjugated with either Manα-6Man or a mixture of Lewis^a and Lewis^x oligosaccharides (not shown), which targeted both the MR and DC-SIGN receptor. Romero *et al.* specifically chose a melanoma antigen, A27L Melan-A/Mart-1 epitope,[120] to evaluate if the synthetic glycocluster could facilitate the cross-presentation of this antigen to elicit a CTL response, with the aim of developing a vaccine. Interestingly, their synthetic molecule was not only able to bind to both MR and DC-SIGN in SPR experiments, but it was also able to induce internalization into DCs, and to stimulate Melan-A-specific $CD8^+$ T-cells to induce IFN-γ production *in vitro*. This work serves as an interesting example of how synthetic mimics of biomacromolecules can elicit receptor-specific, targeted pro-inflammatory responses.

1.3 Synthetic Systems for Controlling Cell–Cell Communication: Cytokine Mimetics

The immune system relies heavily on intercellular communication to regulate critical surveillance and defense functions. Cytokines are a diverse family of soluble molecules that have been divided into subtypes based on their function, secreting cell type, and presumed target. These subtypes include chemokines, which trigger cell movement or chemotaxis; interleukins, which are key players in transferring information between leukocytes; and interferons, named for their ability to interfere with viral replication. Because of their roles in co-ordinating both pro- and anti-inflammatory processes, cytokines are critical for maintaining immune homeostasis. Small molecules capable of replicating cytokine functions have been widely sought.[121] Here we discuss several examples of synthetic molecules capable of functioning as cytokine mimics.

Chemokines are a class of small proteins (8–12 kDa) that are primarily responsible for stimulating leukocyte migration, which is critical for their activation, differentiation, and survival.[122] Chemokine receptors are members of the class of G-protein coupled receptors (GPCRs), which contain seven transmembrane-spanning domains and elicit a complex downstream signaling cascade.[122] Here we focus on a few examples of how synthetic approaches to mimic the natural chemokines have led to novel ligands for several chemokine receptors, including CCR1, CCR3, and CXCR3.

CCR1 interacts with natural peptide ligands CCL3 and CCL5, and triggers chemotaxis of monocytes, macrophages, and T-cells.[123] Jensen *et al.* have recently identified two small molecule metal chelators – 2,2-bipyridine and 1,10-phenanthroline (Figure 1.5, **26**) – which can serve as chemokine mimics. When these compounds are complexed to Cu^{2+} or Zn^{2+}, they are capable of inducing G-protein signaling by binding CCR1. Interestingly, the mechanism of this effect was found to involve binding of the metal–chelator complex deep within a cleft in CCR1, leading to antagonism of the CCL5 binding interaction and enhancement of CCL3 association.[124]

CCR3 is found primarily on eosinophils, and is involved in modulating allergic reactions when activated by its natural ligand CCL11/eotaxin.[125] Using a high-throughput cell-based small-molecule screen involving FACS-based detection of changes in cellular morphology, researchers have identified the chemokine mimic CH0076989 (**27**).[126,127] Mechanistic studies have revealed that **27** can promote chemotaxis of a pre-B lymphoma cell line in a CCR3-specific manner, albeit not as potently as CCL11. Follow-up homology modeling studies have suggested that the binding site for **27** on CCR3 is similar to that for the natural ligand (CCL11), yet appears to overlap with prototypical

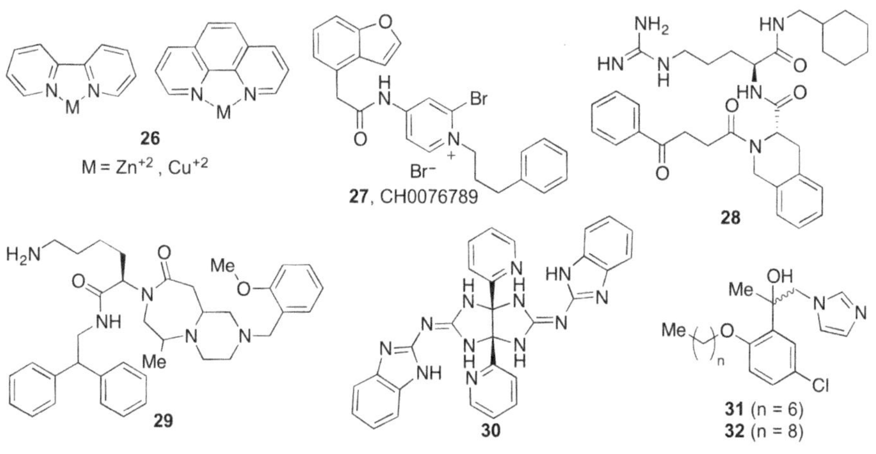

Figure 1.5 Molecules that influence immune cell communication.

antagonist binding regions.[126] Taken together, these findings have indicated that minor changes in the orientation of synthetic ligands can result in major changes in the receptor response. Therefore, these studies have the potential to aid in future structure-based design efforts of both agonists and antagonists of CCR3.

CXCR3 is expressed primarily on T-cells and regulates chemotactic responses to natural ligands CXCL9, CXCL10, and CXCL11.[128] CXCR3 agonists have shown benefits in therapeutic applications such as wound healing and anti-tumor activity through enhancement leukocyte chemotaxis.[129–131] Using a calcium mobilization assay in CXCR3-transfected HEK293 T-cells, Stroke *et al.* screened an encoded combinatorial library of tetrahydro-isoquinolines and piperidinyl diazepanones to search for chemokine mimics. These studies have led to the discovery of several CXCR3 agonists (**28** and **29**) capable both of inhibiting CXCL10 and CXCL11 binding, and promoting T-cell chemotaxis.[132] Follow-up experiments have revealed these small molecule agonists can bind three critical residues (D112, D195, and E196) within an intrahelical pocket in CXCR3, thus mimicking the Pro-Arg-Val sequence (positions 37–39) in CXCL10.[133] These molecules represent both structural and functional mimics of chemokines, and may provide effective alternatives to recombinant CXCL10 and CXCL11 tumor therapies.[130,131]

The cytokine Granulocyte-Colony Stimulating Factor (**G-CSF**) is a potent regulator of survival, proliferation, differentiation, and activation of granulocytes. G-CSF binds to the G-CSF receptor (CD114), which results in activation of the JAK/STAT signaling pathway, transcriptional activation of interferon responsive genes, and enhancement of cell proliferation and differentiation.[134,135] Indeed, recombinant G-CSF or filgrastim (marketed by Amgen under the brand name Neupogen®) is used clinically to replenish the loss of leukocytes in response to chemotherapy or bone-marrow transplantation.[136] Synthetic mimics of G-CSF have the potential to provide effective, low cost, and non-immunogenic alternatives to such treatments. To identify G-CSF mimics, one group has developed a high-throughput screening assay utilizing a myeloid-derived murine cell line with a JAK/STAT pathway-dependent luciferase reporter construct. Execution of this screen has led to the identification of the non-peptidic small molecule SB247464 (**30**); this compound has been shown to elicit JAK/STAT signaling, and increase the neutrophil count in an immunosuppressed neutropenic mouse model (BDF-1).[137] Notably, the two-fold rotational symmetry in **30** is believed to be critical to its ability to mimic G-CSF function in inducing oligmerization and activation of CD114.[137–139] Interestingly, although human G-CSF can activate both murine and human G-CSFR, SB247464 has only been found to stimulate murine, but not human, CD114. To address this deficiency, another group performed a high-throughput cellular screen of 10,000 synthetic compounds for their ability to stimulate proliferation of a human G-CSF sensitive BAF/B03 cell line (hematopoetic B-cell line).[140] These studies have identified imidazole derivatives SSCL02446 (**31**) and SSCL02448 (**32**), which have been shown to stimulate STAT3 signaling and increase neutrophil counts *in vivo* in neutropenic rats.[140]

1.4 Synthetic Molecules that Modulate the Complement System

The complement system is an integral part of humoral immunity, serving as a cytotoxic effector system against a variety of pathological stimuli. As shown in Figure 1.6, the complement pathway involves a network of proteins that act in concert to regulate immune cell responses, clear infections and cellular debris, and lyse antibody-opsonized targets.[141] The activation of complement, either through the classical, lectin, or alternative pathways,[142,143] leads to a cascade of enzymatic reactions, culminating in complement-mediated pro-inflammatory signaling, phagocytosis, or lysis of targets.[144] As shown in Figure 1.6, the responses of the complement system are also negatively regulated by various control proteins, including C1-INH, factor H, C4BP, CD46, and CD55. Dysregulation of any of these components can lead to the development of diseases including glomerulonephritis, pancreatitis, psoriasis, rheumatoid arthritis, and asthma.[141,145]

A molecular-level understanding of the complement cascade has facilitated the development of synthetic molecules capable of mimicking the endogenous regulators. For example, in order to develop mimics of complement factor H–related protein 1 (CFHR1) – a human plasma protein that inhibits the activity of C5 convertase – researchers have developed an assay to monitor the C5 cleavage in the presence of a library of aptamers.[146] This screening protocol has led to the identification of C5C6 (Figure 1.7, **32**), a specific inhibitor of C5 esterase that can prevent the formation of C5a and

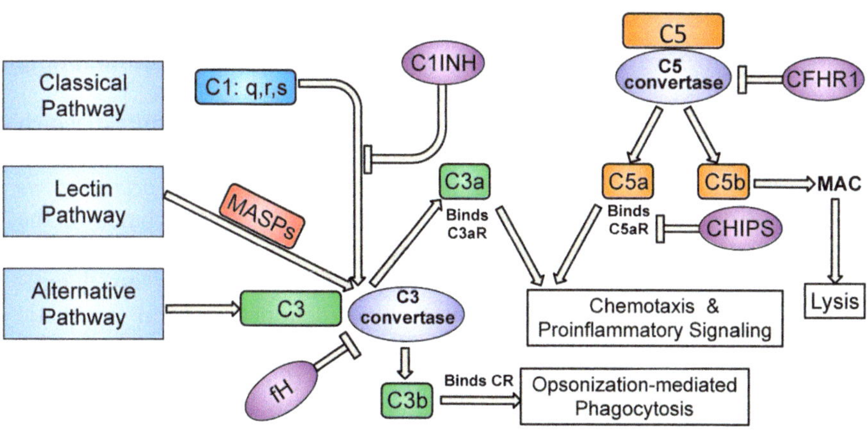

Figure 1.6 Schematic representation of the complement cascade. The endogenous complement inhibitors are colored in purple. MASPs: mannan-binding lectin serine proteases; C1: complement component 1; C3: complement component 3, C5: complement component 5; CFHR1: complement factor H–related protein 1; CHIPS: chemotaxis inhibitory protein Staphylococcus aureus; fH: factor H.

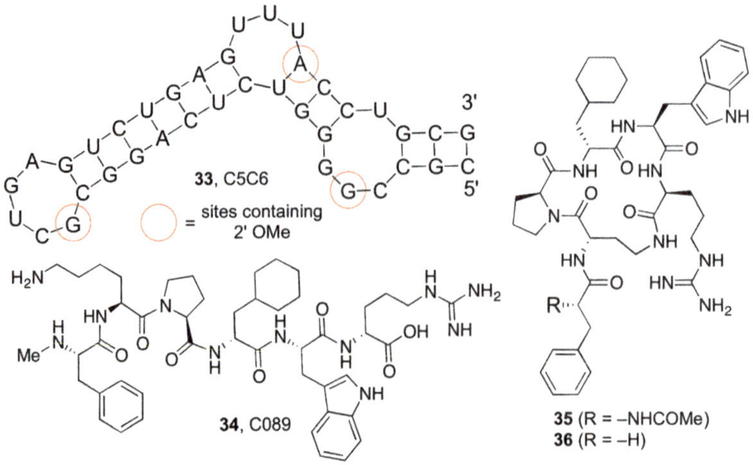

Figure 1.7 Synthetic complement modulators. C5C6 aptamer (32) inhibits C5 convertase (sites containing 2'-OMe indicated with red circles). C089 (34) and cyclic derivatives 35 and 36 inhibit C5.

C5b, much like CFHR1. The three 2'-OMe substitutions in **33** have proven critical for increasing its serum stability and enabling maximal inhibition of complement-mediated hemolytic activity of sheep erythrocytes.[147] Subsequent truncation of C5C6, followed by conjugation to PEG and capping with an inverted nucleotide at the 3' end, have provided ARC1905, an aptamer derivative with improved pharmacokinetic properties compared to **33**.[148] This molecule has shown clinical promise as a treatment for age-related macular degeneration (AMD) and is an excellent example of a fully synthetic, modified aptamer capable of emulating a natural complement-inhibiting protein.

Several naturally-occurring peptides and proteins have been identified, which are capable of enabling pathogens to evade the human complement response.[149] For example, the Gram-positive bacteria *Staphylococcus aureus* secretes a chemotaxis inhibitory protein called CHIPS. CHIPS binds to the N-terminus of C5a and abrogates the binding of C5a to the C5a receptor (Figure 1.6, C5aR or CD88,).[150] With the goal of developing novel anti-inflammatory compounds, researchers have identified several synthetic molecules that mimic the function of CHIPS. In one such example – called the chemotaxis inhibitory construct protein of *S. aureus* (CHOPS)[151] – researchers took into account critical structural features of CHIPS (an α-helix and three β-strands), as well as the spatial orientation of its binding to C5aR. To this end, a D-Pro-Gly was incorporated into CHOPS, which helped maintain the helical structural features, resembling the motifs found in CHIPS. Using isothermal titration calorimetry, researchers then found CHOPS to have an affinity for C5aR of 3–4 μM. Although this value is three-orders of magnitude lower than the affinity of CHIPS for C5aR, CHOPS has served as an important lead compound for further ligand-design efforts.[151]

An alternative mechanism for synthetically mimicking the action of CHIPs has been to develop C5a mimetics that block the C5a–C5aR interaction.[152] C5a has two binding sites that enable it to bind and activate the C5aR; one of these resides at the core inter-helical loops of C5a and contains positively charged residues,[153,154] and the other comprises an eight-residue motif at the protein's C-terminus, which itself is sufficient to agonize C5aR.[155] Researchers have utilized this octapeptide agonist as a basis for developing linear (**34**, C089) and cyclic (**35**, **36**) synthetic antagonists of the C5a–C5aR interaction.[156] These molecules have exhibited inhibitory effects on neutrophil chemotaxis and cytokine production from macrophages both *in vitro* and *in vivo*.[157] These non-immunogenic, synthetic molecules have the potential to serve as starting points for effective anti-complement therapeutics, capable of treating a range of diseases including arthritis, ischemia-reperfusion injuries, and sepsis.[146,158]

Factor H (fH) is an inhibitory protein that is highly abundant in human plasma, and prevents C3 convertase formation by binding to components C3b and C3d.[159] Factor H is recruited to endothelial cells by polyanionic ligands such as sulfated heparin, dermatan sulfate, and glycosaminoglycan, thus protecting these cells from the alternative complement pathway.[160] Therefore, functionalization of biomaterials with synthetic structures capable of recruiting fH has served as a useful strategy for preparing complement-compatible materials.[145] For example, using phage-display-based screening technologies, Wu *et al.* have identified a hexapeptide termed 5C6, which is capable of binding fH without interfering with its complement-inhibitory properties.[161] Immobilization of 5C6 on pegylated polystyrene or glass surfaces, using thiol-maleimide bond-forming reactions, prevented complement fixation in erythrocyte hemolytic assays.[162] This research has provided an interesting and potentially useful approach to the production of biocompatible materials.

1.5 Mimicking and Hijacking Antibody Function

Antibody proteins bridge the innate and adaptive wings of the immune system; they are produced in response to immunogenic epitopes, yet they function by activating innate cytotoxic processes. Antibody-based therapeutic approaches have blossomed over the past decade, in part because they exploit both innate and adaptive features; they can be generated against wide-ranging disease-relevant epitopes, and they can also activate endogenous immune effector mechanisms, such as complement-dependent cytotoxicity (CDC) and antibody-dependent cellular cytotoxicity (ADCC). The development of synthetic systems capable of emulating – and improving upon – the function of antibody molecules has represented an exciting focus of synthetic immunology research in recent years.

For example, with the goal of emulating the adaptive immune system's ability to rapidly identify highly selective epitope-targeting motifs, Kodadek and colleagues have developed a systematic small-molecule synthesis and screening protocol. This process involves construction of oligomeric, bead-based small molecule libraries using split-and-mix synthesis methods, followed by execution of protein- and/or cell-binding selections directly on synthesis resins.[163–166] Because small molecule ligands identified through this approach are both potent and selective – in analogy to natural antibody proteins – they have been termed "antibody mimics."[167,168] Indeed, using their streamlined synthesis-screening protocol, the Kodadek lab has identified nanomolar "antibody mimics" against both soluble and cell-surface-bound targets such as cholera toxin,[169] VEGF receptor,[170,171] orexin receptor,[172] phosphoproteins,[173] and other systems.[174–179]

Researchers have also explored the development of novel cytotoxic agents that are capable of exploiting the innate effector functions of antibodies. For example, attachment of synthetic targeting motifs to intact antibodies or immunoglobulin constant (Fc) domains has led to the development of novel anticancer agents for targeting the endothelin receptor,[180] or $\alpha_v\beta_3$ and $\alpha_v\beta_5$ integrins.[181–184] Bioconjugation reactions leading to the assembly of these chimeric agents can also be performed *in vivo,* potentially providing new approaches for *in situ* self-assembly of targeted chemotherapeutic agents.[185] Another strategy has been to develop small molecules that hijack antibodies already present in the human bloodstream, and recruit them to disease-relevant cell-surface targets. Indeed, humans possess endogenous antibodies that recognize simple, low-molecular weight haptens such as 2,4-dinitrophenyl, α-gal trisaccharide, and others. By attaching these antibody recognition motifs to high-affinity chemical groups that recognize disease-associated targets, it has been possible to develop bifunctional molecules that exploit the innate cytotoxic properties of antibodies to clear pathologic cells (Figure 1.8). These synthetic agents – called antibody-recruiting molecules (ARMs) – have been used to target cancers,[186–196] bacteria,[197–201] and viruses,[202–204] and are currently in clinical development. A review summarizing ARM-based strategies has recently appeared in the literature.[184]

Overall, by mimicking and/or exploiting the properties of antibody proteins, researchers hope to develop more effective, safer and less costly alternatives to current immunotherapeutic strategies.

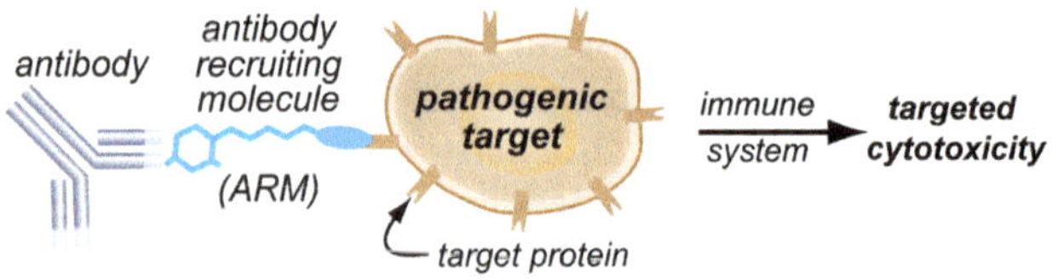

Figure 1.8 Schematic diagram representing the concept of antibody recruiting molecules (ARMs). ARMs are bifunctional molecules capable of bindig to pathogenic target and simultaneously recruiting antibody molecules, this eliciting antibody-dependent immune responses.

1.6 Synthetic Molecules that Modulate T- and B-cell Responses

Immune responses to foreign invaders frequently result in a series of molecular and cellular events that endow a host with long-term immunity (Figure 1.9). This process – termed adaptive immunity – begins with the uptake and processing of non-self molecular motifs by DCs, macrophages, B-cells, and other professional antigen presenting cells (APCs). APCs then present antigens to T-cells, and when accompanied by the appropriate pro-inflammatory signals, the activation and proliferation of antigen-specific B- and T-cells

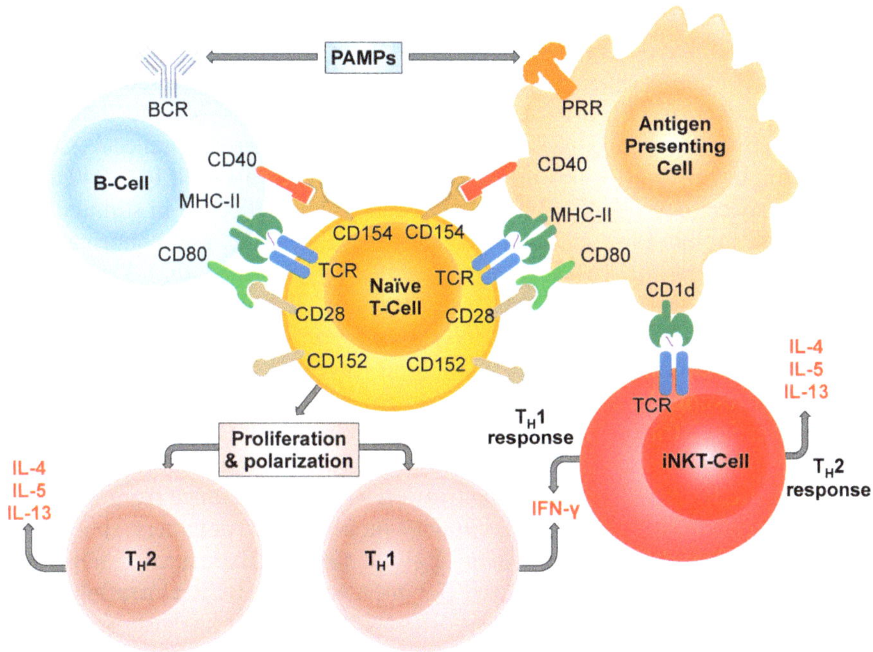

Figure 1.9 Summary of components of the adaptive immune system. Following an infection, pathogen associated molecular patters (PAMPs) are recognized and internalized by pattern recognition receptors (PRRs) on the antigen presenting cells, for subsequent presentation of antigenic epitopes on major histocompatibility complex class II (MHC II) proteins. These MHC II complexes are recognized by specific naïve T-cell receptors (TCRs). Costimulatory signals mediated by CD40 and CD80 of the antigen presenting cells, and CD154 and CD28 on the CD4+ T-cells lead to T-cell proliferation and polarization to T helper cells (T_H1 and T_H2), which then secrete inflammatory cytokines (labeled in red). CD152 regulates this proinflammatory response by inhibiting the interaction between CD28 and CD80. Invariant natural killer cells (iNKTs) are a subset of T-cells whose TCRs recognize lipidated molecules presented by CDld on antigen presenting cells to elicit either T_H1 or T_H2 cytokine responses.

ensues. B-cells differentiate into plasma cells, capable of secreting antigen-specific antibody molecules that give rise to humoral immunity, while T-cells can differentiate into variants of helper T-cells or cytotoxic T-lymphocytes (CTLs), responsible for antigen-specific cell-mediated immunity. Synthetic approaches to emulate, stimulate and/or suppress adaptive immune responses have been explored, and we highlight a selected few in this section.

With the goal of developing synthetic molecules to promote adaptive immune responses against *Trypanosoma cruzi* – the causative agent of Chagas disease – Fournel *et al.* have developed synthetic molecules that mimic the CD40 ligand (CD154).[205] CD154 is expressed on activated T-cells, while its receptor is present mainly on APCs. The CD40–CD154 interaction is critical for regulating B-cell proliferation and memory cell development, as well as "licensing" DCs to present antigen to cytotoxic T-cells.[206] To mimic natural CD154, researchers developed synthetic structures called miniCD40Ls (Figure 1.10, Panel A). These compounds contain a rigid, symmetrical, trimeric scaffold comprised of either β^3-tripeptide (**37**) or D,L-α-hexapepetide (**38**) coupled to a peptide sequence derived from CD40 ligand (CD154).[205] This trimeric structure reflects X-ray crystallographic findings, which indicate that CD154 presents itself to CD40 as a trimer.[207] Indeed, compound **37** has proven functionally effective in stimulating mouse B-cell proliferation and maturation of a DC cell line, and confirmatory surface plasmon resonance (SPR) experiments have revealed that **37** and **38** binds CD40 trivalently. Furthermore, exposure of *T. cruzi*-infected mice to miniCD40Ls leads to increases in IFN-γ production by lymphocytes, proliferation of B-cells, maturation of DCs, and reversal of *T. cruzi*-mediated immunosuppression, leading to effective control over infection *in vivo*.[208]

An interesting strategy for modulating B-cell receptor (BCR) activity has recently been disclosed by Kiessling and colleagues. Because crosslinking of cell-surface BCRs can induce B-cell activation without assistance from T-cells, these researchers have developed hapten-functionalized polymers that can display BCR ligands in a multivalent fashion (**39**).[209] Exposure of hapten-specific B-cells to these functionalized polymers leads to BCR clustering and production of anti-hapten antibodies *in vitro*. This research has also provided insights into the molecular requirements for B-cell activation and lays a foundation for a novel immunization protocol that could prove useful when extended to living organisms.[210,211]

Invariant natural killer cells (iNKTs) are a class of T-cells involved in the recognition of lipid antigens, which are presented on APCs by CD1 family members – a class of APC-expressed glycoproteins related to MHC I.[212] Derivatives of the natural product agelasphin (Figure 1.10, Panel B, **40**), have proven useful in understanding and mimicking CD1-mediated antigen presentation. Agelasphin and derivatives such as KRN7000 (α-GalCer, **41**) can be taken up by scavenger receptors (SR) on macrophages and DCs, which are then presented to iNKT-cells bound to CD1d. This activation event can lead to the production of both T_H1 (IFN-γ) and T_H2 (IL-4, IL-5, and IL-13) cytokine responses *in vitro*.[213–215] More recently, synthetic, crystallographic,[216] and computational studies[217] have led to a more detailed understanding of the

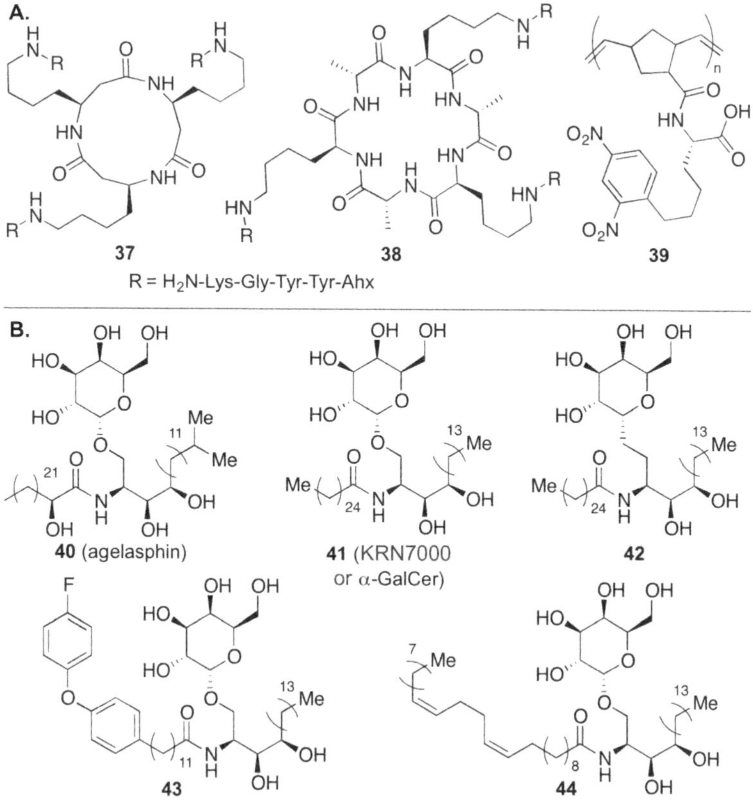

Figure 1.10 Molecules that modulate the function of B-cells (Panel A) and iNKT-cells (Panel B).

CD1d-glycolipid interaction.[218] Importantly, synthetic studies have resulted in the generation of α-GalCer analogs capable of polarizing iNKT-cells to produce either T_H1 or T_H2 cytokine responses.[218–222] In one example, altering only the connectivity between the glycan and lipid portions of ligands, as in C-linked compound **42**, has led to increases in T_H1 cytokine production, and enhanced activity *in vivo* against melanoma metastases in mice.[223] T_H1 responses can be further enhanced by functionalization of one acyl chain with arenes, leading to the optimized structure **43**. Compound **43** has been found to elicit protection against bacterial (*Sphingomonas casulata* and *S. aureus*) and viral (Japanese encephalitis virus) infections in mouse models.[224] Conversely, alterations in the length and saturation of an acyl chain in α-GalCer derivatives were sufficient to elicit potent T_H2 responses.[225] For example, the molecule C20:2 (**44**) was found to reduce IFN-γ levels and increase IL-4 and IL-13 production in mice, conferring protection from type 1 diabetes.[226] These examples illustrate the ability to utilize synthetic derivatives of a natural molecule to fine-tune the adaptive immune response.

In certain situations, immune tolerance to self-antigens can be permissive to pathological processes such as cancer and autoimmune diseases in the host. Therefore, strategies to overcome self-tolerance may have useful therapeutic applications. For example, the Lerner and Schultz laboratories have disclosed a novel approach for breaking immunological self-tolerance to TNF-α, thus generating long-lasting immune responses that can curb TNF-α-mediated pro-inflammatory responses. To this end, these researchers have utilized amber codon suppression mutagenesis strategies to replace a single tyrosine (Tyr-86) in murine TNF-α (mTNF-α) with the non-natural amino acid *p*-nitrophenylalanine (pNO_2Phe).[227] These researchers selected pNO_2Phe for incorporation, as it is known that nitroaryl groups are highly immunogenic haptens,[228] and disruption of the native conformation of the protein would be minimal due to the structural similarities between pNO_2Phe and Tyr. Indeed, it was found that vaccination of mice with this mutated protein led to a high titer antibody response against *natural* mTNF-α, capable of protecting mice against challenges with LPS.[229] Further mechanistic studies suggested the involvement of T-cells to mediate Ig class switching to generate anti-mTNF IgG antibodies, demonstrating that a single mutation was enough to break T-cell self-tolerance to non-mutated regions of a protein.[230] This technique has further been utilized to incorporate other naturally and non-naturally occurring post-translational modifications, such as sulfotyrosine and 3-nitrotyrosine, and it has been successfully employed to break self-tolerance to other endogenous proteins, including retinol binding protein 4 (RBP4),[230] and epidermal growth factor (EGF).[231] This approach has provided valuable insight into the mechanism of autoimmunity as well as interesting leads for novel vaccination strategies.

Advances in synthetic organic chemistry over the last two decades have made possible the synthesis of complex carbohydrates and glycopeptides in both solution- and solid-phase formats.[232–236] These molecules have important applications as synthetic antigens for synthetic vaccine development,[237–240] carbohydrate-based drugs,[241] drug-delivery systems,[242,243] glycan screening,[244,245] and understanding of basic immunology.[246] Carbohydrate mimics of antigenic epitopes have been developed in the search for vaccines against cancer,[1,236,241,247,248] viruses,[1,236,247,249] bacteria,[236,249] and parasites.[1,236,249,250] Interested readers are directed to the chapter in this book devoted to synthetic vaccines for a more detailed discussion of this exciting area of synthetic immunology.

In order to survey the antigen repertoire within hosts, T- and B-cells must constantly traffic between the periphery and primary (bone marrow and thymus) and secondary (lymph nodes) lymphoid organs.[251] Thus, a useful strategy for modulating T- and B-cell function has been to alter lymphocyte localization within the lymphatic system by preventing egress from lymph nodes.[252,253] Fingolimod (Gilenya®, Figure 1.11, Panel A, **45**), a derivative of the fungal metabolite ISP-1 (**46**),[254] has been found to inhibit lymphocyte egression from secondary lymphoid organs by way of interactions with the sphingosine1-phosphate receptors (S1PRs). Sphingosine1-phosphate (**47**, S1P) is a critical mediator of both T-cell[255,256] and B-cell[257] egression from the lymphoid tissue, and it is believed that fingolimod serves as an S1P mimic after being phosphorylated by sphingosine kinases.[258] Agonism of S1PRs induced by fingolimod leads to rapid receptor internalizion and degradation on lymphocytes, resulting in inhibition of lymphoid egression.[259,260] Fingolimod has been shown to be an effective therapeutic for multiple sclerosis, and was approved by the FDA in 2010 as the first oral treatment for the disease.[261] Fingolimod is a unique example of a synthetic mimic of a natural receptor ligand, which has led to an effective therapeutic.

Strategies to prevent the interactions of TCRs with cognate peptide–MHC II complexes have also proven useful for immune modulation. One example of such an approach has come from Khosla and colleagues, inspired by an understanding of the molecular basis for celiac sprue – a disease characterized by inflammatory responses upon exposure to dietary gluten.[262] Initially, these researchers found that protease digestion of α-gliadin, a member of the gluten protein family, leads to the production of a protease-resistant 33-amino-acid peptide fragment (Figure 1.11, Panel B, **48**). Exposure of this stable peptide to

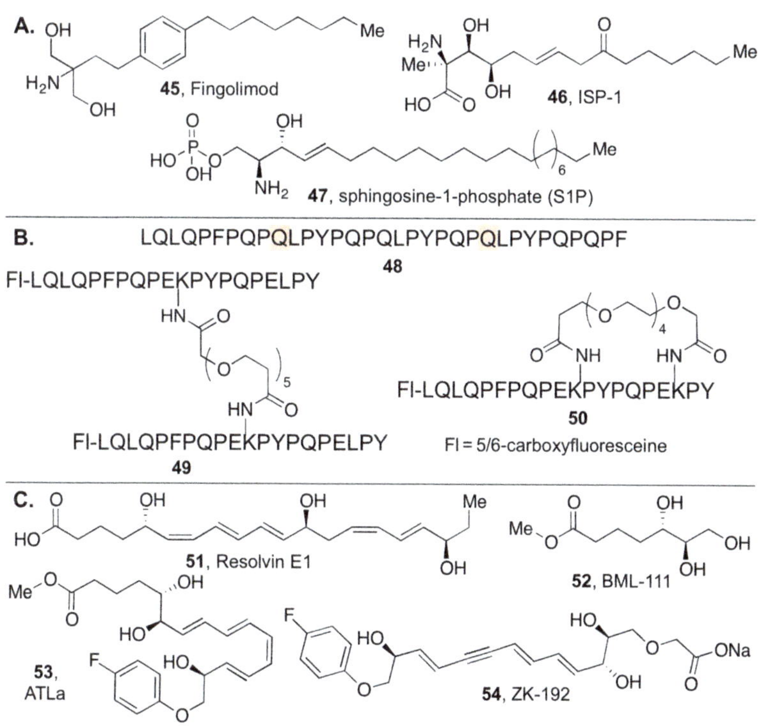

Figure 1.11 (A) Fingolimod (45) is a synthetic derivative of the natural product ISP-1 (46) . Fingolimod was found to modulate S1PR, whose natural ligand is sphingosine-1-phosphate (47). (B) Peptide sequence (48) is derived from α-gliadin, and deaminated at two glutamine residues (rd boxes) by transglutaminase 2 (TG2). Cyclic peptide 50 and linear dimer 49, have been shown to bind HLA-DQ2, yet inhibit recognition by autoimmune T-cells. (C) Specialized pro-resolving mediators (SPMs), including the endogenous resolvin E1 (51) and synthetic derivatives BML-111 (52), ATLa (53), and ZK-192 (54).

the enzyme transglutaminase 2 (TG2), which can convert glutamine residues to glutamic acids, then leads to the accumulation of mono-deamidated products; these products are presented to auto-reactive T-cells through MHC II proteins of the HLA-DQ2 allelic variant.[263] Thus, the authors have designed dimeric (**49**)[264,265] and cyclic (**50**)[265] analogs of a minimal deamidated binding motif derived from **48**; indeed, these synthetic constructs have been shown to bind HLA-DQ2, and inhibit T-cell proliferation *in vitro*, even in the presence of deamidated **48**. These examples highlight how an understanding of molecular immunology and pathophysiology can enable the development of synthetic immunomodulators with exciting therapeutic potential.

Another strategy for modulating T-cell-mediated immunity has involved the development of immunosuppressive lipid byproducts, derived from a family of naturally occurring materials. This class of natural metabolites – termed proresolving mediators (SPMs) – includes resolvin E1 (Figure 1.11, Panel C, **51**), protectins, lipoxins, and maresins.[266] These anti-inflammatory compounds are produced in the host during the resolution phase of the inflammatory process, and regulate leukocyte trafficking, complement stimulation, and cytokine release.[266,267] Synthetic analogs of SPMs have been developed – such as BML-111 (**52**),[268] ATLa (**53**),[269] and ZK-192 (**54**)[270] – that can suppress immune responses in animal models for arthritis, haemorrhagic-shock-induced lung injury, and inflammatory colitis, respectively. Because, unlike corticosteroids, these molecules are not thought to elicit broad-spectrum immunosuppression, they have the potential to give rise to effective anti-inflammatory treatments while minimizing side-effect profile.[267]

1.7 Concluding Remarks

The diverse research efforts covered in this chapter bring to light a few important themes. First, as the immune system's ability to distinguish "self" from "non-self" fundamentally depends on molecular recognition, the field of immunology is at once both a chemical and a biological science. This duality is underscored by the many organic chemists who have made important contributions to immunological science, and presents numerous opportunities in Synthetic Immunology moving forward. Indeed, despite the diversity of scientific contributions covered in this chapter, the field of synthetic immunology is still in its infancy. Opportunities for organic chemists, chemical biologists, and other molecular scientists are continuously increasing as the barriers between the "traditional" disciplines of biology and chemistry continue to blur. Hopefully this chapter will help inspire interdisciplinary researchers to start pursuing these exciting directions, which we believe will have important ramifications in both academic and clinical settings.

References

1. R. D. Astronomo and D. R. Burton, *Nat Rev Drug Discov*, 2010, **9**, 308–324.

2. T. D. Gilmore and M. R. Garbati, *Current Topics in Microbiology and Immunology*, 2011, **349**, 245–263.
3. J. D. Powell, K. N. Pollizzi, E. B. Heikamp and M. R. Horton, *Anni Rev Immunol*, 2012, **30**, 39–68.
4. A. Kontzias, A. Laurence, M. Gadina and J. J. O'Shea, *F1000 Medicine Reports*, 2012, **4**, 5.
5. J. J. O'Shea and R. Plenge, *Immunity*, 2012, **36**, 542–550.
6. E. T. Boder and W. Jiang, *Annu Rev Chem. Biomol Eng*, 2011, **2**, 53–75.
7. A. C. Chan and P. J. Carter, *Nat Rev Immunol*, 2010, **10**, 301–316.
8. M. Dougan and G. Dranoff, *Annu Rev Immunol*, 2009, **27**, 83–117.
9. R. M. Steinman, *Annu Rev Immunol*, 2012, **30**, 1–22.
10. M. A. Dobrovolskaia, P. Aggarwal, J. B. Hall and S. E. McNeil, *Mol Pharm*, 2008, **5**, 487–495.
11. G. T. Jennings and M. F. Bachmann, *Ann Rev Pharmacol Toxicol*, 2009, **49**, 303–326.
12. J. De Souza Reboucas, I. Esparza, M. Ferrer, M. L. Sanz, J. M. Irache and C. Gamazo, *J Biomed Biotechnol*, 2012, **2012**, 474605, 1–13.
13. M. Look, A. Bandyopadhyay, J. S. Blum and T. M. Fahmy, *Adv Drug Deliv Rev*, 2010, **62**, 378–393.
14. A. M. Bodles-Brakhop, R. Heller and R. Draghia-Akli, *Mol Ther*, 2009, **17**, 585–592.
15. M. A. Liu, *Immunol Rev*, 2011, **239**, 62–84.
16. A. Judge and I. MacLachlan, *Hum Gene Ther*, 2008, **19**, 111–124.
17. S. Zhu, H. Wurdak and P. G. Schultz, *Future Med Chem*, 2010, **2**, 965–973.
18. C. Cobaleda, W. Jochum and M. Busslinger, *Nature*, 2007, **449**, 473–477.
19. L. Bruno and M. Merkenschlager, *Cell Cycle*, 2008, **7**, 2296–2298.
20. K. Araki, A. P. Turner, V. O. Shaffer, S. Gangappa, S. A. Keller, M. F. Bachmann, C. P. Larsen and R. Ahmed, *Nature*, 2009, **460**, 108–112.
21. K. Takeda and S. Akira, *Cell Microbiol*, 2003, **5**, 143–153.
22. K. Takeda, T. Kaisho and S. Akira, *Annu Rev Immunol*, 2003, **21**, 335–376.
23. E. J. Hennessy, A. E. Parker and L. A. J. O'Neill, *Nat Rev Drug Discov*, 2010, **9**, 293–307.
24. M. Hedayat, K. Takeda and N. Rezaei, *Med Res Rev*, 2012, **32**, 294–325.
25. E. Lin, J. E. Freedman and L. M. Beaulieu, *Cardiovasc Ther*, 2009, **27**, 117–123.
26. C. Leon, R. Tory, J. Jia, O. Sivak and K. Wasan, *Pharm Res*, 2008, **25**, 1751–1761.
27. H. Kanzler, F. J. Barrat, E. M. Hessel and R. L. Coffman, *Nat Med*, 2007, **13**, 552–559.
28. M. R. Hutchinson, A. L. Northcutt, T. Hiranita, X. Wang, S. S. Lewis, J. Thomas, K. van Steeg, T. A. Kopajtic, L. C. Loram, C. Sfregola, E. Galer, N. E. Miles, S. T. Bland, J. Amat, R. R. Rozeske, T. Maslanik, T. R. Chapman, K. A. Strand, M. Fleshner, R. K. Bachtell,

A. A. Somogyi, H. Yin, J. L. Katz, K. C. Rice, S. F. Maier and L. R. Watkins, *J Neurosci*, 2012, **32**, 11187–11200.
29. M. R. Hutchinson, Y. Zhang, M. Shridhar, J. H. Evans, M. M. Buchanan, T. X. Zhao, P. F. Slivka, B. D. Coats, N. Rezvani, J. Wieseler, T. S. Hughes, K. E. Landgraf, S. Chan, S. Fong, S. Phipps, J. J. Falke, L. A. Leinwand, S. F. Maier, H. Yin, K. C. Rice and L. R. Watkins, *Brain, Behav, and Immun*, 2010, **24**, 83–95.
30. K. Cheng, X. Wang, S. Zhang and H. Yin, *Angew Chem Int Ed*, 2012, **51**, 12246–12249.
31. K. Yoshiizumi, F. Nakajima, R. Dobashi, N. Nishimura and S. Ikeda, *Bioorg Med Chem*, 2002, **10**, 2445–2460.
32. A. Geloen, L. Helin, B. Geeraert, E. Malaud, P. Holvoet and G. Marguerie, *PLoS One*, 2012, **7**, e37633.
33. P. G. Lysko, J. Weinstock, C. L. Webb, M. E. Brawner and N. A. Elshourbagy, *J Pharmacol Exp\ Ther*, 1999, **289**, 1277–1285.
34. C. Hashimoto, K. L. Hudson and K. V. Anderson, *Cell*, 1988, **52**, 269–279.
35. C. A. Janeway, Jr. and R. Medzhitov, *Annu Rev Immunol*, 2002, **20**, 197–216.
36. X. Z. West, N. L. Malinin, A. A. Merkulova, M. Tischenko, B. A. Kerr, E. C. Borden, E. A. Podrez, R. G. Salomon and T. V. Byzova, *Nature*, 2010, **467**, 972–976.
37. C. Erridge, *J Leukocyte Biol*, 2010, **87**, 989–999.
38. K. Farhat, S. Riekenberg, H. Heine, J. Debarry, R. Lang, J. Mages, U. Buwitt-Beckmann, K. Roschmann, G. Jung, K. H. Wiesmuller and A. J. Ulmer, *J Leukocyte Biol*, 2007, **83**, 692–701.
39. W. G. Bessler, M. Cox, A. Lex, B. Suhr, K. H. Wiesmuller and G. Jung, *J Immunol*, 1985, **135**, 1900–1905.
40. O. Takeuchi, T. Kawai, P. F. Muhlradt, M. Morr, J. D. Radolf, A. Zychlinsky, K. Takeda and S. Akira, *Int Immunol*, 2001, **13**, 933–940.
41. F. Reutter, G. Jung, W. Baier, B. Treyer, W. G. Bessler and K.-H. Wiesmüller, *J Peptide Res*, 2005, **65**, 375–383.
42. C. Galanos, M. Gumenscheimer, P. Muhlradt, E. Jirillo and M. Freudenberg, *J Endotoxin Res*, 2000, **6**, 471–476.
43. M. S. Jin, S. E. Kim, J. Y. Heo, M. E. Lee, H. M. Kim, S.-G. Paik, H. Lee and J.-O. Lee, *Cell*, 2007, **130**, 1071–1082.
44. Q. Chen, T. S. Davidson, E. N. Huter and E. M. Shevach, *J Immunol*, 2009, **183**, 4458–4466.
45. K. Cheng, X. Wang, S. Zhang and H. Yin, *Angew Chem*, 2012, **51**, 12246–12249.
46. S. Kim and M. Karin, *Ann N Y Acad Sci*, 2011, **1217**, 191–206.
47. S. T. Ishizaka and L. D. Hawkins, *Expert Rev Vaccines*, 2007, **6**, 773–784.
48. C. Alexander and E. T. Rietschel, *J Endotoxin Res*, 2001, **7**, 167–202.
49. F. Peri and M. Piazza, *Biotechnol Adv*, 2012, **30**, 251–260.
50. D. A. Johnson, *Curr Top Med Chem*, 2008, **8**, 64–79.

51. J. R. Baldridge, C. W. Cluff, J. T. Evans, M. J. Lacy, J. R. Stephens, V. G. Brookshire, R. Wang, J. R. Ward, Y. M. Yorgensen, D. H. Persing and D. A. Johnson, *J Endotoxin Res*, 2002, **8**, 453–458.
52. A. G. Stover, *J Biol Chem*, 2003, **279**, 4440–4449.
53. H. M. Kim, B. S. Park, J.-I. Kim, S. E. Kim, J. Lee, S. C. Oh, P. Enkhbayar, N. Matsushima, H. Lee, O. J. Yoo and J.-O. Lee, *Cell*, 2007, **130**, 906–917.
54. P. Rallabhandi, A. Awomoyi, K. E. Thomas, A. Phalipon, Y. Fujimoto, K. Fukase, S. Kusumoto, N. Qureshi, M. B. Sztein and S. N. Vogel, *J Immunol*, 2008, **180**, 1139–1147.
55. D. P. Rossignol and M. Lynn, *J Endotoxin Res*, 2002, **8**, 483–488.
56. W. J. Christ, O. Asano, A. L. Robidoux, M. Perez, Y. Wang, G. R. Dubuc, W. E. Gavin, L. D. Hawkins, P. D. McGuinness and M. A. Mullarkey, *Science*, 1995, **268**, 80.
57. C. W. Cluff, J. R. Baldridge, A. G. Stöver, J. T. Evans, D. A. Johnson, M. J. Lacy, V. G. Clawson, V. M. Yorgensen, C. L. Johnson, M. T. Livesay, R. M. Hershberg and D. H. Persing, *Infect Immun*, 2005, **73**, 3044–3052.
58. L. D. Hawkins, S. T. Ishizaka, P. McGuinness, H. Zhang, W. Gavin, B. DeCosta, Z. Meng, H. Yang, M. Mullarkey, D. W. Young, D. P. Rossignol, A. Nault, J. Rose, M. Przetak, J. C. Chow and F. Gusovsky, *J Pharmacol Exp Ther*, 2002, **300**, 655–661.
59. M. Przetak, J. Chow, H. Cheng, J. Rose, L. D. Hawkins and S. T. Ishizaka, *Vaccine*, 2003, **21**, 961–970.
60. M. Przetak, J. Chow, H. Cheng, J. Rose, L. D. Hawkins and S. T. Ishizaka, *Vaccine*, 2003, **21**, 961–970.
61. M. Ortega-Munoz, J. Morales-Sanfrutos, F. Perez-Balderas, F. Hernandez-Mateo, M. Dolores Giron-Gonzalez, N. Sevillano-Tripero, R. Salto-Gonzalez and F. Santoyo-Gonzalez, *Org Biomol Chem*, 2007, **5**, 2291–2301.
62. M. Piazza, M. Colombo, I. Zanoni, F. Granucci, P. Tortora, J. Weiss, T. Gioannini, D. Prosperi and F. Peri, *Angew Chem Int Ed Engl*, 2011, **50**, 622–626.
63. A. Pantel, C. Cheong, D. Dandamudi, E. Shrestha, S. Mehandru, L. Brane, D. Ruane, A. Teixeira, L. Bozzacco, R. M. Steinman and M. P. Longhi, *Eur J Immunol*, 2011, **42**, 101–109.
64. M. Piazza, V. Calabrese, G. Damore, R. Cighetti, T. Gioannini, J. Weiss and F. Peri, *ChemMedChem*, 2012, **7**, 213–217.
65. R. N. Coler, S. Bertholet, M. Moutaftsi, J. A. Guderian, H. P. Windish, S. L. Baldwin, E. M. Laughlin, M. S. Duthie, C. B. Fox, D. Carter, M. Friede, T. S. Vedvick and S. G. Reed, *PLoS One*, 2011, **6**, e16333.
66. F. Heil, H. Hemmi, H. Hochrein, F. Ampenberger, C. Kirschning, S. Akira, G. Lipford, H. Wagner and S. Bauer, *Science*, 2004, **303**, 1526–1529.
67. J. F. Gerster, K. J. Lindstrom, R. L. Miller, M. A. Tomai, W. Birmachu, S. N. Bomersine, S. J. Gibson, L. M. Imbertson, J. R. Jacobson,

R. T. Knafla, P. V. Maye, N. Nikolaides, F. Y. Oneyemi, G. J. Parkhurst, S. E. Pecore, M. J. Reiter, L. S. Scribner, T. L. Testerman, N. J. Thompson, T. L. Wagner, C. E. Weeks, J. D. Andre, D. Lagain, Y. Bastard and M. Lupu, *J Med Chem*, 2005, **48**, 3481–3491.
68. N. J. Horscroft, D. C. Pryde and H. Bright, *J Antimicrob Chemother*, 2012.
69. H. Hemmi, T. Kaisho, O. Takeuchi, S. Sato, H. Sanjo, K. Hoshino, T. Horiuchi, H. Tomizawa, K. Takeda and S. Akira, *Nat Immunol*, 2002, **3**, 196–200.
70. M. Fogel, J. A. Long, P. J. Thompson and J. W. Upham, *J Leukoc Biol*, 2002, **72**, 932–938.
71. B. Drobits, M. Holcmann, N. Amberg, M. Swiecki, R. Grundtner, M. Hammer, M. Colonna and M. Sibilia, *J Clin Invest*, 2012, **122**, 575–585.
72. E. L. Smits, P. Ponsaerts, Z. N. Berneman and V. F. Van Tendeloo, *Oncologist*, 2008, **13**, 859–875.
73. T. Wei, J. Gong, F. Jamitzky, W. M. Heckl, R. W. Stark and S. C. Rossle, *Protein Sci*, 2009, **18**, 1684–1691.
74. R. G. Govindaraj, B. Manavalan, S. Basith and S. Choi, *PLoS One*, 2011, **6**, e25118.
75. L. Deml, R. Schirmbeck, J. Reimann, H. Wolf and R. Wagner, *Clin Chem Lab Med*, 1999, **37**, 199–204.
76. D. Yu, Q. Zhao, E. R. Kandimalla and S. Agrawal, *Bioorg Med Chem Lett*, 2000, **10**, 2585–2588.
77. D. Yu, E. R. Kandimalla, L. Bhagat, J. Y. Tang, Y. Cong, J. Tang and S. Agrawal, *Nucleic Acids Res*, 2002, **30**, 4460–4469.
78. E. R. Kandimalla, L. Bhagat, D. Wang, D. Yu, F. G. Zhu, J. Tang, H. Wang, P. Huang, R. Zhang and S. Agrawal, *Nucleic Acids Res*, 2003, **31**, 2393–2400.
79. E. R. Kandimalla, D. Yu, Q. Zhao and S. Agrawal, *Bioorg Med Chem*, 2001, **9**, 807–813.
80. D. Yu, E. R. Kandimalla, Y. Cong, J. Tang, J. Y. Tang, Q. Zhao and S. Agrawal, *J Med Chem*, 2002, **45**, 4540–4548.
81. L. Bhagat, F. G. Zhu, D. Yu, J. Tang, H. Wang, E. R. Kandimalla, R. Zhang and S. Agrawal, *Biochem Biophys Res Commun*, 2003, **300**, 853–861.
82. D. P. Sester, S. Naik, S. J. Beasley, D. A. Hume and K. J. Stacey, *J Immunol*, 2000, **165**, 4165–4173.
83. D. Wang, Y. Li, D. Yu, S. S. Song, E. R. Kandimalla and S. Agrawal, *Int J Oncol*, 2004, **24**, 901–908.
84. A. M. Krieg, *Nucleic Acid Ther*, 2012, **22**, 77–89.
85. S. R. Jacobs and B. Damania, *J Leukoc Biol*, 2012, **92**, 469–477.
86. M. Chamaillard, M. Hashimoto, Y. Horie, J. Masumoto, S. Qiu, L. Saab, Y. Ogura, A. Kawasaki, K. Fukase, S. Kusumoto, M. A. Valvano, S. J. Foster, T. W. Mak, G. Nunez and N. Inohara, *Nat Immunol*, 2003, **4**, 702–707.

87. N. Inohara, Y. Ogura, F. F. Chen, A. Muto and G. Nunez, *J Biol Chem*, 2001, **276**, 2551–2554.
88. M. Hasegawa, A. Kawasaki, K. Yang, Y. Fujimoto, J. Masumoto, E. Breukink, G. Nuñez, K. Fukase and N. Inohara, *J Biol Chem*, 2007, **282**, 11757–11764.
89. Y. Fujimoto, A. R. Pradipta, N. Inohara and K. Fukase, *Nat Prod Rep*, 2012, **29**, 568–579.
90. O. Gutierrez, C. Pipaon, N. Inohara, A. Fontalba, Y. Ogura, F. Prosper, G. Nunez and J. L. Fernandez-Luna, *J Biol Chem*, 2002, **277**, 41701–41705.
91. N. Inohara, Y. Ogura, A. Fontalba, O. Gutierrez, F. Pons, J. Crespo, K. Fukase, S. Inamura, S. Kusumoto, M. Hashimoto, S. J. Foster, A. P. Moran, J. L. Fernandez-Luna and G. Nunez, *J Biol Chem*, 2003, **278**, 5509–5512.
92. C. Ogawa, Y. J. Liu and K. S. Kobayashi, *Curr Bioact Compd*, 2011, **7**, 180–197.
93. L. Kager, U. Potschger and S. Bielack, *Ther Clin Risk Manage*, 2010, **6**, 279–286.
94. K. Geddes, J. G. Magalhaes and S. E. Girardin, *Nat Rev Drug Discov*, 2009, **8**, 465–479.
95. S.-D. Zhang, G. Liu, S.-Q. Xia, P. Wu and L. Zhang, *J Comb Chem*, 2002, **4**, 131–137.
96. H. Z. Yang, S. Xu, X. Y. Liao, S. D. Zhang, Z. L. Liang, B. H. Liu, J. Y. Bai, C. Jiang, J. Ding, G. F. Cheng and G. Liu, *J Med Chem*, 2005, **48**, 5112–5122.
97. H. Tada, S. Aiba, K. Shibata, T. Ohteki and H. Takada, *Infect Immun*, 2005, **73**, 7967–7976.
98. G. Trinchieri, *Nat Rev Immunol*, 2003, **3**, 133–146.
99. L. Klotz, S. Burgdorf, I. Dani, K. Saijo, J. Flossdorf, S. Hucke, J. Alferink, N. Novak, M. Beyer, G. Mayer, B. Langhans, T. Klockgether, A. Waisman, G. Eberl, J. Schultze, M. Famulok, W. Kolanus, C. Glass, C. Kurts and P. A. Knolle, *J Exp Med*, 2009, **206**, 2079–2089.
100. M. G. Belvisi, D. J. Hele and M. A. Birrell, *Eur J Pharmacol*, 2006, **533**, 101–109.
101. T. M. Willson, P. J. Brown, D. D. Sternbach and B. R. Henke, *J Med Chem*, 2000, **43**, 527–550.
102. Y. J. Surh, H. K. Na, J. M. Park, H. N. Lee, W. Kim, I. S. Yoon and D. D. Kim, *Biochem Pharmacol*, 2011, **82**, 1335–1351.
103. J. M. Lehmann, L. B. Moore, T. A. Smith-Oliver, W. O. Wilkison, T. M. Willson and S. A. Kliewer, *J Biol Chem*, 1995, **270**, 12953–12956.
104. R. T. Nolte, G. B. Wisely, S. Westin, J. E. Cobb, M. H. Lambert, R. Kurokawa, M. G. Rosenfeld, T. M. Willson, C. K. Glass and M. V. Milburn, *Nature*, 1998, **395**, 137–143.
105. J. Uppenberg, C. Svensson, M. Jaki, G. Bertilsson, L. Jendeberg and A. Berkenstam, *J Biol Chem*, 1998, **273**, 31108–31112.

106. D. J. Parks, N. C. Tomkinson, M. S. Villeneuve, S. G. Blanchard and T. M. Willson, *Bioorg Med Chem Lett*, 1998, **8**, 3657–3658.
107. B. R. Henke, S. G. Blanchard, M. F. Brackeen, K. K. Brown, J. E. Cobb, J. L. Collins, W. W. Harrington, Jr., M. A. Hashim, E. A. Hull-Ryde, I. Kaldor, S. A. Kliewer, D. H. Lake, L. M. Leesnitzer, J. M. Lehmann, J. M. Lenhard, L. A. Orband-Miller, J. F. Miller, R. A. Mook, Jr., S. A. Noble, W. Oliver, Jr., D. J. Parks, K. D. Plunket, J. R. Szewczyk and T. M. Willson, *J Med Chem*, 1998, **41**, 5020–5036.
108. B. R. Henke, K. K. Adkison, S. G. Blanchard, L. M. Leesnitzer, R. A. Mook, Jr., K. D. Plunket, J. A. Ray, C. Roberson, R. Unwalla and T. M. Willson, *Bioorg Med Chem Lett*, 1999, **9**, 3329–3334.
109. B. G. Shearer and A. N. Billin, *Biochim Biophys Acta*, 2007, **1771**, 1082–1093.
110. J. J. García-Vallejo and Y. van Kooyk, *Immunol Rev*, 2009, **230**, 22–37.
111. P. R. Taylor, S. Gordon and L. Martinez-Pomares, *Trends Immunol*, 2005, **26**, 104–110.
112. T. Geijtenbeek, S. Van Vliet, A. Engering, B. 'T. Hart and Y. Van Kooyk, *Annu Rev Immunol*, 2004, **22**, 33–54.
113. A. Engering, T. B. Geijtenbeek, S. J. van Vliet, M. Wijers, E. van Liempt, N. Demaurex, A. Lanzavecchia, J. Fransen, C. G. Figdor, V. Piguet and Y. van Kooyk, *J Immunol*, 2002, **168**, 2118–2126.
114. I. Mellman and R. M. Steinman, *Cell*, 2001, **106**, 255–258.
115. B. J. Appelmelk, I. van Die, S. J. van Vliet, C. M. Vandenbroucke-Grauls, T. B. Geijtenbeek and Y. van Kooyk, *J Immunol*, 2003, **170**, 1635–1639.
116. L. R. Prost, J. C. Grim, M. Tonelli and L. L. Kiessling, *ACS Chem Biol*, 2012, **7**, 1603–1608.
117. M. J. Borrok and L. Kiessling, *J Am Chem Soc*, 2007, **129**, 12780–12785.
118. S. L. Mangold, L. R. Prost and L. L. Kiessling, *Chem Sci*, 2012, **3**, 772.
119. O. Srinivas, P. Larrieu, E. Duverger, C. Boccaccio, M. T. Bousser, M. Monsigny, J. F. Fonteneau, F. Jotereau and A. C. Roche, *Bioconjugate Chem*, 2007, **18**, 1547–1554.
120. P. Romero, N. Gervois, J. Schneider, P. Escobar, D. Valmori, C. Pannetier, A. Steinle, T. Wolfel, D. Lienard, V. Brichard, A. van Pel, F. Jotereau and J. C. Cerottini, *J Immunol*, 1997, **159**, 2366–2374.
121. A. Whitty and C. Borysenko, *Chem Biol*, 1999, **6**, R107–R118.
122. A. Viola and A. D. Luster, *Annu Rev Pharmacol Toxicol*, 2008, **48**, 171–197.
123. T. J. Schall and A. E. Proudfoot, *Nat Rev Immunol*, 2011, **11**, 355–363.
124. P. C. Jensen, S. Thiele, T. Ulven, T. W. Schwartz and M. M. Rosenkilde, *J Biol Chem*, 2008, **283**, 23121–23128.
125. T. Natori, M. Morita, K. Akimoto and Y. Koezuka, *Tetrahedron*, 1994, **50**, 2771–2784.
126. E. L. Wise, C. Duchesnes, P. C. da Fonseca, R. A. Allen, T. J. Williams and J. E. Pease, *J Biol Chem*, 2007, **282**, 27935–27943.
127. I. Sabroe, A. Hartnell, L. A. Jopling, S. Bel, P. D. Ponath, J. E. Pease, P. D. Collins and T. J. Williams, *J Immunol*, 1999, **162**, 2946–2955.

128. M. Wijtmans, D. Verzijl, R. Leurs, I. J. de Esch and M. J. Smit, *ChemMedChem*, 2008, **3**, 861–872.
129. C. C. Yates, D. Whaley, P. Kulasekeran, W. W. Hancock, B. Lu, R. Bodnar, J. Newsome, P. A. Hebda and A. Wells, *Am J Pathol*, 2007, **171**, 484–495.
130. A. D. Luster and P. Leder, *J Exp Med*, 1993, **178**, 1057–1065.
131. P. J. Hensbergen, P. G. Wijnands, M. W. Schreurs, R. J. Scheper, R. Willemze and C. P. Tensen, *J Immunother*, 2005, **28**, 343–351.
132. I. L. Stroke, A. G. Cole, S. Simhadri, M.-R. Brescia, M. Desai, J. J. Zhang, J. R. Merritt, K. C. Appell, I. Henderson and M. L. Webb, *Biochem Biophys Res Commun*, 2006, **349**, 221–228.
133. B. Nedjai, H. Li, I. L. Stroke, E. L. Wise, M. L. Webb, J. R. Merritt, I. Henderson, A. E. Klon, A. G. Cole, R. Horuk, N. Vaidehi and J. E. Pease, *Br J Pharmacol*, 2012, **166**, 912–923.
134. L. M. Budel, I. P. Touw, R. Delwel and B. Lowenberg, *Blood*, 1989, **74**, 2668–2673.
135. Y. Hanazono, T. Hosoi, T. Kuwaki, S. Matsuki, K. Miyazono, K. Miyagawa and F. Takaku, *Exp Hematol*, 1990, **18**, 1097–1103.
136. I. A. Tabbara, C. D. Ghazal and H. H. Ghazal, *Cancer Invest*, 1997, **15**, 353–357.
137. S.-S. Tian, P. Lamb, A. G. King, S. G. Miller, L. Kessler, J. I. Luengo, L. Averill, R. K. Johnson, J. G. Gleason, L. M. Pelus, S. B. Dillon and J. Rosen, *Science*, 1998, **281**, 257–259.
138. S. G. Miller, *Drug Discov Today*, 2000, **5**, 77–83.
139. M. L. Doyle, S. S. Tian, S. G. Miller, L. Kessler, A. E. Baker, M. R. Brigham-Burke, S. B. Dillon, K. J. Duffy, R. M. Keenan, R. Lehr, J. Rosen, L. A. Schneeweis, J. Trill, P. R. Young, J. I. Luengo and P. Lamb, *J Biol Chem*, 2003, **278**, 9426–9434.
140. K. Kusano, S. Ebara, K. Tachibana, T. Nishimura, S. Sato, T. Kuwaki and T. Taniyama, *Blood*, 2004, **103**, 836–842.
141. A. Sahu and J. D. Lambris, *Immunopharmacology*, 2000, **49**, 133–148.
142. M. J. Walport, *N Engl J Med*, 2001, **344**, 1140–1144.
143. M. J. Walport, *N Engl J Med*, 2001, **344**, 1058–1066.
144. J. V. Sarma and P. A. Ward, *Cell Tissue Res*, 2011, **343**, 227–235.
145. D. Ricklin and J. D. Lambris, *Adv Exp Med Biol*, 2013, **734**, 1–22.
146. T. M. Woodruff, K. S. Nandakumar and F. Tedesco, *Mol Immunol*, 2011, **48**, 1631–1642.
147. G. Biesecker, L. Dihel, K. Enney and R. A. Bendele, *Immunopharmacology*, 1999, **42**, 219–230.
148. D. Ricklin and J. D. Lambris, *Nat Biotechnol*, 2007, **25**, 1265–1275.
149. A. Laarman, F. Milder, J. van Strijp and S. Rooijakkers, *J Mol Med (Berl)*, 2010, **88**, 115–120.
150. B. Postma, M. J. Poppelier, J. C. van Galen, E. R. Prossnitz, J. A. van Strijp, C. J. de Haas and K. P. van Kessel, *J Immunol*, 2004, **172**, 6994–7001.
151. A. Bunschoten, J. H. Ippel, J. A. Kruijtzer, L. Feitsma, C. J. de Haas, R. M. Liskamp and J. Kemmink, *Amino Acids*, 2011, **40**, 731–740.

152. H. Qu, D. Ricklin and J. D. Lambris, *Mol Immunol*, 2009, **47**, 185–195.
153. K. W. Mollison, W. Mandecki, E. R. Zuiderweg, L. Fayer, T. A. Fey, R. A. Krause, R. G. Conway, L. Miller, R. P. Edalji and M. A. Shallcross, *Proc Natl Acad Sci USA*, 1989, **86**, 292–296.
154. A. M. Finch, A. K. Wong, N. J. Paczkowski, S. K. Wadi, D. J. Craik, D. P. Fairlie and S. M. Taylor, *J Med Chem*, 1999, **42**, 1965–1974.
155. M. Kawai, D. A. Quincy, B. Lane, K. W. Mollison, J. R. Luly and G. W. Carter, *J Med Chem*, 1991, **34**, 2068–2071.
156. Z. D. Konteatis, S. J. Siciliano, G. Van Riper, C. J. Molineaux, S. Pandya, P. Fischer, H. Rosen, R. A. Mumford and M. S. Springer, *J Immunol*, 1994, **153**, 4200–4205.
157. P. N. Monk, A. M. Scola, P. Madala and D. P. Fairlie, *Br J Pharmacol*, 2007, **152**, 429–448.
158. M. Otto, H. Hawlisch, P. N. Monk, M. Muller, A. Klos, C. L. Karp and J. Kohl, *J Biol Chem*, 2004, **279**, 142–151.
159. D. Ricklin, G. Hajishengallis, K. Yang and J. D. Lambris, *Nat Immunol*, 2010, **11**, 785–797.
160. V. P. Ferreira, M. K. Pangburn and C. Cortes, *Mol Immunol*, 2010, **47**, 2187–2197.
161. Y. Q. Wu, H. Qu, G. Sfyroera, A. Tzekou, B. K. Kay, B. Nilsson, K. Nilsson Ekdahl, D. Ricklin and J. D. Lambris, *J Immunol*, 2011, **186**, 4269–4277.
162. P. H. Nilsson, K. N. Ekdahl, P. U. Magnusson, H. Qu, H. Iwata, D. Ricklin, J. Hong, J. D. Lambris, B. Nilsson and Y. Teramura, *Biomaterials*, 2013, **34**, 985–994.
163. J. M. Astle, L. S. Simpson, Y. Huang, M. M. Reddy, R. Wilson, S. Connell, J. Wilson and T. Kodadek, *Chem Biol*, 2010, **17**, 38–45.
164. B. K. Sarma and T. Kodadek, *ACS Comb Sci*, 2012, **14**, 558–564.
165. A. Aditya and T. Kodadek, *ACS Comb Sci*, 2012, **14**, 164–169.
166. C. Aquino, M. Sarkar, M. J. Chalmers, K. Mendes, T. Kodadek and G. C. Micalizio, *Nat Chem*, 2012, **4**, 99–104.
167. T. Kodadek, *Curr Opin Chem Biol*, 2010, available online, DOI: 10.1016/j.cbpa.2010.1007.1001, and references contained therein.
168. T. Kodadek, *Nat Chem*, 2009, **1**, 183–185.
169. L. S. Simpson, L. Burdine, A. K. Dutta, A. P. Feranchak and T. Kodadek, *J Am Chem Soc*, 2009, **131**, 5760–5762.
170. J. M. Astle, D. G. Udugamasooriya, J. E. Smallshaw and T. Kodadek, *Int J Pept Res Ther*, 2008, **14**, 223–227.
171. D. G. Udugamasooriya, S. P. Dineen, R. A. Brekken and T. Kodadek, *J Am Chem Soc*, 2008, **130**, 5744–5752.
172. X. Qi, J. Astle and T. Kodadek, *Mol Biosyst*, 2010, **6**, 102.
173. D. Cai, A.-Y. Lee, C.-M. Chiang and T. Kodadek, *Bioorg Med Chem Lett*, 2011, 1–5.
174. M. M. Reddy, R. Wilson, J. Wilson, S. Connell, A. Gocke, L. Hynan, D. German and T. Kodadek, *Cell*, 2011, **144**, 132–142.

175. A. R. Gocke, D. G. Udugamasooriya, C. T. Archer, J. Lee and T. Kodadek, *Chem Biol*, 2009, **16**, 1133–1139.
176. P. Alluri, M. Reddy, K. Bachhawat–Sikder, H. Olivos and T. Kodadek, *J Am Chem Soc*, 2003, **125**, 13995–14004.
177. K. S. Shores, D. G. Udugamasooriya, T. Kodadek and D. R. Knapp, *J Proteome Res*, 2008, **7**, 1922–1931.
178. H.-S. Lim, D. Cai, C. T. Archer and T. Kodadek, *J Am Chem Soc*, 2007, **129**, 12936–12937.
179. T. Kodadek, M. M. Reddy, H. J. Olivos, K. Bachhawat-Sikder and P. G. Alluri, *Acc Chem Res*, 2004, **37**, 711–718.
180. V. R. Doppalapudi, N. Tryder, L. Li, T. Aja, D. Griffith, F.-F. Liao, G. Roxas, M. P. Ramprasad, C. Bradshaw and C. F. Barbas Iii, *Bioorg Med Chem Lett*, 2007, **17**, 501–506.
181. J. Xiao, R. Chen, M. A. Pawlicki and T. J. Tolbert, *J Am Chem Soc*, 2009, **131**, 13616–13618.
182. C. Rader, S. C. Sinha, M. Popkov, R. A. Lerner and C. F. Barbas, *Proc Natl Acad Sci USA*, 2003, **100**, 5396–5400.
183. J. Gavrilyuk, U. Wuellner, S. Salahuddin, R. Goswami, S. Sinha and C. Barbas, *Bioorg Med Chem Lett*, 2009, **19**, 3716–3720.
184. P. J. McEnaney, C. G. Parker, A. X. Zhang and D. A. Spiegel, *ACS Chem Biol*, 2012, **7**, 1139–1151.
185. M. Popkov, B. Gonzalez, S. C. Sinha and C. F. Barbas, *Proc Natl Acad Sci*, 2009, **106**, 4378–4383.
186. Y. Lu and P. S. Low, *Cancer Immunol Immunother*, 2002, **51**, 153–162.
187. Y. Lu, E. Sega and S. Low Philip, *Int J Cancer*, 2005, **116**, 710–719.
188. Y. Lu, F. You, I. Vlahov, E. Westrick, M. Fan, P. S. Low and C. P. Leamon, *Mol Pharmaceutics*, 2007, **4**, 695–706.
189. E. H. Holmes, T. G. Greene, W. T. Tino, A. L. Boynton, H. C. Aldape, S. L. Misrock and G. P. Murphy, *The Prostate Suppl*, 1996, **7**, 25–29.
190. A. Dubrovska, C. Kim, J. Elliott, W. Shen, T.-H. Kuo, D.-I. Koo, C. Li, T. Tuntland, J. Chang, T. Groessl, X. Wu, V. Gorney, T. Ramirez Montagut, D. Spiegel, C. Cho and P. Schultz, *ACS Chem Biol*, 2011, **6**, 1223–1231.
191. R. P. Murelli, A. X. Zhang, J. Michel, W. L. Jorgensen and D. A. Spiegel, *J Am Chem Soc*, 2009, **131**, 17090–17092.
192. A. X. Zhang, R. P. Murelli, C. Barinka, J. Michel, A. Cocleaza, W. L. Jorgensen, J. Lubkowski and D. A. Spiegel, *J Am Chem Soc*, 2010, **132**, 12711–12716.
193. C. E. Jakobsche, P. J. McEnaney, A. X. Zhang and D. A. Spiegel, *ACS Chem Biol*, 2011.
194. P. R. Crocker, J. C. Paulson and A. Varki, *Nat Rev Immunol*, 2007, **7**, 255–266.
195. R. M. Owen, C. B. Carlson, J. Xu, P. Mowery, E. Fasella and L. L. Kiessling, *ChemBioChem*, 2007, **8**, 68–82.
196. C. B. Carlson, P. Mowery, R. M. Owen, E. C. Dykhuizen and L. L. Kiessling, *ACS Chem Biol*, 2007, **2**, 119–127.

197. C. R. Bertozzi and M. D. Bednarski, *J Am Chem Soc*, 1992, **114**, 2242–2245.
198. C. R. Bertozzi and M. D. Bednarski, *J Am Chem Soc*, 1992, **114**, 5543–5546.
199. J. Li, S. Zacharek, X. Chen, J. Wang, W. Zhang, A. Janczuk and P. G. Wang, *Bioorg Med Chem*, 1999, **7**, 1549–1558.
200. S. J. Metallo, R. S. Kane, R. E. Holmlin and G. M. Whitesides, *J Am Chem Soc*, 2003, **125**, 4534–4540.
201. V. M. Krishnamurthy, L. J. Quinton, L. A. Estroff, S. J. Metallo, J. M. Isaacs, J. P. Mizgerd and G. M. Whitesides, *Biomaterials*, 2006, **27**, 3663–3674.
202. K. P. Naicker, H. Li, A. Heredia, H. Song and L.-X. Wang, *Org Biomol Chem*, 2004, **2**, 660–664.
203. M. F. Perdomo, M. Levi, M. Saellberg and A. Vahlne, *Proc Natl Acad Sci USA*, 2008, **105**, 12515–12520.
204. C. G. Parker, R. A. Domaoal, K. S. Anderson and D. A. Spiegel, *J Am Chem Soc*, 2009, **131**, 16392–16394.
205. S. Fournel, S. Wieckowski, W. Sun, N. Trouche, H. Dumortier, A. Bianco, O. Chaloin, M. Habib, J. C. Peter, P. Schneider, B. Vray, R. E. Toes, R. Offringa, C. J. Melief, J. Hoebeke and G. Guichard, *Nat Chem Biol*, 2005, **1**, 377–382.
206. C. van Kooten and J. Banchereau, *J Leukoc Biol*, 2000, **67**, 2–17.
207. M. Karpusas, Y. M. Hsu, J. H. Wang, J. Thompson, S. Lederman, L. Chess and D. Thomas, *Structure*, 1995, **3**, 1426.
208. M. Habib, M. Noval Rivas, M. Chamekh, S. Wieckowski, W. Sun, A. Bianco, N. Trouche, O. Chaloin, H. Dumortier, M. Goldman, G. Guichard, S. Fournel and B. Vray, *J Immunol*, 2007, **178**, 6700–6704.
209. E. B. Puffer, J. K. Pontrello, J. J. Hollenbeck, J. A. Kink and L. L. Kiessling, *ACS Chem Biol*, 2007, **2**, 252–262.
210. M. C. Dalakas, *Pharmacol Ther*, 2006, **112**, 57–70.
211. M. J. Townsend, J. G. Monroe and A. C. Chan, *Immunol Rev*, 2010, **237**, 264–283.
212. D. C. Barral and M. B. Brenner, *Nat Rev Immunol*, 2007, **7**, 929–941.
213. V. Cerundolo, P. Barral and F. D. Batista, *Curr Opin Immunol*, 2010, **22**, 417–424.
214. T. Kawano, J. Cui, Y. Koezuka, I. Toura, Y. Kaneko, K. Motoki, H. Ueno, R. Nakagawa, H. Sato, E. Kondo, H. Koseki and M. Taniguchi, *Science*, 1997, **278**, 1626–1629.
215. M. Morita, T. Natori, K. Akimoto, T. Osawa, H. Fukushima and Y. Koezuka, *Bioorg Med Chem Lett*, 1995, **5**, 699–704.
216. M. Koch, V. S. Stronge, D. Shepherd, S. D. Gadola, B. Mathew, G. Ritter, A. R. Fersht, G. S. Besra, R. R. Schmidt and E. Y. Jones, *Nat Immunol*, 2005, **6**, 819–826.
217. J. Nadas, C. Li and P. G. Wang, *J Chem Inf Model*, 2009, **49**, 410–423.
218. T. Tashiro, *Bioscience, Biotechnol Biochem*, 2012, **76**, 1055–1067.

219. A. Bendelac, P. B. Savage and L. Teyton, *Annu Rev Immunol*, 2007, **25**, 297–336.
220. J. Wojno, J.-P. Jukes, H. Ghadbane, D. Shepherd, G. S. Besra, V. Cerundolo and L. R. Cox, *ACS Chem Biol*, 2012, **7**, 847–855.
221. H. J. Blumenfeld, R. Tohn, S. M. Haeryfar, Y. Liu, P. B. Savage and T. L. Delovitch, *Clin Exp Immunol*, 2011, **166**, 121–133.
222. X. Li, M. Fujio, M. Imamura, D. Wu, S. Vasan, C. H. Wong, D. D. Ho and M. Tsuji, *Proc Natl Acad Sci*, 2010, **107**, 13010–13015.
223. G. Yang, J. Schmieg, M. Tsuji and R. W. Franck, *Angew Chem Int Ed Engl*, 2004, **116**, 3906–3910.
224. K. H. Lin, J. J. Liang, W. I. Huang, S. Y. Lin-Chu, C. Y. Su, Y. L. Lee, J. T. Jan, Y. L. Lin, Y. S. E. Cheng and C. H. Wong, *Antimicrob Agents Chemother*, 2010, **54**, 4129–4136.
225. K. O. A. Yu, J. S. Im, A. Molano, Y. Dutronc, P. A. Illarionov, C. Forestier, N. Fujiwara, I. Arias, S. Miyake, T. Yamamura, Y.-T. Chang, G. S. Besra and S. A. Porcelli, *Proc Natl Acad Sci USA*, 2005, **102**, 3383–3388.
226. D. Ly, R. Tohn, B. Rubin, H. Blumenfeld, G. S. Besra, N. Veerapen, S. A. Porcelli and T. L. Delovitch, *Clin Exp Immunol*, 2010, **160**, 185–198.
227. J. Xie and P. G. Schultz, *Curr Opin Chem Biol*, 2005, **9**, 548–554.
228. E. Keinan, *Catalytic Antibodies*, John Wiley and Sons, New York, 2006.
229. J. Grünewald, M.-L. Tsao, R. Perera, L. Dong, F. Niessen, B. G. Wen, D. M. Kubitz, V. V. Smider, W. Ruf, M. Nasoff, R. A. Lerner and P. G. Schultz, *Proc Natl Acad Sci*, 2008, **105**, 11276–11280.
230. J. Grünewald, G. S. Hunt, L. Dong, F. Niessen, B. G. Wen, M.-L. Tsao, R. Perera, M. Kang, B. A. Laffitte, S. Azarian, W. Ruf, M. Nasoff, R. A. Lerner, P. G. Schultz and V. V. Smider, *Proc Natl Acad Sci*, 2009, **106**, 4337–4342.
231. V. Gauba, J. Grünewald, V. Gorney, L. M. Deaton, M. Kang, B. Bursulaya, W. Ou, R. A. Lerner, C. Schmedt, B. H. Geierstanger, P. G. Schultz and T. Ramirez-Montagut, *Proc Natl Acad Sci*, 2011, **108**, 12821–12826.
232. J. Mlynarski and B. Gut, *Chem Soc Rev*, 2012, **41**, 587–596.
233. G. J. L. Bernardes, L. Linderoth, K. J. Doores, O. Boutureira and B. G. Davis, *ChemBioChem*, 2011, **12**, 1383–1386.
234. P. Wang and S. J. Danishefsky, *J Am Chem Soc*, 2010, **132**, 17045–17051.
235. P. H. Seeberger, *Chem Soc Rev*, 2008, **37**, 19–28.
236. P. H. Seeberger and D. B. Werz, *Nature*, 2007, **446**, 1046–1051.
237. P. Wang, Y.-J. Kim, M. Navarro-Villalobos, B. D. Rohde and D. Y. Gin, *J Am Chem Soc*, 2005, **127**, 3256–3257.
238. K. Deng, M. M. Adams and D. Y. Gin, *J Am Chem Soc*, 2008, **130**, 5860.
239. N. Petrovsky and P. D. Cooper, *Expert review of vaccines*, 2011, **10**, 523–537.
240. M. M. Adams, P. Damani, N. R. Perl, A. Won, F. Hong, P. O. Livingston, G. Ragupathi and D. Y. Gin, *J Am Chem Soc*, 2010, **132**, 1939–1945.

241. B. Lepenies, J. Yin and P. H. Seeberger, *Curr Opin Chem Biol*, 2010, **14**, 404–411.
242. D. A. Tomalia and Y. Cheng, *Dendrimer-Based Drug Delivery Systems: From Theory to Practice*, Wiley, New York, 2012.
243. I. R. Vlahov, H. K. R. Santhapuram, F. You, Y. Wang, P. J. Kleindl, S. J. Hahn, J. F. Vaughn, D. S. Reno and C. P. Leamon, *J Orgc Chemy*, 2010, **75**, 3685–3691.
244. P.-H. Liang, C.-Y. Wu, W. A. Greenberg and C.-H. Wong, *Curr Opin Chem Biol*, 2008, **12**, 86–92.
245. H. Lönnberg, *Bioconjugate Chem*, 2009, **20**, 1065–1094.
246. M. A. Oberli, M. Tamborrini, Y.-H. Tsai, D. B. Werz, T. Horlacher, A. Adibekian, D. Gauss, H. M. Möller, G. Pluschke and P. H. Seeberger, *J Am Chem Soc*, 2010, **132**, 10239–10241.
247. J. Warren, X. Geng and S. Danishefsky, *Glycopeptides and Glycoproteins*, 2007, 109–141.
248. E. J. Grayson, G. J. L. Bernardes, J. M. Chalker, O. Boutureira, J. R. Koeppe and B. G. Davis, *Angew Chem Int Ed*, 2011, **50**, 4127–4132.
249. L. Morelli, L. Poletti and L. Lay, *Eur J Org Chem*, 2011, **2011**, 5723–5777.
250. M. C. Hewitt and P. H. Seeberger, *J Org Chem*, 2001, **66**, 4233–4243.
251. J. G. Cyster and S. R. Schwab, *Annu Rev Immunol*, 2012, **30**, 69–94.
252. N. Kawakami, I. Bartholomaus, M. Pesic and M. Mues, *Immunol Rev*, 2012, **248**, 140–155.
253. A. Vossenkamper, P. M. Lutalo and J. Spencer, *Clin Exp Immunol*, 2012, **167**, 7–14.
254. T. Fujita, K. Inoue, S. Yamamoto, T. Ikumoto, S. Sasaki, R. Toyama, K. Chiba, Y. Hoshino and T. Okumoto, *J Antibiot (Tokyo)*, 1994, **47**, 208–215.
255. M. Matloubian, C. G. Lo, G. Cinamon, M. J. Lesneski, Y. Xu, V. Brinkmann, M. L. Allende, R. L. Proia and J. G. Cyster, *Nature*, 2004, **427**, 355–360.
256. M. L. Allende, J. L. Dreier, S. Mandala and R. L. Proia, *J Biol Chem*, 2004, **279**, 15396–15401.
257. G. Cinamon, M. Matloubian, M. J. Lesneski, Y. Xu, C. Low, T. Lu, R. L. Proia and J. G. Cyster, *Nat Immunol*, 2004, **5**, 713–720.
258. K. Adachi and K. Chiba, *Perspect Med Chem*, 2008, **1**, 11–23.
259. M. Mehling, R. Lindberg, F. Raulf, J. Kuhle, C. Hess, L. Kappos and V. Brinkmann, *Neurology*, 2010, **75**, 403–410.
260. Z. J. Luo, T. Tanaka, F. Kimura and M. Miyasaka, *Immunopharmacology*, 1999, **41**, 199–207.
261. V. Brinkmann, *Br J Pharmacol*, 2009, **158**, 1173–1182.
262. L. Shan, Ø. Molberg, I. Parrot, F. Hausch, F. Filiz, G. M. Gray, L. M. Sollid and C. Khosla, *Science*, 2002, **297**, 2275–2279.
263. L. Shan, S.-W. Qiao, H. Arentz-Hansen, Ø. Molberg, G. M. Gray, L. M. Sollid and C. Khosla, *J Proteome Res*, 2005, **4**, 1732–1741.

264. J. Xia, M. Siegel, E. Bergseng, L. M. Sollid and C. Khosla, *J Am Chem Soc*, 2006, **128**, 1859–1867.
265. J. Xia, E. Bergseng, B. Fleckenstein, M. Siegel, C. Y. Kim, C. Khosla and L. M. Sollid, *Bioorg Med Chem*, 2007, **15**, 6565–6573.
266. C. N. Serhan, *Annu Rev Immunol*, 2007, **25**, 101–137.
267. M. J. Zhang and M. Spite, *Annu Rev Nutr*, 2012, **32**, 203–227.
268. L. Zhang, X. Zhang, P. Wu, H. Li, S. Jin, X. Zhou, Y. Li, D. Ye, B. Chen and J. Wan, *Inflamm Res*, 2008, **57**, 157–162.
269. A. J. Schottelius, C. Giesen, K. Asadullah, I. M. Fierro, S. P. Colgan, J. Bauman, W. Guilford, H. D. Perez and J. F. Parkinson, *J Immunol*, 2002, **169**, 7063–7070.
270. S. Fiorucci, J. L. Wallace, A. Mencarelli, E. Distrutti, G. Rizzo, S. Farneti, A. Morelli, J. L. Tseng, B. Suramanyam, W. J. Guilford and J. F. Parkinson, *Proc Natl Acad Sci USA*, 2004, **101**, 15736–15741.

CHAPTER 2

Immunopharmacotherapy for Nicotine Addiction

JONATHAN W. LOCKNER AND KIM D. JANDA*

Department of Chemistry, The Scripps Research Institute, 10550 North Torrey Pines Road, La Jolla, CA 92037, USA
*Email: kdjanda@scripps.edu

2.1 Introduction

Harnessing the immune system for defense against nicotine (Figure 2.1) is a research area that has received significant attention over the past 20 years. Nicotine plays a central role in precipitating addiction to smoking tobacco, a vice that possesses millions globally. In fact, according to the World Health Organization, there are nearly 1.3 billion smokers worldwide, and smoking is responsible for 5 million deaths annually.[1] The economic impact is also sobering: in the United States alone, smoking costs \$96 billion in medical expenses and \$97 billion in lost productivity each year.[2] The epidemiological link between chronic tobacco use and myriad diseases is well understood, and many smokers indeed wish to quit. However, currently available cessation aids do not help much. Pharmacotherapeutic interventions, whether agonists or antagonists, target brain receptors implicated in the reinforcement of nicotine dependence.[3] By acting centrally, these medications produce a variety of side effects.[4]

Alternatively, a nicotine vaccine stimulates the immune system to identify nicotine as an antigen, eliciting antibodies that alter nicotine pharmacokinetics, disrupting the positive reinforcement circuitry that attends each cigarette that is lit up. Anti-nicotine antibodies reduce the concentration of free nicotine in the blood and prevent it from entering the central nervous system (Figure 2.2).

RSC Drug Discovery Series No. 36
Biotherapeutics: Recent Developments using Chemical and Molecular Biology
Edited by Lyn H. Jones and Andrew J. McKnight

Published by the Royal Society of Chemistry, www.rsc.org

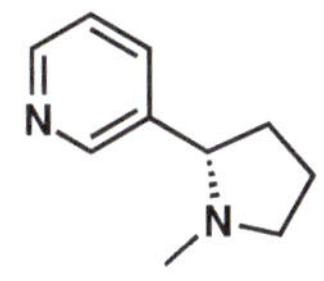

Figure 2.1 (*S*)-(−)-nicotine, the central mediator of tobacco addiction and a target for immunopharmacotherapy.

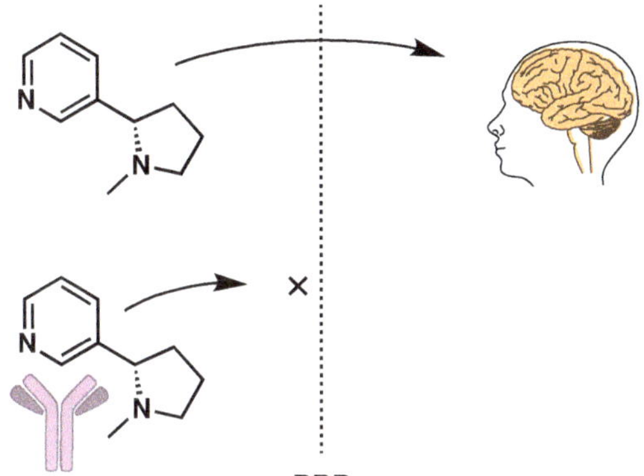

Figure 2.2 Normally, nicotine enters brain unimpeded, exerting psychoactive effects. Vaccination elicits antibodies that preclude drug entry into brain, blunting its reward capacity.
Adapted from Bachmann and Jennings with permission from the Royal Society.[5]

Blocking the activation of brain reward systems can facilitate extinction of the addictive behavior, leading to better smoking cessation outcomes. A clinically licensed nicotine vaccine would be a complementary addition to the armamentarium of tools, which, when leveraged appropriately, promises to afford significantly better rates of sustained smoking abstinence.

Of course, withdrawal symptoms may very well be exacerbated by the preclusion of nicotine from the brain, and adjunct therapies (counseling, antidepressants) are still needed to provide broadly successful outcomes.

2.2 Nicotine and the Body

Nicotine is the most widely used addictive drug in the world. It is a plant alkaloid found in tobacco leaves and it is legally available in cigarettes, cigars, pipe tobacco, and chewing tobacco. Nicotine is intimately linked with cigarette smoking, the leading preventable cause of death in the United States.[6,7] Nicotine itself is known to have numerous cardiovascular, endocrine, and metabolic effects, and this small molecule is a neuroteratogen.[8–10] Smoking contributes to coronary heart disease, stroke, vascular disease, peptic ulcers, chronic lung diseases, lung cancer, fetal brain damage, and morbidity. Despite the vast body of knowledge that links nicotine use with myriad diseases, people continue to smoke.[11]

2.3 Nicotine and the Brain

Many research reports underscore the notion that people continue to smoke because of the addictive effects of nicotine.[8,11,12] While nicotine itself imparts physiological effects on the central nervous system that reinforce its continued consumption, there is also the behavioral component of chronic tobacco use, in which the act of smoking another cigarette is brought on by particular social or emotional cues, or merely force of habit. While such behavioral reinforcement may be opposed through appropriate forms of behavioral therapy, the physical dependence on nicotine must be disrupted through pharmacotherapy, immunopharmacotherapy, or both.

Because nicotine is legally and widely available, there is relatively little stigma associated with its use (unlike cocaine, for instance). Although many chronic smokers have expressed a desire to stop smoking, and most who quit do so without treatment, less than 5% of unaided attempts result in sustained, long-term abstinence.[13] The high rate of relapse in smokers who try to quit emphasizes the strong effect of nicotine dependence.[14]

Two of the most popular smoking cessation therapies are nicotine gum and transdermal nicotine patches. As substitutes for cigarettes, these replacement products deliver small amounts of nicotine to the user over a period of time. Because nicotine is still being delivered to the physically dependent brain, the user is less inclined to reach for a cigarette. Meanwhile, counseling, changes in routines, and sheer willpower might be sufficient to eventually achieve prolonged abstinence. The amount of nicotine delivered through these products can be tapered down over time, thereby weaning the user off the drug. (Interestingly, active immunization could achieve the same effect, as a gradual increase in anti-nicotine antibodies would be joined by a reciprocal decrease in the rate and extent of nicotine entering the brain.)

Based on several placebo-controlled studies, nicotine chewing gum increased the rates of smoking abstinence 0–20%.[15] Nicotine patches have typically afforded long-term success rates of 10–30% with minimal interventive support.[16] Thus, nicotine replacement products are inadequate cessation aids, and the user is still exposed to nicotine and its adverse physiological effects.

Other pharmacotherapies, such as mecamylamine and clonidine, serotonergic agents such as buspirone, and antidepressants such as bupropion, have also been used.[12] Bupropion has perhaps been the most effective in aiding smoking cessation efforts, although the results are limited.[17] However, all of these pharmacotherapies exert their effects within the central nervous system (CNS), wherein numerous untoward side effects present a substantial counterweight to the potential therapeutic benefit.

It is important to appreciate the particular means by which nicotine, a plant alkaloid with a molecular weight of only 162 Dalton, exerts its effects. Within seconds of inhalation of tobacco smoke, nicotine begins to accumulate in the

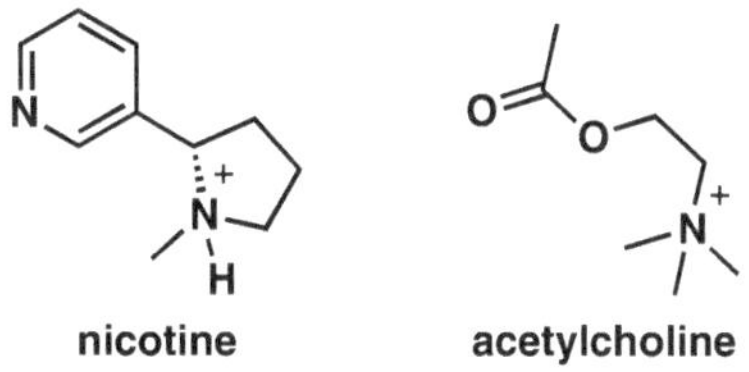

Figure 2.3 Nicotine and acetylcholine are similar pharmacophores.

brain.[18] Nicotine binds to nicotinic acetylcholine receptors (nAChRs), resulting in release of dopamine.[19] Acetylcholine is the endogenous ligand for nicotinic receptors, and it is worthwhile to note the physicochemical similarities between acetylcholine and nicotine (Figure 2.3). Several studies of nicotine–nAChR binding interactions verify that the nicotinic pharmacophore is analogous to acetylcholine: strong cation-pi interaction between nicotine's pyrrolidinium cation and a tryptophan residue,[20,21] hydrogen-bonding between nicotine's pyridine nitrogen through a water molecule to a protein backbone NH,[22] and presence of a hydrogen-bond network.[23] A recent review by Russo[24] emphasizes that genetic variability in nAChRs plays a substantial role in nicotine addiction and its treatment.

Nicotine binding to nAChRs induces the release of dopamine, which in turn induces pleasure, reduces stress and anxiety, improves mood, and improves concentration. Chronic exposure to nicotine results in adaptive changes, including upregulation of these receptors. Lack of exposure to nicotine (i.e. during a cessation attempt) leads to unpleasant withdrawal symptoms: depression, irritability, anxiety, and anhedonia.[25,26] For a more detailed account of the molecular and behavioral aspects of nicotine addiction, a recent review by Benowitz[27] is recommended.

Very commonly, withdrawal symptoms perpetuate continual use of nicotine products, so that the addiction becomes governed by a need to manage negative affect. The pathophysiology of nicotine dependence includes tolerance development, receptor upregulation, physiological dependence, and reinforcing effects, as well as the many other effects of nicotine on behavior and physiological functioning.[28]

Within the brain lies the mesolimbic dopaminergic system, the main target through which nicotine exerts its rewarding effects.[29,30] Nicotine activates nicotinic acetylcholine receptors on dopaminergic neurons located in the ventral tegmental area. These neurons project to several brain regions, including the nucleus accumbens.[31] Increased dopamine release in the nucleus accumbens,[32] following nAChR activation by nicotine,[33] reinforces the reward-seeking aspect of nicotine consumption. Ultimately, the clinical definition of drug addiction takes hold, in which the user is characterized by compulsive and relapsing dependence on nicotine.[34]

Currently available medications for nicotine addiction have provided limited help.[8] First-line therapies include nicotine replacement products, bupropion, and varenicline. Second-line therapies include clonidine and nortriptyline.

Beyond these, there is a litany of potential medications under development: MAOIs, SSRIs, nicotinic receptor antagonists, opioid receptor antagonists, bromocriptine, anxiolytics, CYP inhibitors, and cannabinoid-1 receptor antagonists.[35] In a multi-faceted approach, it is clear that pharmacotherapies targeting negative aspects of withdrawal, such as depression and anxiety, would be of major benefit to one who is motivated to abstain from smoking. However, in order to offer more complete protection against relapse, the positive aspects of nicotine consumption must also be mitigated. It has been appropriately stated that "further developments in pharmacotherapy for smoking cessation should focus on reducing the rewarding effects of smoking."[36] Existing therapeutics have had limited success in curbing nicotine dependence, and these drugs all possess the liability of acting within the brain, with untoward side effects.[37,38] Therefore, the alternative tactic of active immunization against nicotine holds great appeal.

2.4 Nicotine and the Immune System

Two recent reports by the US surgeon general have delineated the numerous health consequences of smoking.[39,40] Among these health consequences, several impinge on the immune system.[41,42] Work by Nouri-Shirazi *et al.* has demonstrated that chronic nicotine exposure can have lasting effects on the immune system. In a mouse model, nicDCs were generated, which exhibit Th2 commitment at the expense of Th1 protection. As a result, there is suppression of the immune system's response to vaccines attempting to use Th1 adjuvants for immunopotentiation.[43]

A depressed, compromised immune system brought on by chronic tobacco exposure means that the user is weakening his or her own protective shield against infection, let alone the numerous malignancies and cardiovascular and respiratory[44] diseases. Delayed wound healing[45] and increased susceptibility to infections (including the common cold, influenza, and tuberculosis) are associated with chronic tobacco exposure.[46] There is evidence that chronic exposure to nicotine can lead to metabolic disease, as the nicotine is metabolized to nornicotine, the latter of which can form covalent adducts with materials *in vivo*.[47]

Nicotine itself is non-immunogenic, but when attached to a carrier (e.g. protein) it can elicit an immune response,[48] leading to anti-nicotine antibody production. As will be emphasized in later sections of this chapter, wide variation in the magnitude and quality of antibody response has been observed in several clinical trials of nicotine vaccines. Variable responses to nicotine vaccines may be due to genetic variations in expression of human leukocyte antigen (HLA) complexes, cytokines, and T-cell receptors, ultimately defining the repertoire of available monocytes and lymphocytes that are pivotal to mounting an immune response during nicotine immunogen challenge. For instance, both HLA complex and immunoregulatory cytokine gene polymorphisms correlate with variable immune response to recombinant hepatitis B virus (HBV) vaccines.[49] Furthermore, individuals with an immunodeficiency brought on by chronic illness or infection may not respond optimally to an administered vaccine. Indeed, such conditions are often exclusion criteria utilized during selection of subjects for nicotine vaccine clinical trials. Additionally, it has been suggested that immunologic tolerance to nicotine via a T-cell independent response[50] might also account for lackluster response to a given nicotine vaccine.[51]

2.5 Nicotine and the Antibody

> *"Nicotine vaccines empty the traffic on the highway of pleasure because the antibodies binding to nicotine provide an efficient roadblock already outside the brain"*[52]

The notion of using an antibody to bind nicotine has its roots in the early 1970s. At the time, nicotine-specific antibodies were generated for the purpose of immunoassays (ELISA and RIA) then being developed for detection and quantification.[53,54] Following these two earliest examples of nicotine–protein immunoconjugates by Langone and Matsushita, numerous publications emerged. Researchers devised and optimized antibody-based methods for detecting nanogram quantities of nicotine and its metabolites in urine, blood, and smoke. However, not until the 1990s did it become clear that nicotine-specific antibodies could be used for therapeutic, not merely diagnostic, purposes.

Early on in the quest for antibody-based diagnostic assays for nicotine, Langone *et al.* found that when using 2- or 6-aminonicotine derivatives as haptens (Figure 2.4), the elicited nicotine antibodies exhibited significant cross-reaction with *N*-methylpyrrolidine, another constituent of tobacco smoke.[53] However, using similar haptens, others observed only limited binding of *N*-methylpyrrolidine.[55,56] Using 2- and 6-aminonicotine derivatives, Castro

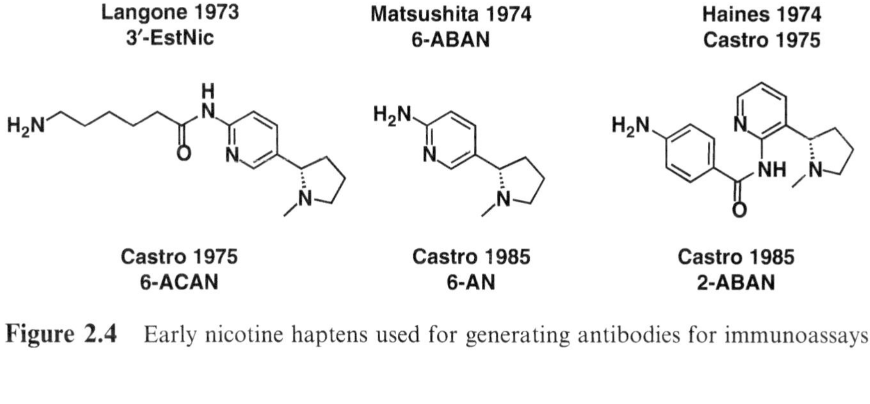

Figure 2.4 Early nicotine haptens used for generating antibodies for immunoassays.

et al. examined the effects of linker length and composition on affinity and specificity, and somewhat surprisingly observed no significant differences.[56] Matsushita *et al.* continued to evaluate pyridine-linked haptens, reasoning that unhindered exposure of the pyrrolidine ring of a nicotine hapten would lead to antibodies with superior specificity.[54] Meanwhile, Langone *et al.* preferred a different hapten design, in which linkage to carrier protein is made through the pyrrolidine ring of nicotine.[53,57] Using the **3′-EstNic** hapten (perhaps the most widely recognized, studied, and accepted hapten for anti-nicotine antibody production), they prepared anti-nicotine antisera and monoclonal antibodies.[57,58] Affinities on the order of $K_d \sim 10$ nM were reported for various mAbs, although other workers using the same hapten isolated mAbs with only micromolar K_d values for nicotine and increased cross-reaction for several nicotine metabolites and tobacco components.[59]

As early as 1972, the proof-of-principle of immunizing an animal against a drug substance had been demonstrated. However, it seems that this fledgling field of immunopharmacotherapy lay dormant for another 10–20 years while attention was diverted to treating drug abuse through small-molecule agonists and antagonists. Possibly, this dormant period was caused by the conclusion of Bonese in his 1974 report of heroin self-administration in a rhesus monkey that had been vaccinated with a morphine-BSA conjugate: "This blockade has been shown to be dose dependent, and it can be overcome by high doses of the drug."[60] Who knows where this field would be if research had carried on, unmitigated, during the 1980s?

The first report of a *therapeutic* use of an immunoconjugate – composed of a nicotine hapten and a protein carrier – for the stimulation of an anti-nicotine antibody response came in 1997.[61] Inspired by work against morphine,[62] heroin,[60] and cocaine,[63] Hieda leveraged the first immunopharmacotherapeutic attack on nicotine.[61] They used a **6-CMUNic**-KLH conjugate, and observed a modest anti-nicotine antibody response in rats. Their serum nicotine-specific antibody binding capacity was determined to be on the order of 210 ng/mL in rats. In the fifteen years since this seminal publication, significant gains have been achieved, owing to fervent participation by a large number of researchers, clinicians, and funding agencies. Now, the best performing formulations elicit serum nicotine binding capacities on the order of 280 μg/mL in rats,[64] three orders of magnitude improvement. Even so, further optimization is required. Several research groups have been pursuing active immunization strategies for nicotine dependence in recent years, with varying degrees of success. It did not take long for researchers to identify hapten designs (Figure 2.5) that permitted the generation of antibodies with high selectivity (low cross-reactivity to acetylcholine, cotinine, etc.) and high affinity (e.g. nanomolar K_d values). However, the Achilles heel remains the lack of uniformly high immunogenicity in preclinical studies (***Section 2.6***) and in clinical trials (***Section 2.7***).

Anti-nicotine antibodies extend the half-life of serum nicotine (not cleared as rapidly), impacting nicotine pharmacokinetics in rats. These antibodies may also alter nicotine distribution in various tissues other than the brain and

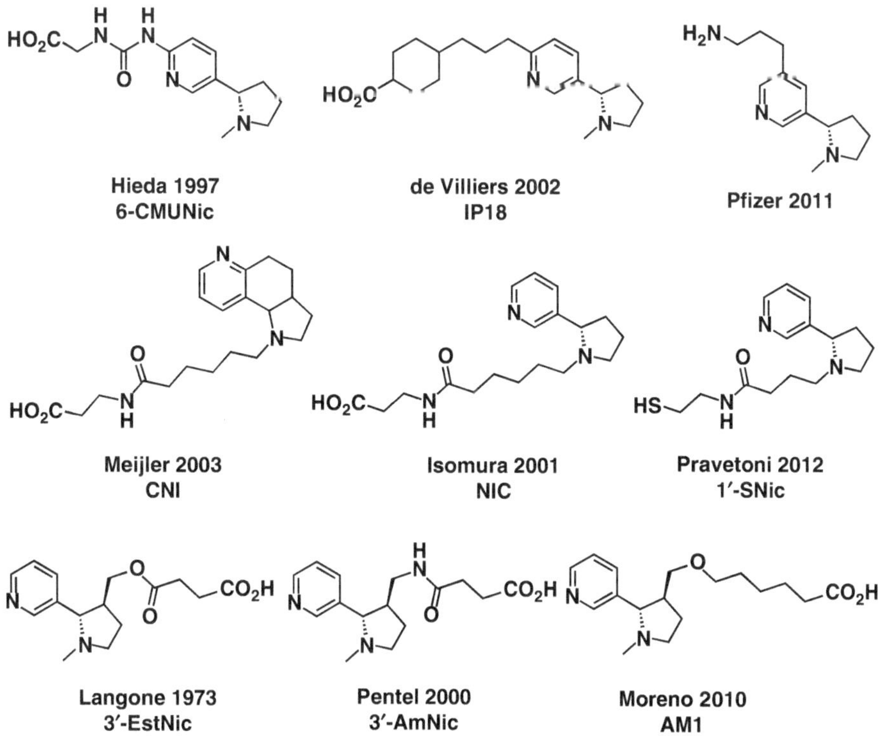

Figure 2.5 Nicotine haptens used for immunopharmacotherapy development.

serum.[65] For instance, there is a significant amount of extravascular nicotine-specific IgG in fat. Thus, it could be that vascular patrolling IgG could bind nicotine in the bloodstream, eventually handing it off to extravascular IgG, such as in fat tissue. This phenomenon may help explain how a 33-fold excess of nicotine (relative to serum nicotine-specific IgG) is nevertheless thwarted from exerting its psychoactive effects in a rat model, further refuting the misconception that efficacy for immunopharmacotherapy requires that there be an equivalent molar ratio between the antibody binding capacity and the drug dose.[66] In fact, this is not the case: substoichiometric antibody doses (see previous, for example) effectively antagonize these drugs, disproving the notion that antibody capacity is easily surmounted.[67]

2.6 Preclinical Studies of Nicotine Immunopharmacotherapy

Many of the earliest studies on nicotine immunopharmacotherapy utilized polyclonal antibodies from rabbits or goats. In a few cases, murine monoclonal antibodies were prepared. These efforts were important in that they established proof-of-principle that anti-nicotine antibodies could alter nicotine pharmacokinetics. The following pages trace much of the history of preclinical nicotine vaccine research, with emphasis on active immunization (Table 2.1).

2.6.1 Janda

In 2001, the Janda laboratory reported the use of **NIC** hapten in mice.[68] Even in this, our foray into anti-nicotine immunization, we emphasized the importance of respecting the stereochemistry of the (*S*)-nicotine molecule. Preservation of stereochemical integrity would elicit an antibody response better able to specifically bind and sequester natural (*S*)-nicotine from tobacco smoke; a minor portion is (*R*). To that end, we designed a synthetic route for our inaugural nicotine hapten (**NIC**), whereby a classical resolution would permit access to pure (*S*)-**NIC**. Another aspect explored in this study was the use of a less immunogenic linker, because it was shown by Hieda that the urea-based linker in their **6-CMUNic** hapten elicited anti-linker antibodies. As such, the linker used in **NIC** is characterized by a lack of immunogenic heteroatoms (oxygens, nitrogens) near the nicotine "epitope." Immunization of mice with **NIC**-KLH yielded several mAbs with good affinity ($K_d \sim 0.2\,\mu M$) and selectivity for (*S*)-(−)-nicotine. No cross-reactivity with cotinine, nornicotine, or *N*-methylpyrrolidine was detected at 100- to 1000-fold molar excess of these compounds using competition ELISA.

Further study of this **NIC**-KLH went in multiple directions. Active immunization using **NIC**-KLH led to ~45% decrease in locomotor activity. Like others in this field, we have also investigated passive immunization using nicotine-specific monoclonal antibodies. We found that NIC9D9, a mAb derived from **NIC**-KLH challenge, was effective (~67% decrease) in abrogating the psychoactive effects of nicotine in a rat locomotor assay.[69]

Table 2.1 Impact of vaccination against nicotine on pharmacokinetics, brain function, and behavior. Adapted from Raupach, Hoogsteder and Onno van Schayck.[36]

Parameter	*Effect*
Serum [nicotine] following acute nicotine administration	↑
Serum [nicotine] following chronic nicotine administration	↑
Brain [nicotine] following acute nicotine administration	↓
Brain [nicotine] following chronic nicotine administration	↓
Serum nicotine half-life	↑
Dopamine release in NAcc following acute nicotine administration	×
Locomotor hyperactivity following acute nicotine administration	↓
Acquisition of nicotine self-administration	↓
Maintenance of nicotine self-administration	↓
Nicotine intake in self-administration setting	↑
Reinstatement of nicotine self-administration following extinction	×
Withdrawal signs in presence of nicotine	~
Withdrawal sign in absence of nicotine	↓
Reduction of withdrawal signs following acute nicotine administration	↓

Another design consideration that came into play was the appreciation of conformational restriction as a useful means for limiting the heterogeneity of epitopic presentation during antibody production.[70] Building in conformational rigidity is a common tactic in structure-based drug design (SBDD) of small molecule ligands for binding sites in medicinal chemistry campaigns. In the present context, our lab designed nicotine haptens in which molecular constraint prohibits free rotation about the sigma bond joining the pyridine and *N*-methylpyrrolidine rings. From this campaign emerged **CNI**, a constrained nicotine hapten[71,72] that mimics one of the energetically most stable conformations of nicotine at physiological pH.

Immunization of mice with **NIC**-KLH provided relatively low serum anti-nicotine antibody titers (~1 : 3200). Serum affinity for (*S*)-(−)-nicotine was $K_{d\text{-}avg}$ ~1.7 μM. On the other hand, immunization with **CNI**-KLH resulted in antisera with greatly increased titers (~1 : 25 000) and improved affinity ($K_{d\text{-}avg}$ ~0.60 μM). Also, **CNI**-KLH antisera showed >10 : 1 specificity for (*S*)-(−)-nicotine versus the major metabolite (*S*)-(−)-cotinine, similar to **NIC**-KLH antisera. Further studies using both **NIC** and **CNI** in self-administration models have been reported.[73]

In another instance of hapten modification, we sought to improve upon existing 3′-linker nicotine haptens (i.e. Langone's **3′-EstNic** and Pentel's **3′-AmNic**). We reasoned that a 3′-ether linker would be less immunogenic and more hydrolytically stable than an ester or amide. This hapten, **AM1**, was examined in mice and in rats, with notable observation of high affinity (K_d 5.68 nM) and impact on nicotine IVSA in rats.[74,75] Additionally, the **AM1** hapten has also been featured in adenovirus-based vaccines by the Crystal group.[76–79]

2.6.2 Pentel

In 1997, the Pentel group described the use of urea derivative **6-CMUNic** coupled to KLH. Rats were immunized with **6-CMUNic**-KLH, and although serum antibodies bound nicotine with good affinity, and plasma nicotine concentrations were elevated in immunized rats compared to non-immunized rats upon nicotine administration, no significant alteration in brain nicotine concentration was observed.[61] However, these measurements of brain nicotine levels were taken 40 minutes after administration of nicotine. In follow-up studies,[80,81] significant differences in brain nicotine levels were detected when measurements were taken between 30 seconds and 6 minutes after injection. Using a more suitable time-point for brain nicotine measurement, it was found that rats immunized with **6-CMUNic**-KLH showed 28–48% reduction in brain nicotine concentration compared to rats immunized with only KLH.

Subsequent to these early reports on **6-CMUNic**-KLH, the Pentel group focused much of their attention on a different immunoconjugate, **3′-AmNic**-rEPA. In this case, nicotine is linked through the pyrrolidine ring to recombinant *Pseudomonas aeruginosa* exoprotein A (rEPA). Rats were immunized with **3′-AmNic**-rEPA or treated with polyclonal antibodies obtained from rabbits immunized with the same immunoconjugate.[82–84] Active immunization generated a 64% reduction in brain nicotine levels upon injection of nicotine.

In a separate study, the effects of continuous nicotine administration during 11 weeks of active immunization were examined.[85] At the end of this period, serum anti-nicotine antibody titers reached values of over 1 : 10 000 in all animals. A single dose of labeled nicotine was administered, and vaccinated rats showed 40–60% lower brain nicotine levels than control rats. Continuous exposure to nicotine, at concentrations comparable to those found in heavy smokers, does not saturate the capacity of anti-nicotine antibodies to bind nicotine. Further studies confirmed these results and also showed that vaccinated rats were less prone to seizures induced by high doses of nicotine.[86]

Thus, in the first several years of effort by the Pentel group, significant strides were made towards producing a viable candidate to move into clinical trials. By this point, K_d values for anti-nicotine antibodies were 20–40 nM, and serum anti-nicotine antibody concentrations in rats were 50–200 μg/mL. Assuming that total serum IgG is 10 mg/mL, this represents 0.5–2% of total IgG.[87] Considering that the typical serum nicotine concentration in a smoker is 15–50 ng/mL,[88] and taking into account the relative size of nicotine (162 Dalton) versus IgG (~150 000 Dalton), one may reasonably anticipate that serum anti-nicotine antibody concentrations on the order of 50–200 μg/mL would be clinically sufficient to abrogate the drug's psychoactive effects.

Because active immunization takes some time to reach sufficient titer levels, combined active and passive strategies have been evaluated. For instance, an anti-nicotine mAb Nic311 has been used in conjunction with the immunoconjugate **3′-AmNic**-rEPA.[89,90]

A major focus of the Pentel group in the wake of the NicVAX failure is the evaluation of various types of combination vaccines, the components of which develop distinct B-cell populations that synergistically afford greater protection. They have evaluated the potential usefulness of "bivalent" (**6′-CMUNic**-KLH and **3′-AmNic**-rEPA)[91] and "trivalent" (**6′-CMUNic**-BSA and **3′-AmNic**-rEPA and **1′-SNic**-KLH)[92] preparations. Each immunogen elicited antibodies that did not appreciably cross-react with the other immunogens, presumably because each immunogen recruited separate populations of B-cells. The cumulative impact of the individual components led to overall improved vaccine immunogenicity.

Mecamylamine is a very effective nAChR antagonist, but causes many peripheral side effects. Therefore, it was reasoned that the combination of mAb Nic311 and mecamylamine would work synergistically to attenuate the impact of nicotine.[93] Low doses of mecamylamine that were ineffective alone completely blocked nicotine discrimination when combined with Nic311. The benefit of nAChR antagonism was retained while peripheral toxicity was largely eliminated. Thus, immunotherapy can be augmented with a nAChR antagonist.

2.6.3 Sanderson

In 2003, a departure from the normal hapten–protein conjugate approach for nicotine vaccine generation was reported.[94] Instead of a protein, a mere

19-residue peptide, a MUC1-C5a dual agonist, served as the "carrier" for a nicotine hapten. The hapten in this case is **3′-AmNic**, the YKQGGFLGL fragment is a B-cell epitope, and the YSFKPMPLaR fragment interacts with C5a receptors on antigen presenting cells. It had been previously shown that when YKQGGFLGL is attached specifically to the N-terminus of YSFKPMPLaR, it is effectively delivered to C5aR-bearing APCs, processed, and presented, invoking the T-cell help required for epitope-specific antibody production.[95] That this 19-mer elicits a Th1-type (humoral) immune response was verified by antibody isotype analysis (IgG2a, IgG2b). Adding a nicotine hapten to the N-terminus of YKQGGFLGL constitutes a trivial, yet highly effective, modification of this B-cell epitope.

This peptide-based nicotine vaccine, NicYKQGGFLGLYSFKPMPLaR, given in weekly 200 µg doses for five weeks in the absence of any additional adjuvant, was able to abrogate the impact of nicotine in a rat behavioral assay. Immunized rats were significantly less sensitive to high concentrations (0.4 mg/kg) of nicotine than their non-vaccinated counterparts. Serum anti-nicotine antibody titers were rather low (1 : 223 in vaccinated animals), but these measurements were performed after completion of behavioral studies, one month after the final boost and after extensive exposure to nicotine.

It should be appreciated that such a peptide-based vaccine formulation as exemplified by Sanderson would not require refrigeration, as is the case with many of the nicotine hapten–protein conjugates (rEPA, rCTB, etc.) explored by others. Furthermore, many of the immunoconjugate vaccines described earlier were formulated with an adjuvant to enhance antibody titers. The direction taken here by Sanderson reflects the careful design of a self-adjuvanting peptide carrier that elicits an antibody response while not triggering an inflammatory response by the immune system, a feature that could be beneficial for a vaccine expected to receive broad human use.

2.6.4 Bunce/Bond/Aker

There does not appear to be much in the way of *preclinical* information for TA-NIC. The company behind TA-NIC, Xenova/Celtic, had perhaps (at least initially) been placing a majority of its emphasis on TA-CD, its cocaine vaccine clinical candidate.[96] In any case, there is some indication that its performance mirrored that of other nicotine vaccine candidates that have undergone clinical evaluation thus far. In a 2004 column[97] describing clinical trials to watch, it was stated that the phase I trial of TA-NIC resulted in 43% of participants quitting smoking or at least indicating that smoking was no longer as pleasurable. So, while early indicators may have generated some excitement, further clinical evaluation produced disappointing results, and development of this product has been abandoned. Additional details can be found in ***Section 2.7.2***.

2.6.5 Maurer/Bachmann

There does not appear to be much in the way of readily accessible *preclinical* information for NicQb/NIC002. It may be that preclinical findings were never shared via peer-reviewed journals. Additionally, in this case, the immunogenic platform (virus-like particles, VLPs) used in NicQb/NIC002 had already been the subject of intense research for a number of years, with applications to targets other than drugs of abuse. Quite possibly, the researchers behind NicQb/NIC002 seized the opportunity to combine their preclinical findings of VLP efficacy in other therapeutic contexts with others' preclinical findings on nicotine vaccines. Further details can be found in ***Section 2.7.1***.

2.6.6 Cerny

In 1990, a Swiss patent was issued to E. H. Cerny.[98] It is claimed by Cerny that this was the very first instance of anti-drug vaccination, including vaccination against nicotine. Cerny and coworkers[99] reported the preparation of a nicotine vaccine based on the pyrrolidine-linked hapten (**3′-EstNic**) that was used by the groups of Langone and Pentel. This hapten was conjugated to the carrier protein cholera toxin B, and mice were immunized with the resulting immunoconjugate. Both intranasal and subcutaneous administration routes were evaluated. After one month, serum anti-nicotine antibody titers were in the range of 1 : 1000 to 1 : 2000 – relatively low by today's standards. Moderate affinity ($K_{d\text{-avg}} \sim 60\,\mu M$) for nicotine was reported, and brain nicotine levels (following injection of nicotine, 600 ng) were 90% lower in vaccinated mice compared to their non-vaccinated counterparts. With this murine study, Cerny demonstrated that intranasal immunizations elicit significant anti-nicotine IgA antibodies in serum and saliva. This is a useful observation because nicotine enters the body through the lungs, and the superficial lining of the lungs contains significant amounts of IgA antibodies.

2.6.7 Svensson/de Villiers

Reported in 2002, de Villiers *et al.* immunized male Wistar rats with an immunoconjugate (**IP18**-KLH) that contains a hapten similar to the one originally used by the group of Pentel (**6-CMUNic**), in which a linker is attached to the pyridine moiety of nicotine. Using *in vivo* voltammetry, the authors examined the effect of nicotine on dopamine release in the shell of the nucleus accumbens, a region of the brain heavily implicated in the neurobiology of drug addiction. It was observed that immunized rats showed no significant dopamine release, whereas control animals showed dopamine release that was dose-dependent on administered nicotine.[100] In other studies, the nicotine-seeking behavior among immunized rats was examined. Rats were trained to self-administer nicotine by pressing a lever for intravenous injection.[101] They were then extinguished from this nicotine seeking behavior, and subsequently administered small amounts of nicotine that are known to induce nicotine seeking. Rats that had developed high titers of serum anti-nicotine antibodies (>1 : 10 000) did not respond to this reinforcing primer, whereas control animals showed reinstatement of nicotine self-administration. Recently, an additional report by this group describes their evaluation of a series of nicotine immunogens in rats, with serological analyses of antibody affinity and selectivity for nicotine, as well as further use of *in vivo* voltammetry to measure nicotine-induced dopamine release. One of these immunogens, **IP31**-KLH, is similar to **IP18**-KLH but bears non-racemic (*S*)-(−)-nicotine haptens. Immunization with **IP31**-KLH elicited antibodies with superior selectivity for (*S*)-(−)-nicotine, compared to all other immunogens tested.[102]

2.6.8 Crystal

In collaboration with the Janda laboratory, the Crystal group has been exploring the use of adenovirus, both in the protein carrier sense, and as a vector for gene expression. While this latter tactic is outside the scope of the present chapter (but see ***Section 2.8.6***), its very existence bears witness to the fact that researchers are leveraging cutting edge tactics to tackle the problem of nicotine addiction. As for the conventional hapten–protein immunoconjugate tactic, nicotine hapten **AM1** has been conjugated to adenovirus capsid protein (serotype 5 hexon), which is known to be highly immunogenic.[79]

With all of the preclinical studies, it appears that serum nicotine-specific antibody concentrations of 100–200 μg/mL are effective in altering nicotine pharmacokinetics and behavioral effects in rats. However, for humans, it is not yet known what threshold will be broadly efficacious. The maximum antibody concentration measured during a phase II clinical trial of NicVAX was 45 μg/mL.[103] However, serum hapten-specific antibody concentrations of up to 100 μg/mL have been reported in humans.[104,105]

2.7 Clinical Trials of Nicotine Vaccines

To date, there have been at least six different nicotine vaccines in the clinic (Table 2.2). As per usual, clinical trials of nicotine vaccine candidates first seek to evaluate the safety and immunogenicity of the vaccines, then more advanced trials evaluate the impact of the vaccines on smoking cessation and/or relapse prevention, compared to placebo.

Table 2.2 Nicotine vaccines in clinical development.

Vaccine	*Development*	*Carrier*	*Adjuvant*
Chilka Ltd	Phase II (halted)	Recomb. cholera toxin B (rCTB)	Alum
TA-NIC (Xenova/Celtic)	Phase II (halted)	Recomb. cholera toxin B (rCTB)	Alum
Niccine (Ind. Pharmaceutica)	Phase II (halted)	Tetanus toxoid (TT)	Alum
NicQb/NIC002 (Cytos/Novartis)	Phase II (halted)	Virus-like particle (VLP) of bacteriophage Qb	Alum
NicVAX (Nabi/GSK)	Phase III (halted)	*P aeruginosa* recomb. exoprotein A (rEPA)	Alum
SEL-068 (Selecta Biosciences)	Phase I (on-going)	Synthetic vaccine particle (SVP)	Resiquimod, Th epitope

In sum, early evaluations of the vaccines verified their safety, tolerability, and immunogenicity. Unfortunately, in advanced trials involving larger cohorts and having primary endpoints of sustained abstinence for a period of months post-immunization, the vaccines imparted little or no advantage over placebo. However, in many cases, if one performs subgroup analyses of the clinical outcomes, a solid correlation between high anti-nicotine antibody titers and greater likelihood of sustained abstinence is observed. Studies thus far have focused on smoking cessation, the ability to achieve sustained abstinence during the designated trial period. The closely linked aspect of relapse prevention, though not studied as much, would seek to evaluate the level of protection afforded by a vaccine, such that a lapse (i.e. one cigarette) would not give way to full-blown relapse. A recent Cochrane review of several clinical trials has been published,[106] with the following summary statement: *"Despite encouraging preclinical data, none of the four included studies detected a statistically significant difference in long-term cessation"*

2.7.1 Cytos Biotechnology AG

NIC002 (formerly NicQb) is comprised of a nicotine hapten attached to a virus-like particle (VLP), which is 25–30 nm diameter, formulated with alum adjuvant, and administered IM. The VLP is a uniform assembly of bacteriophage Qb coat protein expressed in *Escherichia coli*. NicQb/NIC002 has an excellent safety profile, modest immunogenicity, and there is a strong correlation between antibody concentration and smoking cessation. In a phase I study,[107] NicQb (100 μg) was administered with saline or with alum, and it was found that alum increased titers 1.5-fold and average serum nicotine-specific antibody concentration was ~10 μg/mL. In a subsequent phase II study, smokers were given NicQb (100 μg) with alum, in a regimen of five injections rather than only two. Antibody titers increased with each boost. There was a small but statistically insignificant difference in outcomes between vaccine placebo groups. However, it was observed in several studies that long-term abstinence among trial participants correlated with antibody titer. In this phase II trial of NIC002, study participants were categorized as low, medium, or high responders based on their nicotine antibody levels. The high responders had higher abstinence rates at 6- and 12-month time points.[108] No evidence of compensatory smoking was observed among clinical trial participants. In another study, it was found that a higher dose of NIC002 (300 μg instead of 100 μg) afforded four-fold higher antibody levels. Thus, by merely adjusting the dose, significant gains were observed.

2.7.2 ImmuLogic/Cantab/Xenova/Celtic Pharma

TA-NIC consists of a nicotine hapten, recombinant cholera toxin B (rCTB), formulated with alum, and administered IM. A phase I study (2001–2002) in 60 subjects verified the vaccine's immunogenicity and tolerability; 10 μg and 50 μg doses, given four or five times at 2-week intervals, produced anti-nicotine antibody titers that persisted for up to 12 weeks.[109] A subsequent phase I/II study (2003–2004) found that the 50 µg and 250 µg doses were well tolerated, but the 1000 µg dose had an unacceptable side effect profile. Study participants who received the 250 µg dose had a 25% abstinence rate compared with 9% in the placebo group.

2.7.3 Nabi Biopharmaceuticals/Biota Pharmaceuticals

NicVAX consists of a nicotine hapten, recombinant exoprotein A (rEPA) from *P aeruginosa*, formulated with alum, and administered IM. Phase I clinical trials were initiated in 2002 by Nabi Biopharmaceuticals to evaluate the safety and immunogenicity of NicVAX developed by the group of Pentel. As exciting results funneled in during the early evaluations of NicVAX, GlaxoSmithKline signed a $540 million deal with Nabi. At least five phase I/II clinical trials have been conducted, confirming that NicVAX is well tolerated and highly immunogenic.[103,110,111]

An early phase I trial (2002) of NicVAX showed that a single dose of this vaccine elicited a rapid immune response in non-smokers, and the vaccine was well tolerated. A subsequent phase I/II trial (2003) validated two additional factors: (a) the vaccine generates substantial antibodies even in smokers, and (b) multiple injections are well tolerated. Another phase II study (2003) examined the effect of NicVAX in smokers who expressed a desire to quit. Promising results emerged from this study, further fueling interest in this particular nicotine vaccine candidate.

Indeed, as was observed with NIC002 and TA-NIC, a dose-dependent increase in antibody concentration was observed. And again, a correlation between antibody concentration and cessation rate was observed.[103] For NicVAX, refinements in the dose (to 400 µg) and in the frequency (to six injections) led to gains in immunogenicity and continued to reaffirm the correlation between antibody concentration and cessation rate.

Ultimately, the desired end point of phase III trials was not met. There was no statistical difference between NicVAX and placebo groups in abstinence rates at 8–12 months. Others have pointed out that a potential pitfall of the NicVAX phase III study designs was that "treatment efficacy is assessed several months after the active agent responsible for the therapeutic effect (namely, high antibody levels to nicotine) has subsided."[51]

2.7.4 Chilka Limited

This nicotine vaccine consists of a nicotine hapten, recombinant cholera toxin B (rCTB), formulated with alum, and administered IM.[99] Cerny has a few papers that vaguely describe their immunogen, and he has even provided a timeline for clinical evaluation of their nicotine vaccine, with regulatory approval repeatedly being pushed back. There is some indication that it went into phases I/II. Interestingly, Cerny has recently partnered with Burkhard (at University of Connecticut) to explore new tactics for nicotine vaccine development. They have been awarded a \$2.5 million, five-year grant from the National Institutes of Health, presumably to apply Burkhard's self-assembling peptide nanoparticle (SAPN) platform to the nicotine vaccine problem.

2.7.5 Independent Pharmaceutica AB

Niccine consists of the nicotine hapten **IP18** (or non-racemic **IP31**) and carrier protein TT in formulation with alum. During a recent phase II clinical trial focusing on relapse prevention, participants received a regimen of six doses (of vaccine or placebo) over seven months, and mean anti-nicotine antibody levels (in the vaccine group) reached 1.34 μg/mL.[112] (This is much lower than the 45 μg/mL maximum antibody concentration observed in a NicVAX phase II study.[103]) According to the authors of the Niccine clinical trial report, "It is estimated that 1 μg/mL of nicotine antibodies bind 2 ng/mL of nicotine." A smoker will typically achieve a steady-state concentration upwards of 15–30 ng/mL[113] of nicotine during his/her daily routine, and it is therefore not all that surprising that Niccine failed to produce better outcomes than placebo. This vaccine is not being developed any further.

2.7.6 Selecta Biosciences

SEL-068 consists of a nicotine hapten incorporated into proprietary synthetic vaccine particles (SVPs), and is in phase I clinical evaluation. Based on publically available patent applications with Selecta Biosciences as the assignee, the constituents of SEL-068 are presumably the following: a polylactide-block polyethylene glycol (PLA-PEG) co-polymer, Resiquimod (a small-molecule agonist of Toll-like receptors 7 and 8), an MHC class II binding peptide (ILMQYIKANSKFIGIPMGLPQSIALSSLMVAQ) to elicit T-cell help, and, of course, a nicotine hapten, attachment likely at position 1′-, 3′-, or 4′- on the pyrrolidine ring, or 2-, or 6- on the pyridine ring. PLA is the core, and PEG is the shell; the latter is exploited to attenuate/modulate the uptake and persistence of the nanoparticles. Subcutaneous injections were administered on days 0, 14, and 28.

2.7.7 Pfizer Vaccines Ltd

At the time of this writing, there is some indication that Pfizer is actively evaluating a nicotine vaccine in the clinic. This is substantiated by the existence of at least one active clinical trial to evaluate **NIC7** immunoconjugates: a phase I trial to evaluate the safety of four ascending dose levels. There is also a 2012 patent application (with Pfizer Vaccines Ltd as assignee) that describes nicotine haptens, prominently featuring attachment at positions 2-, 5-, and 6- on the pyridine ring. At a recent conference of the Society for Research on Nicotine and Tobacco, preclinical results of **NIC7-DT** (diphtheria toxoid bearing nicotine haptens) formulated with alum (Alhydrogel-85) and CpG 24555 were disclosed.[114] Intramuscular injections were administered on days 0, 28, and 42. They identified several haptens (including **NIC7**, presumably) that afford greater reduction in brain nicotine in vaccinated mice, compared to the hapten used in NicVAX.[115] Also, it was asserted that this vaccine is more immunogenic than a NicQb mimetic. What should not go unnoticed from these recent entries (SEL-068 and **NIC7**-DT) is the careful combination of adjuvants aimed at inducing a fulminant immune response.

The overall picture of nicotine vaccine research, though fraught with clinical disappointments, is one in which a successful nicotine vaccine is nevertheless close at hand. To be sure, there remains significant room for improvement. Among the first wave of clinical candidates, some featured sub-optimal nicotine haptens, and most featured rather conventional (protein-based) carriers and/or adjuvants (alum). At present, efficacy is only good for ~30% of trial participants, but improvements in immunogenicity and concomitant increases in serum nicotine binding capacity should result in better nicotine vaccines. So far, serum anti-nicotine antibody concentrations as high as 45 μg/mL[103] have been achieved, but it is known that higher antigen-specific antibody concentrations can be generated. However, is there an upper limit on the amount of antibodies (and mature B-cells that produce them) we can, so to speak, request the immune system to allocate for defense against nicotine? If we do reach an asymptote with regard to nicotine-specific antibody concentration that can be achieved, may we enhance other components to make a more lasting, enduring (i.e. memory B-cells) response?

2.8 New Directions to Improve Immunogenicity

"The remaining challenge is to induce sufficient levels of antibody with sufficient quality in the majority of smokers"[5]

On the one hand, proof-of-principle has been repeatedly borne out from clinical trials conducted thus far: active immunization against nicotine can improve smoking cessation outcomes. On the other hand, it is clear that in order to make a nicotine vaccine work well for the masses, the vaccine must be highly immunogenic. Antibody concentration is critical. It takes 2–3 months for active immunization to produce sufficient titers of antibodies, and efforts to shorten this induction period would be beneficial. The first generation (NicVAX, NIC002, TA-NIC, Niccine, and Chilka Ltd's candidate) of nicotine vaccines tested so far in the clinic all utilized alum, but a veritable plethora of adjuvants now exist.[116] Knowledge of the immune system and of vaccine engineering[117] must be leveraged to generate follow-up candidates capable of stimulating fulminant yet focused T- and B-cell activation, leading to persistent immunological memory characterized by a long-lasting presence of anti-nicotine antibodies in circulation. In the following sections, we describe several emerging developments, some of which are already in use for nicotine vaccine research, and some of which may indeed play a role in the coming years.

2.8.1 Carrier

In addition to those protein carriers already examined in clinical trials, *E coli* outer membrane protein C (OmpC) and keyhole limpet hemocyanin (KLH) have been proposed as potentially superior alternatives. OmpC is already featured in a flu vaccine,[118] and KLH is present in vaccines for hypertension as well as several cancers. However, much attention is currently being devoted to carriers that are more rationally and purposefully designed than the conventional protein carrier manifold.

The next wave of nicotine vaccine clinical candidates will likely be characterized by the exploitation of various nanoparticle carrier designs, rather than the relatively smaller protein carriers (e.g. rEPA, rCTB). It is worth noting that the VLPs featured in NicQb/NIC002 represent a potent type of nanoparticle delivery system. Given their size (diameters of 25–30 nm) and structural (epitopic) homogeneity, they ably induced fulminant B- and T-cell responses, with induction of primarily IgG2a antibodies. Like viruses, many nanoparticle designs are often highly symmetric and highly repetitive.[117,119] The latest incarnation to reach the clinic is SEL-068, which utilizes Selecta's proprietary Synthetic Vaccine Particles (SVPs), represents a shift from a biological to fully synthetic carrier.

In addition to VLPs and SVPs, there exists self-assembling polypeptide nanoparticles (SAPNs), nanofibers, cubosomes, liposomes, polymeric beads, PLGA, PEG, gold nanoparticles, and α-Gal liposomes, to name a few (Table 2.3). Size definitely matters for governing immune tolerance versus immunogenicity. Antigen presenting cells cross-present antigen 1000-fold more efficiently if it is particulate rather than soluble.[120] Particulate size and surface properties can be adjusted to modulate kinetics of antigen delivery to APCs in draining lymph nodes. Particles under 20 nm in size can arrive at the lymph node within hours, draining directly through the lymphatics, whereas particles above 200 nm in size arrive within days, and must be transported by cells.[121,122] This can dictate whether an antigen will be tolerated or an immune reaction will be generated. There is an evolutionary shift from empirical study to rational design[123,124] of therapeutic vaccines, including within the nicotine vaccine field. More and more, conventional 'dirty' carrier proteins are being jettisoned in favor of novel, carefully engineered, self-adjuvanting, particulate carriers.

As mentioned earlier (***Section 2.7.4***), there is evidence that Burkhard at University of Connecticut is teaming up with Cerny to evaluate novel nicotine vaccine formulations.[125] Burkhard's research focuses on SAPNs, while Cerny (recall Chilka Ltd) has had a long history of interest in vaccines against drugs of abuse, including nicotine. In 2011, the Burkhard–Cerny team was awarded $2.5 million over five years, presumably to leverage SAPN technology for improving nicotine vaccine efficacy.

2.8.2 Delivery Method

Might we gain inroads through unconventional delivery mechanisms? Novel delivery methods could be preferable to intramuscular injections, which require a licensed clinician to administer. Also, patient adherence may be improved by replacing painful intramuscular injections with subcutaneous injections, especially if as many as six injections are required. Alternatively, it may be advantageous to engage the mucosal arm of the immune system, since it is through the mouth, throat, and lungs that nicotine travels prior to entering the bloodstream. One potential limitation may be that IgA found in mucosa tends not to be as high-affinity as IgG found in the bloodstream. Newer routes of vaccine administration include inhalation,[126] intranasal,[99,127] intradermal,[128] and transcutaneous[129] – some of which have already been evaluated in the context of drug of abuse vaccine research.

The particular route of administration can also dramatically influence the vaccination schedule, possibly minimizing the number of doses required to achieve functional titers over several months. Additionally, the quantity of immunogen per dose may be moderated. In nicotine vaccine clinical trials, relevant doses have ranged from 100 to 400 μg per injection, and in at least one study, 1000 μg was not well tolerated.

2.8.3 Adjuvants

> *"Recent advances in the development of biologics, including gene therapy, bioengineer[ing], and nanotechnology methods are providing a wealth of opportunities to discover and develop new biologics that are more potent and specific and with longer duration of action against drugs of abuse"*[130]

Adjuvant development continues to flourish. Alum, MF59, and MPL are in approved vaccines.[131] Also, some of GlaxoSmithKline's Adjuvant Systems (AS01, AS03) are approved for use in human vaccines. Adjuvant research is being conducted at an accelerating pace, and many vaccines, including those for nicotine, stand to benefit from the tidal wave of new insight and technologies on this front.[116] Furthermore, as the mechanisms of immune response are being elucidated at the molecular level, we look forward to very soon being able to customize immunogenicity, stability, and distribution of an injected vaccine.[132]

Recruitment of specific cell surface receptors or natural antibodies may help improve anti-nicotine immunogenicity. Mannosylated vaccine constructs allow for targeting payload via interaction with lectin binding elements on the surface of dendritic cells (DCs). By decorating nanoparticulate surfaces with sugars, such as α-Gal or rhamnose, natural antibodies for these glycotopes can be readily co-opted for the purpose of more efficient antigen processing by APCs. Instead of co-opting natural antibodies, one can also decorate a vaccine construct with monoclonal antibodies (mAbs) that are specific for surface molecules, such as Clec9A, on DCs. Tactics such as these allow for targeting DCs, leading to enhanced antigen uptake and activation of T- and B-cells.[150]

In sum, a carrier that is immunologically conspicuous, combined with adjuvants that potently activate specific signal transduction pathways, will increasingly become the theme of future vaccine formulations. Greater understanding of the mechanisms by which synthetic particles enhance

Table 2.3 New generation vaccines will be increasingly characterized by the incorporation of carefully designed particulate immunogens.

Platform	*Reference*	*Particle diameter*	*Application*
VLPs	Bachmann[117]	~25–100 nm	Nicotine
SAPNs	Kaba[133]	~25 nm	Malaria
Nanofibers	Rudra[134]	~15 nm (width)	Ovalbumin
SVPs	Pittet[135]; Gu[136]	~150–250 nm	Nicotine
Cubosomes	Rizwan[137]	~260–350 nm	Ovalbumin
Liposomes	Alving[138]; Fujii[139]	~50–500 nm	Malaria, HIV
ICMVs	Moon[140,141]	~200 nm	Malaria
α-Gal liposomes	Abdel-Motal[142]; Galili[143]	0.5–10 μm	HIV, influenza
Ab-coated liposomes	Mangat[144]	~1–10 μm	
Lipid-coated PLGA particles	Bershteyn[145]	~100–1000 nm	
Gold nanoparticles	Cruz[146]; Moyano[147]	~10–20 nm	
Viral vectors	Crystal[78,148]; Tatsis[149]	N/A	Nicotine

immunity will be critical to maximize the potential of nanoparticle vaccine strategies.[141]

2.8.4 Hapten Design

"Further studies are in progress in our laboratory to produce specific L-nicotine compounds that can be used as antigens to produce anti-nicotine antibodies"[151]

This statement was made, more than 35 years ago, during pursuit of higher sensitivity in nicotine immunoassays. It is well known that antibodies can exhibit exquisite substrate selectivity.[58,152–154] Bjercke found that stereospecific monoclonal antibodies to nicotine could be generated. In the case of their anti-nicotine mAb 402C10, only 0.25 ng of (*S*)-(−)-nicotine produced 50% inhibition, whereas 6.4 ng of (*R*)-(+)-nicotine were needed to achieve the same effect. Put another way, a single molecule of natural (*S*)-(−)-nicotine had the inhibitory potency of ~26 molecules of (*R*)-(+)-nicotine. (Elsewhere,[154] it was found that the relative binding affinities of (−)-cocaine and (+)-cocaine for mAb 2E2 were 1.00 and $3.4 \pm 1.3 \times 10^2$) One way to galvanize this point experimentally would be to intentionally make an (*R*)-(+)-nicotine-specific mAb, then see if its ability to bind (*S*)-(−)-nicotine is inferior to its ability to bind (*R*)-(+)-nicotine. This has important implications for nicotine haptens that are stereoisomeric mixtures, as similar "training" of polyclonal sera may end up wasting antibody. Thus, it behooves us to design haptens with respect for stereochemical integrity, as this feature is wrapped up in the larger context of epitopic homogeneity.

Through a comprehensive summary of linker attachment, it appears that nearly every position on nicotine has been exploited (Figure 2.6). Among the three exceptions, linkage via either the 3-position of the pyridine ring or the

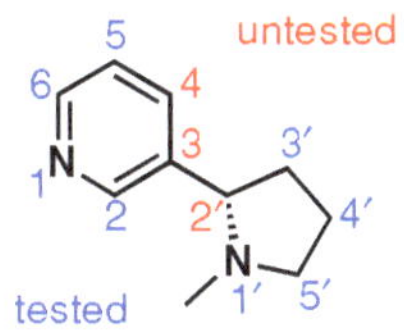

Figure 2.6 Nicotine hapten designs: linkages tested (blue) and untested (red).

2′-position of the pyrrolidine ring would seem to be impossible and/or offer no benefit over existing hapten designs. However, the remaining exception, linkage via the 4-position on the pyridine ring, might furnish as-yet untested hapten designs with desirable attributes (preclusion of stereoisomeric mixtures, and minimal obstruction of heteratom determinants during clonal selection).

In any case, linker position should be distant from the predominant sites of nicotine metabolism in order to allow the immune system (B-cells) to elicit antibodies that recognize nicotine in preference to its metabolites. Furthermore, linker composition should preferably be stable and non-immunogenic *per se*. Conjugation chemistry (i.e. the means by which the haptens are attached to a protein or other type of immunogenic carrier) should preferably be reliable, robust, and consistently reproducible.

2.8.5 Caveat: No Nicotine Vaccine Should Be Viewed as a "Magic Bullet" Cure

In reading the literature, rather than a balanced view of substance abuse as both nature (genetics, brain disease) and nurture (peer pressure, family history, lack of personal willpower, and restraint) there seems to be evidence of polarization to one or the other extreme. Both extremes are detrimental. Those who view substance abuse as a "moral failure rather than a brain disease" tend to minimize the potential benefit afforded by pharmacotherapeutic and immunotherapeutic interventions. At the other end of the spectrum is the camp of individuals who regard addiction as purely biological. We caution against a myopic view of a nicotine vaccine as a "magic bullet" cure and instead urge all of us to keep in mind that winning the war on tobacco addiction is more than just winning the battle against a 162 Dalton small molecule.[36,155] Such "reductionist" thinking erroneously oversimplifies the problem, inviting absolution as decreased personal responsibility (e.g. exercising self-discipline, seeking counseling) and increased resignation that "it's in my genes" undermine one's efforts to kick the habit.

It must be remembered that a nicotine vaccine will never provide complete blockade of the drug. The vaccine is not a "cure-all." While it can (and should) be reasonably expected to increase the odds a motivated treatment seeker will not relapse, it cannot *per se* guarantee abstinence (or non-relapse) as the outcome. Vaccination might help stave off full-blown relapse brought about by

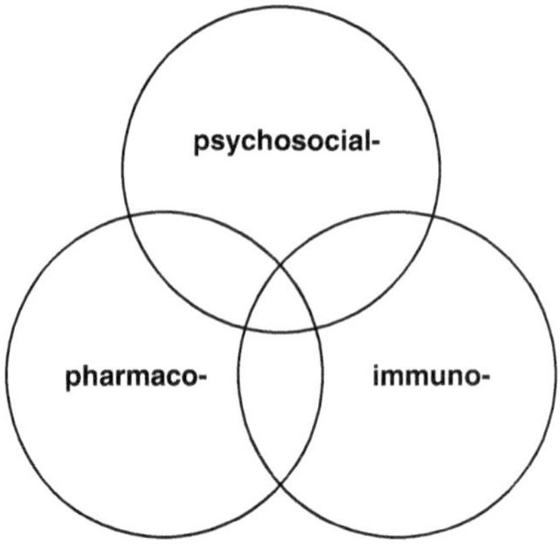

Figure 2.7 Treatment outcomes for smoking cessation are enhanced when all therapeutic arms are brought to bear on the condition.

a moment of weakness (e.g. lapse of a single cigarette), but it does not eliminate other triggers to relapse (e.g. anxiety, stress, or environmental cues). The fact that there has thus far been a disconnect between robust success in animal models and only modest success in human clinical trials could reasonably be attributed to "an incomplete understanding of the complex physiological and behavioural processes contributing to tobacco addiction."[36] To be sure, further research is needed to unravel these complexities, and it is important to view each incremental stride forward as a means toward the overarching aim of facilitating tobacco use reduction. More sophisticated and nuanced understanding of individual differences, whether gender,[156,157] genetic,[158,159] hormonal,[160] environmental, or otherwise, will continue to propel the fight against smoking, through individualized treatments.[161]

Chronic tobacco use is a multi-faceted problem requiring a multi-faceted solution.[36] For tobacco abuse reduction programs to be broadly successful, a combination of psychosocial-, pharmaco-, and immunopharmacotherapies will be needed (Figure 2.7).[130] As one case in point, a recent phase II study[162] evaluated the combined impact of varenicline and vaccination on smoking cessation. Although the study design omitted any deliberate counseling component, it is easy enough to conceive of a follow-up study in which all three components are indeed being leveraged.

2.8.6 Miscellany

The half-life of nicotine is two hours, and nicotine is primarily metabolized in the liver by CYP2A6 to cotinine.[163] In principle, a selective CYP2A6 inhibitor could delay metabolism of nicotine and artificially increase the half-life of nicotine, which would assuage the need to smoke another cigarette. (This tactic for pharmacokinetic enhancement[164] has been exploited for extending the half-life of viral protease inhibitors via co-administration of ritonavir,[165] an inhibitor of CYP3A4.) Through the combination of a nicotine vaccine with a

CYP2A6 inhibitor[166] (wherein the latter extends the serum half-life of parent molecule nicotine), anti-nicotine antibodies (elicited by the nicotine vaccine) would have more opportunity to exert their pharmacokinetic antagonism of nicotine's CNS effects.

Alternatively, cotinine-specific antibodies would prohibit cotinine from entering the CNS, meanwhile permitting nicotine (in a replacement product such as gum or patch) to more effectively mitigate withdrawal symptoms during abstinence from smoking. Applying this rationale, GlaxoSmithKein (GSK) has evaluated a cotinine vaccine[167] to reduce receptor antagonism by cotinine[168,169] in the CNS. By ameliorating the impact of cotinine, a nicotine replacement product might have greater efficacy as a smoking cessation aid.

Adenoviral vectors offer a means by which the genetic instructions for protein production can be inserted into host cells. Research by the Crystal group has focused on soliciting the host cells to express the antigen-binding fragment (Fab) of a high affinity anti-nicotine monoclonal antibody (NIC9D9) *in vivo*, representing an entirely different means for arming the host immune system with capacity to bind and sequester nicotine molecules in the bloodstream.[78]

2.9 Conclusions and Future Prospects

> *"The development and clinical implementation of drug vaccination is expensive; however, with an expenditure greater than US$467.7 billion seen in the United States in 2005 alone, the cost of drug abuse to federal, state, and local governments is staggering"*[132]

> *"We believe that much more programmatic and parametric research is needed on vaccine design, formulation, and immunization schedules, as well as their effect on nicotine-specific antibody production and any accompanying changes in the pharmacokinetic and pharmacodynamics effects of nicotine"*[170]

Over the last twenty years, tremendous progress has been made toward the goal of a broadly efficacious nicotine vaccine. The appeal of this type of smoking cessation aid resides in its unique mode of action. Whereas all currently used medications for the treatment of dependence disorders consist of drugs that act within the CNS, antibodies act by binding the drug of abuse in the bloodstream before it can reach the brain. Because these antibodies operate outside the CNS, there are minimal side effects. A large body of data has accumulated, which substantiates the premise that active immunization against nicotine works not only in animal models, but also in humans.

Proof-of-principle has clearly been established, but significant improvement in vaccine immunogenicity through tailored delivery vehicles and adjuvant(s) will surely characterize the next wave of nicotine vaccine candidates in the clinic. With continued robust research in the areas of novel carriers and adjuvants, we are optimistic about the prospects for nicotine immunopharmacotherapy in the 21st century.

Acknowledgements

The authors wish to acknowledge financial support from the Tobacco-Related Disease Research Program (20XT-0156).

References

1. K. M. Esson and S. R. Leeder, *The Millennium Development Goals and Tobacco Control: An Opportunity for Global Partnership*, World Health Organization, Geneva, Switzerland, 2004.
2. *MMWR Morb Mortal Wkly Rep*, 2008, **57**, 1226–1228.
3. H. J. Aubin, L. Karila and M. Reynaud, *Curr Pharm Des*, 2011, **17**, 1343–1350.
4. J. T. Hays and J. O. Ebbert, *Drugs*, 2010, **70**, 2357–2372.
5. M. F. Bachmann and G. T. Jennings, *Philos Trans R Soc, B*, 2011, **366**, 2815–2822.
6. D. E. Nelson, R. S. Kirkendall, R. L. Lawton, J. H. Chrismon, R. K. Merritt, D. A. Arday and G. A. Giovino, *MMWR CDC Surveillance Summaries*, 1994, **43**, 1–8.
7. U.S. Dept. of Health and Human Services, Public Health Service, Office of the Surgeon General. *The Surgeon General's Report on Reducing the Health Consequences of Smoking: 25 Years of Progress*. Rockville, MD: DHHS, 1989.
8. N. L. Benowitz, *Annu Rev Pharmacol Toxicol*, 1996, **36**, 597–613.
9. K. O. Haustein, *Int J Clin Pharmacol Ther*, 1999, **37**, 417–427.
10. T. A. Slotkin, *J Pharmacol Exp Ther*, 1998, **285**, 931–945.
11. K. A. Perkins and M. Stitzer, in *Handbook of Substance Abuse: Neurobehavioral Pharmacology*, eds. R. E. Tarter, R. T. Ammerman and P. J. Ott, Plenum, New York, 1998, pp. 299–317.
12. J. E. Rose, *Annu Rev Med*, 1996, **47**, 493–507.
13. J. E. Henningfield, *N Engl J Med*, 1995, **333**, 1196–1203.
14. C. P. O'Brien and A. T. McLellan, *Lancet*, 1996, **347**, 237–240.
15. J. R. Hughes, M. G. Goldstein, R. D. Hurt and S. Shiffman, *J Am Med Assoc*, 1999, **281**, 72–76.
16. E. C. Westman, E. D. Levin and J. E. Rose, *Arch Intern Med*, 1993, **153**, 1917–1923.
17. B. D. Jamerson, M. Nides, D. E. Jorenby, R. Donahue, P. Garrett, J. A. Johnston, M. C. Fiore, S. I. Rennard and S. J. Leischow, *Clin Ther*, 2001, **23**, 744–752.
18. J. E. Rose, A. G. Mukhin, S. J. Lokitz, T. G. Turkington, J. Herskovic, F. M. Behm, S. Garg and P. K. Garg, *Proc Natl Acad Sci USA*, 2010, **107**, 5190–5195.
19. J. P. Changeux, *Nat Rev Neurosci*, 2010, **11**, 389–401.

20. D. L. Beene, G. S. Brandt, W. Zhong, N. M. Zacharias, H. A. Lester and D. A. Dougherty, *Biochemistry (Mosc)*, 2002, **41**, 10262–10269.
21. X. Xiu, N. L. Puskar, J. A. Shanata, H. A. Lester and D. A. Dougherty, *Nature*, 2009, **458**, 534–537.
22. A. P. Blum, H. A. Lester and D. A. Dougherty, *Proc Natl Acad Sci USA*, 2010, **107**, 13206–13211.
23. A. P. Blum, K. R. Gleitsman, H. A. Lester and D. A. Dougherty, *J Biol Chem*, 2011, **286**, 32251–32258.
24. P. Russo, A. Cesario, S. Rutella, G. Veronesi, L. Spaggiari, D. Galetta, S. Margaritora, P. Granone and D. S. Greenberg, *Curr Med Chem*, 2011, **18**, 91–112.
25. G. F. Koob, *Ann N Y Acad Sci*, 2000, **909**, 170–185.
26. S. S. Watkins, G. F. Koob and A. Markou, *Nicotine Tob Res*, 2000, **2**, 19–37.
27. N. L. Benowitz, *N Engl J Med*, 2010, **362**, 2295–2303.
28. J. E. Henningfield, L. M. Schuh and M. E. Jarvik, in *Neuropsychopharmacology: The Fourth Generation of Progress*, eds. F. E. Bloom and D. J. Kupfer, Raven Press, New York, 2000.
29. W. A. Corrigall, K. B. Franklin, K. M. Coen and P. B. Clarke, *Psychopharmacology (Berl)*, 1992, **107**, 285–289.
30. G. Di Chiara and A. Imperato, *Proc Natl Acad Sci USA*, 1988, **85**, 5274–5278.
31. R. C. Pierce and V. Kumaresan, *Neurosci Biobehav Rev*, 2006, **30**, 215–238.
32. M. F. Suaud-Chagny, K. Chergui, G. Chouvet and F. Gonon, *Neuroscience*, 1992, **49**, 63–72.
33. S. Leonard and D. Bertrand, *Nicotine Tob Res*, 2001, **3**, 203–223.
34. E. J. Nestler, *Nat Neurosci*, 2005, **8**, 1445–1449.
35. C. Cohen, E. Kodas and G. Griebel, *Pharmacol Biochem Behav*, 2005, **81**, 387–395.
36. T. Raupach, P. H. Hoogsteder and C. P. Onno van Schayck, *Drugs*, 2012, **72**, e1–e16.
37. S. Ross and D. Williams, *Expert Opin Drug Saf*, 2005, **4**, 995–1003.
38. D. Gunnell, D. Irvine, L. Wise, C. Davies and R. M. Martin, *BMJ*, 2009, **339**, b3805.
39. U.S. Dept. of Health and Human Services, Public Health Service, Office of the Surgeon General. *The Health Consequences of Smoking: A Report of the Surgeon General*. Rockville, MD: DHHS, 2004.
40. U.S. Dept. of Health and Human Services, Public Health Service, Office of the Surgeon General. *How Tobacco Smoke Causes Disease: The Biology and Behavioral Basis for Smoking-Attributable Disease: A Report of the Surgeon General*. Rockville, MD: DHHS, 2010.
41. M. Sopori, *Nat Rev Immunol*, 2002, **2**, 372–377.
42. M. R. Stampfli and G. P. Anderson, *Nat Rev Immunol*, 2009, **9**, 377–384.
43. M. Nouri-Shirazi and E. Guinet, *J Immunol*, 2012, **188**, 2359–2370.
44. M. Nouri-Shirazi and E. Guinet, *Immunol Lett*, 2006, **103**, 167–176.

45. P. Silverstein, *Am J Med*, 1992, **93**, 22S–24S.
46. L. Arcavi and N. L. Benowitz, *Arch Intern Med*, 2004, **164**, 2206–2216.
47. T. J. Dickerson and K. D. Janda, *Proc Natl Acad Sci USA*, 2002, **99**, 15084–15088.
48. B. F. Erlanger, *Pharmacol Rev*, 1973, **25**, 271–280.
49. C. Wang, J. Tang, W. Song, E. Lobashevsky, C. M. Wilson and R. A. Kaslow, *Hepatology*, 2004, **39**, 978–988.
50. K. Lindroth, E. F. Mastache, I. Roos, A. G. Fernandez and C. Fernandez, *Immunology*, 2004, **112**, 413–419.
51. X. Y. Shen, F. M. Orson and T. R. Kosten, *Clin Pharmacol Ther*, 2012, **91**, 60–70.
52. E. H. Cerny, L. Franzini-Brunner and T. Cerny, in *Alcohol, Tobacco, and Cancer*, eds. C. H. Cho and V. Purohit, Karger, Basel, 2006, pp. 288–304.
53. J. J. Langone, H. B. Gjika and H. Van Vunakis, *Biochemistry (Mosc)*, 1973, **12**, 5025–5030.
54. H. Matsushita, M. Noguchi and E. Tamaki, *Biochem Biophys Res Commun*, 1974, **57**, 1006–1010.
55. A. Castro, N. Monji, H. Ali, E. R. Bowman and H. McKennis, *Biochem Arch*, 1985, **1**, 173–183.
56. A. Castro, H. McKennis, N. Monji and E. R. Bowman, *Biochem Arch*, 1985, **1**, 205–214.
57. J. J. Langone and H. Van Vunakis, *Methods Enzymol*, 1982, **84**, 628–640.
58. R. J. Bjercke, G. Cook, N. Rychlik, H. B. Gjika, H. Van Vunakis and J. J. Langone, *J Immunol Methods*, 1986, **90**, 203–213.
59. A. Abad, J. J. Manclus, C. March and A. Montoya, *Anal Chem*, 1993, **65**, 3227–3231.
60. K. F. Bonese, B. H. Wainer, F. W. Fitch, R. M. Rothberg and C. R. Schuster, *Nature*, 1974, **252**, 708–710.
61. Y. Hieda, D. E. Keyler, J. T. Vandevoort, J. K. Kane, C. A. Ross, D. E. Raphael, R. S. Niedbalas and P. R. Pentel, *J Pharmacol Exp Ther*, 1997, **283**, 1076–1081.
62. B. Berkowitz and S. Spector, *Science*, 1972, **178**, 1290–1292.
63. M. R. Carrera, J. A. Ashley, L. H. Parsons, P. Wirsching, G. F. Koob and K. D. Janda, *Nature*, 1995, **378**, 727–730.
64. M. Pravetoni, D. E. Keyler, M. D. Raleigh, A. C. Harris, M. G. Lesage, C. K. Mattson, S. Pettersson and P. R. Pentel, *Biochem Pharmacol*, 2011, **81**, 1164–1170.
65. S. D. Satoskar, D. E. Keyler, M. G. LeSage, D. E. Raphael, C. A. Ross and P. R. Pentel, *Int Immunopharmacol*, 2003, **3**, 957–970.
66. J. M. Scherrmann, N. Terrien, M. Urtizberea, P. Pierson, H. Denis and J. M. Bourre, *J Toxicol Clin Toxicol*, 1989, **27**, 1–35.
67. G. Pitas, E. M. Laurenzana, D. K. Williams, S. M. Owens and W. B. Gentry, *Drug Metab Dispos*, 2006, **34**, 906–912.
68. S. Isomura, P. Wirsching and K. D. Janda, *J Org Chem*, 2001, **66**, 4115–4121.

69. M. R. Carrera, J. A. Ashley, T. Z. Hoffman, S. Isomura, P. Wirsching, G. F. Koob and K. D. Janda, *Bioorg Med Chem*, 2004, **12**, 563–570.
70. M. M. Meijler, M. Matsushita, L. J. Altobell, 3rd, P. Wirsching and K. D. Janda, *J Am Chem Soc*, 2003, **125**, 7164–7165.
71. *World Pat.*, WO2005040338A2, 2005.
72. *World Pat.*, WO2009120954A2, 2009.
73. A. Y. Moreno, M. R. Azar, G. F. Koob and K. D. Janda, *Vaccine*, 2012.
74. A. Y. Moreno, M. R. Azar, N. A. Warren, T. J. Dickerson, G. F. Koob and K. D. Janda, *Mol Pharmaceutics*, 2010, **7**, 431–441.
75. *US Pat.*, US20120225087A1, 2012.
76. B. P. De, J. L. Boyer, O. Alsaied, A. Y. Moreno, K. Janda, S. Worgall and R. G. Crystal, *Mol Ther*, 2009, **17**, S281–S282.
77. M. J. Hicks, J. B. Rosenberg, B. P. De, O. Pagovich, C. N. Young, J.-P. Qiu, S. M. Kaminsky, N. R. Hackett, S. Worgall, K. D. Janda, R. L. Davisson and R. G. Crystal, *Mol Ther*, 2012, **20**, S43–S43.
78. M. J. Hicks, J. B. Rosenberg, B. P. De, O. E. Pagovich, C. N. Young, J.-P. Qiu, S. M. Kaminsky, N. R. Hackett, S. Worgall, K. D. Janda, R. L. Davisson and R. G. Crystal, *Sci Transl Med*, 2012, **4**.
79. J. B. Rosenberg, B. P. De, M. J. Hicks, A. Y. Moreno, K. D. Janda, S. M. Kaminsky and R. G. Crystal, *Mol Ther*, 2012, **20**, S127–S127.
80. Y. Hieda, D. E. Keyler, J. T. Vandevoort, J. K. Kane, C. A. Ross, D. E. Raphael, P. R. Pentel and R. S. Niedbala, *Addiction*, 1998, **93**, 915–915.
81. Y. Hieda, D. E. Keyler, J. T. VanDeVoort, R. S. Niedbala, D. E. Raphael, C. A. Ross and P. R. Pentel, *Psychopharmacology (Berl)*, 1999, **143**, 150–157.
82. D. H. Malin, J. R. Lake, A. Lin, M. Saldana, L. Balch, M. L. Irvin, H. Chandrasekara, C. L. Alvarado, Y. Hieda, D. E. Keyler, P. R. Pentel, S. Ennifar, L. E. Basham, R. Naso and A. Fattom, *Pharmacol, Biochem Behav*, 2001, **68**, 87–92.
83. D. H. Malin, C. L. Alvarado, K. S. Woodhouse, H. Karp, E. Urdiales, D. Lay, P. Appleby, W. D. Moon, S. Ennifar, L. Basham and A. Fattom, *Life Sci*, 2002, **70**, 2793–2798.
84. P. R. Pentel, D. H. Malin, S. Ennifar, Y. Hieda, D. E. Keyler, J. R. Lake, J. R. Milstein, L. E. Basham, R. T. Coy, J. W. Moon, R. Naso and A. Fattom, *Pharmacol Biochem Behav*, 2000, **65**, 191–198.
85. Y. Hieda, D. E. Keyler, S. Ennifar, A. Fattom and P. R. Pentel, *Int J Immunopharmacol*, 2000, **22**, 809–819.
86. Y. Tuncok, Y. Hieda, D. E. Keyler, S. Brown, S. Ennifar, A. Fattom and P. R. Pentel, *Exp Clin Psychopharmacol*, 2001, **9**, 228–234.
87. P. R. Pentel and D. E. Keyler, in *New Generation Vaccines*, eds. M. M. Levine, J. B. Kaper, R. Rappuoli, M. A. Liu and M. F. Good, Marcel Dekker, New York, 3rd edn., 2004, pp. 1057–1066.
88. J. Hukkanen, P. Jacob, 3rd and N. L. Benowitz, *Pharmacol Rev*, 2005, **57**, 79–115.
89. S. A. Roiko, A. C. Harris, D. E. Keyler, M. G. LeSage, Y. Zhang and P. R. Pentel, *J Pharmacol Exp Ther*, 2008, **325**, 985–993.

90. K. E. Cornish, A. C. Harris, M. G. Lesage, D. E. Keyler, D. Burroughs, C. Earley and P. R. Pentel, *Int Immunopharmacol*, 2011, **11**, 1809–1815.
91. D. E. Keyler, S. A. Roiko, C. A. Earley, M. P. Murtaugh and P. R. Pentel, *Int Immunopharmacol*, 2008, **8**, 1589–1594.
92. M. Pravetoni, D. E. Keyler, R. R. Pidaparthi, F. I. Carroll, S. P. Runyon, M. P. Murtaugh, C. A. Earley and P. R. Pentel, *Biochem Pharmacol*, 2012, **83**, 543–550.
93. M. G. LeSage, D. Shelley, M. Pravetoni and P. R. Pentel, *Pharmacol, Biochem Behav*, 2012, **102**, 157–162.
94. S. D. Sanderson, S. R. Cheruku, M. P. Padmanilayam, J. L. Vennerstrom, G. M. Thiele, M. I. Palmatier and R. A. Bevins, *Int Immunopharmacol*, 2003, **3**, 137–146.
95. R. M. Tempero, M. A. Hollingsworth, M. D. Burdick, A. M. Finch, S. M. Taylor, S. M. Vogen, E. L. Morgan and S. D. Sanderson, *J Immunol*, 1997, **158**, 1377–1382.
96. T. R. Kosten, M. Rosen, J. Bond, M. Settles, J. S. Roberts, J. Shields, L. Jack and B. Fox, *Vaccine*, 2002, **20**, 1196–1204.
97. *Expert Rev Vaccines*, 2004, **3**, 386.
98. *Swiss Pat.*, CH678394, 1990.
99. E. H. Cerny, R. Levy, J. Mauel, M. Mpandi, M. Mutter, C. Henzelin-Nkubana, L. Patiny, G. Tuchscherer and T. Cerny, *Onkologie*, 2002, **25**, 406–411.
100. S. H. de Villiers, N. Lindblom, G. Kalayanov, S. Gordon, A. Malmerfelt, A. M. Johansson and T. H. Svensson, *Respiration*, 2002, **69**, 247–253.
101. N. Lindblom, S. H. de Villiers, G. Kalayanov, S. Gordon, A. M. Johansson and T. H. Svensson, *Respiration*, 2002, **69**, 254–260.
102. S. H. de Villiers, N. Lindblom, G. Kalayanov, S. Gordon, I. Baraznenok, A. Malmerfelt, M. M. Marcus, A. M. Johansson and T. H. Svensson, *Vaccine*, 2010, **28**, 2161–2168.
103. D. K. Hatsukami, D. E. Jorenby, D. Gonzales, N. A. Rigotti, E. D. Glover, C. A. Oncken, D. P. Tashkin, V. I. Reus, R. C. Akhavain, R. E. Fahim, P. D. Kessler, M. Niknian, M. W. Kalnik and S. I. Rennard, *Clin Pharmacol Ther*, 2011, **89**, 392–399.
104. E. N. Janoff, W. D. Hardy, P. D. Smith and S. M. Wahl, *J Immunol*, 1991, **147**, 2130–2135.
105. P. J. Weiss, M. R. Wallace, E. C. Oldfield 3rd, J. O'Brien and E. N. Janoff, *J Infect Dis*, 1995, **171**, 1217–1222.
106. J. Hartmann-Boyce, K. Cahill, D. Hatsukami and J. Cornuz, *Cochrane Database Syst Rev*, 2012, **8**, CD007072.
107. P. Maurer, G. T. Jennings, J. Willers, F. Rohner, Y. Lindman, K. Roubicek, W. A. Renner, P. Muller and M. F. Bachmann, *Eur J Immunol*, 2005, **35**, 2031–2040.
108. J. Cornuz, S. Zwahlen, W. F. Jungi, J. Osterwalder, K. Klingler, G. van Melle, Y. Bangala, I. Guessous, P. Muller, J. Willers, P. Maurer, M. F. Bachmann and T. Cerny, *PLoS One*, 2008, **3**, e2547.

109. C. J. Bunce, P. T. Loudon, C. Akers, J. Dobson and D. M. Wood, *Curr Opin Mol Ther*, 2003, **5**, 58–63.
110. D. K. Hatsukami, S. Rennard, D. Jorenby, M. Fiore, J. Koopmeiners, A. de Vos, G. Horwith and P. R. Pentel, *Clin Pharmacol Ther*, 2005, **78**, 456–467.
111. E. J. Wagena, A. de Vos, G. Horwith and C. P. van Schayck, *Nicotine Tob Res*, 2008, **10**, 213–218.
112. S. Tonstad, E. Heggen, H. Giljam, P.-A. Lagerback, P. Tonnesen, L. D. Wikingsson, N. Lindblom, S. de Villiers, T. H. Svensson and K.-O. Fagerstrom, *Nicotine Tob Res*, 2012.
113. N. L. Benowitz, *Clin Pharmacol Ther*, 2008, **83**, 531–541.
114. H. L. Davis, M. J. McCluskie, N. Zhang, M. Benoit, K. Robertson and D. C. Pryde, presented in part at the 14th Annual Meeting of the Society for Research on Nicotine and Tobacco (SRNT) Europe, Helsinki, August, 2012.
115. M. J. McCluskie, N. Zhang, M. Benoit, K. Robertson, H. L. Davis, D. C. Pryde, D. Blakemore, A. D. Brown, L. H. Jones, D. R. Stead, D. P. Gervais, P. White and J. R. Merson, presented in part at the 14th Annual Meeting of the Society for Research on Nicotine and Tobacco (SRNT) Europe, Helsinki, August, 2012.
116. G. Leroux-Roels, *Vaccine*, 2010, **28 Suppl 3**, C25–36.
117. M. F. Bachmann and G. T. Jennings, *Nat Rev Immunol*, 2010, **10**, 787–796.
118. J. J. Donnelly, R. R. Deck and M. A. Liu, *J Immunol*, 1990, **145**, 3071–3079.
119. M. F. Bachmann, U. H. Rohrer, T. M. Kundig, K. Burki, H. Hengartner and R. M. Zinkernagel, *Science*, 1993, **262**, 1448–1451.
120. M. Kovacsovics-Bankowski, K. Clark, B. Benacerraf and K. L. Rock, *Proc Natl Acad Sci USA*, 1993, **90**, 4942–4946.
121. V. Manolova, A. Flace, M. Bauer, K. Schwarz, P. Saudan and M. F. Bachmann, *Eur J Immunol*, 2008, **38**, 1404–1413.
122. A. Bershteyn, J. P. Chaparro, E. B. Riley, R. S. Yao, R. S. Zachariah and D. J. Irvine, in *IEEE 35th Annual Northeast*, Institute of Electrical and Electronics Engineers, 2009, pp. 1–2.
123. S. M. Levitz and D. T. Golenbock, *Cell*, 2012, **148**, 1284–1292.
124. C. Rueckert and C. A. Guzman, *PLoS Pathog*, 2012, **8**, e1003001.
125. W. Weir, in *The Hartford Courant*, 2011.
126. W. H. Lin, D. E. Griffin, P. A. Rota, M. Papania, S. P. Cape, D. Bennett, B. Quinn, R. E. Sievers, C. Shermer, K. Powell, R. J. Adams, S. Godin and S. Winston, *Proc Natl Acad Sci USA*, 2011, **108**, 2987–2992.
127. K. Hrafnkelsdottir, J. Valgeirsson and S. Gizurarson, *Biol Pharm Bull*, 2005, **28**, 1038–1042.
128. X. Chen, M. Pravetoni, B. Bhayana, P. R. Pentel and M. X. Wu, *Vaccine*, 2012.
129. G. M. Glenn, D. N. Taylor, X. Li, S. Frankel, A. Montemarano and C. R. Alving, *Nat Med*, 2000, **6**, 1403–1406.

130. I. D. Montoya, *Adicciones*, 2012, **24**, 95–103.
131. R. L. Coffman, A. Sher and R. A. Seder, *Immunity*, 2010, **33**, 492–503.
132. K. D. Janda and J. B. Treweek, *Nat Rev Immunol*, 2012, **12**, 67–72.
133. S. A. Kaba, C. Brando, Q. Guo, C. Mittelholzer, S. Raman, D. Tropel, U. Aebi, P. Burkhard and D. E. Lanar, *J Immunol*, 2009, **183**, 7268–7277.
134. J. S. Rudra, Y. F. Tian, J. P. Jung and J. H. Collier, *Proc Natl Acad Sci USA*, 2010, **107**, 622–627.
135. L. Pittet, D. Altreuter, P. Ilyinskii, C. Fraser, Y. Gao, S. Baldwin, M. Keegan, L. Johnston and T. Kishimoto, *J Immunol*, 2012, **188**.
136. F. Gu, L. Zhang, B. A. Teply, N. Mann, A. Wang, A. F. Radovic-Moreno, R. Langer and O. C. Farokhzad, *Proc Natl Acad Sci USA*, 2008, **105**, 2586–2591.
137. S. B. Rizwan, W. T. McBurney, K. Young, T. Hanley, B. J. Boyd, T. Rades and S. Hook, *J Controlled Release*, 2013, **165**, 16–21.
138. C. R. Alving, M. Rao, N. J. Steers, G. R. Matyas and A. V. Mayorov, *Expert Rev Vaccines*, 2012, **11**, 733–744.
139. G. Fujii, W. Ernst and J. Adler-Moore, *Front Biosci*, 2008, **13**, 1968–1980.
140. J. J. Moon, H. Suh, A. Bershteyn, M. T. Stephan, H. Liu, B. Huang, M. Sohail, S. Luo, S. H. Um, H. Khant, J. T. Goodwin, J. Ramos, W. Chiu and D. J. Irvine, *Nat Mater*, 2011, **10**, 243–251.
141. J. J. Moon, H. Suh, A. V. Li, C. F. Ockenhouse, A. Yadava and D. J. Irvine, *Proc Natl Acad Sci USA*, 2012, **109**, 1080–1085.
142. U. M. Abdel-Motal, K. Wigglesworth and U. Galili, *Vaccine*, 2009, **27**, 3072–3082.
143. U. Galili, *Clin Dev Immunol*, 2011, **2011**, 134020.
144. S. Mangat and H. M. Patel, *Life Sci*, 1985, **36**, 1917–1925.
145. A. Bershteyn, M. C. Hanson, M. P. Crespo, J. J. Moon, A. V. Li, H. Suh and D. J. Irvine, *J Controlled Release*, 2011.
146. L. J. Cruz, F. Rueda, B. Cordobilla, L. Simon, L. Hosta, F. Albericio and J. C. Domingo, *Mol Pharm*, 2011, **8**, 104–116.
147. D. F. Moyano, M. Goldsmith, D. J. Solfiell, D. Landesman-Milo, O. R. Miranda, D. Peer and V. M. Rotello, *J Am Chem Soc*, 2012, **134**, 3965–3967.
148. M. J. Hicks, B. P. De, J. B. Rosenberg, J. T. Davidson, A. Y. Moreno, K. D. Janda, S. Wee, G. F. Koob, N. R. Hackett, S. M. Kaminsky, S. Worgall, M. Toth, J. G. Mezey and R. G. Crystal, *Mol Ther*, 2011, **19**, 612–619.
149. N. Tatsis and H. C. Ertl, *Mol Ther*, 2004, **10**, 616–629.
150. I. Caminschi and K. Shortman, *Trends Immunol*, 2012, **33**, 71–77.
151. A. Castro and I. Prieto, *Biochem Biophys Res Commun*, 1975, **67**, 583–589.
152. K. Landsteiner, *The Specificity of Serological Reactions*, C. C. Thomas, Springfield, IL, 1936.
153. K. D. Janda, S. J. Benkovic and R. A. Lerner, *Science*, 1989, **244**, 437–440.

154. S. Paula, M. R. Tabet, C. D. Farr, A. B. Norman and W. J. Ball Jr., *J Med Chem*, 2004, **47**, 133–142.
155. M. J. Dingel, K. Karkazis and B. A. Koenig, *Soc Sci Quart*, 2011, **92**, 1363–1388.
156. K. A. Perkins, *Pharmacol Biochem Behav*, 1999, **64**, 295–299.
157. M. M. Faraday, V. A. O'Donoghue and N. E. Grunberg, *Pharmacol Biochem Behav*, 2003, **74**, 325–333.
158. N. L. Benowitz, G. E. Swan, P. Jacob, 3rd, C. N. Lessov-Schlaggar and R. F. Tyndale, *Clin Pharmacol Ther*, 2006, **80**, 457–467.
159. G. L. Kortmann, C. J. Dobler, L. Bizarro and C. H. Bau, *Am J Med Genet B Neuropsychiatr Genet*, 2010, **153B**, 17–28.
160. N. L. Benowitz, C. N. Lessov-Schlaggar, G. E. Swan and P. Jacob 3rd, *Clin Pharmacol Ther*, 2006, **79**, 480–488.
161. N. D. Volkow, *Addict Sci Clin Pract*, 2010, **5**, 2.
162. *Nabi Biopharmaceuticals Announces Results of NicVAX(R) Phase II Study in Combination With Varenicline*, 2012.
163. C. E. Berkman, S. B. Park, S. A. Wrighton and J. R. Cashman, *Biochem Pharmacol*, 1995, **50**, 565–570.
164. J. R. King, H. Wynn, R. Brundage and E. P. Acosta, *Clin Pharmacokinet*, 2004, **43**, 291–310.
165. A. Hill, S. Khoo, M. Boffito and D. Back, *J Acquired Immune Defic Syndr*, 2011, **58**, e137–138.
166. W. Zhang, T. Kilicarslan, R. F. Tyndale and E. M. Sellers, *Drug Metab Dispos*, 2001, **29**, 897–902.
167. J. L. Oliver, G. Pashmi, P. Barnett, P. Mettens, R. Biemans, P. Monteyne, R. Palmantier, T. Gallagher, S. Ramaya and S. Wonnacott, *Vaccine*, 2007, **25**, 7354–7362.
168. I. Sziraki, H. Sershen, M. Benuck, M. Lipovac, A. Hashim, T. B. Cooper, D. Allen and A. Lajtha, *Neurochem Res*, 1999, **24**, 1471–1478.
169. J. J. Buccafusco, L. C. Shuster and A. V. Terry Jr., *Neurosci Lett*, 2007, **413**, 68–71.
170. R. A. Bevins, J. L. Wilkinson and S. D. Sanderson, *Expert Opin Biol Ther*, 2008, **8**, 379–383.

CHAPTER 3

Carbohydrate Vaccines

BENJAMIN SCHUMANN,[a,b] CHAKKUMKAL ANISH,[a] CLANEY L. PEREIRA[a] AND PETER H. SEEBERGER*[a,b]

[a] Max Planck Institute of Colloids and Interfaces, 14424 Potsdam, Germany;
[b] Freie Universität Berlin, Arnimallee 22, 14195 Berlin, Germany
*Email: Peter.seeberger@mpikg.mpg.de

3.1 Introduction

In the first decades of the 20th century, Heidelberger discovered that the immunodominant type-specific substance of *Streptococcus pneumoniae* is a polysaccharide.[1] A few years later, Avery and Goebel reported that the conjugation of mono- and disaccharides to proteins enables the generation of saccharide-specific antibodies *in vivo*.[2–8] By this time, Landsteiner had already defined the term "hapten" for any molecule that, attached to a protein, can induce a specific immune response.[9] Studies by Francis and Tillett as well as Heidelberger indicated that infection with *S. pneumoniae* led to the production of antibodies against capsular polysaccharides (CPS) from this pathogen.[10,11] As early as 1931, Goebel and Avery chemically attached CPS of type III pneumococci to proteins from horse serum and successfully vaccinated rabbits with these conjugates.[12,13]

Research into the immunogenicity of bacterial polysaccharides continued,[14] but antibiotics soon overshadowed vaccines as the most popular way of fighting infectious diseases. However, it became apparent that antibiotics would not be able to eradicate pathogenic bacteria because resistant strains emerged rapidly.[15–18] Thus, the idea of CPS-based vaccines was reconsidered. After

RSC Drug Discovery Series No. 36
Biotherapeutics: Recent Developments using Chemical and Molecular Biology
Edited by Lyn H. Jones and Andrew J. McKnight

Published by the Royal Society of Chemistry, www.rsc.org

extensive research, polysaccharide vaccines (PSV) were licensed in the 1970s against *Neisseria meningitidis* and *S. pneumoniae*. The first pneumococcal polysaccharide vaccine contained CPS from 14 different serotypes. A 23-valent formulation soon followed, along with further PSV against pathogenic bacteria.[19]

Despite the remarkable success of PSV in preventing invasive disease, polysaccharides were not effective at protecting from pathogens in very young children.[17] To overcome this problem, the classical approach of conjugating polysaccharides to proteins was used. These carbohydrate conjugate vaccines (CCV) are able to raise a potent antibacterial immune response even in infants.[20,21] CCV have helped to significantly decrease the incidence of invasive diseases caused by pathogenic bacteria.[22] Today, glycoconjugate vaccines constitute a multibillion dollar market and are included by default in immunization programs recommended by health authorities.[23,24]

The field of carbohydrate-based vaccines has greatly improved human health and has almost eradicated certain pathogens.[17] The success of vaccines against bacteria has inspired research on preventing infections by fungi, protozoan parasites, helminths and viruses as well as cancer by using similar strategies.[25–27] Defined glycan structures to be used in these vaccines can be generated by synthetic carbohydrate chemistry. In addition, immunology is helping to understand the mechanisms of immunity elicited by CCV. Carbohydrate-based vaccines have been reviewed and the reader is referred to these articles for further insights.[27–36]

3.1.1 Cell Surface Glycans as Vaccine Candidates

There are two classes of vaccines: whole cell and subunit vaccines. Whole cell vaccines were the first vaccines developed and consist of either attenuated or killed organisms.[37] Although whole cell vaccines confer long-lasting immunity against infectious disease, culturing pathogenic organisms in high quantities may be difficult.[37–39] Furthermore, attenuated pathogens might cause disease in immunocompromised individuals in rare cases.[40] Subunit vaccines are preparations of molecularly defined components of the target organisms. These components can be proteins, such as secreted toxins, as well as virus-like particles and cell surface glycans. Subunit vaccines are often considered to be safer than whole cell vaccines because they cannot cause disease.[40] Antigens used in subunit vaccines can be obtained by recombinant expression.[37] Since the immunogenicity of these antigens is often low, subunit vaccines are usually formulated with adjuvants.[37,39,40]

Proteins are the antigens of choice in subunit vaccines due to their ability to raise an efficient immune response. However, protein antigens may fail to protect from disease under certain circumstances: (i) proteins may not be exposed on the surface of a pathogen and may therefore be inaccessible to components of the immune system; (ii) variability of surface proteins may prevent the build-up of a protective immunological memory; and (iii) surface proteins may be very similar to or derived from cellular proteins of the host and hence may not be very immunogenic. Carbohydrate-based vaccines may present an important alternative in these cases.

Several pathogenic bacteria and fungi cover themselves with a layer of capsular polysaccharides, lipopolysaccharides (LPS) and related structures. Microorganisms use these capsules to prevent desiccation and escape from components of the innate immune system by shielding other cell surface antigens.[41,42] Capsules are major virulence factors because they prevent complement activation and opsonization.[43] Most CPS are biosynthesized in a modular manner from oligosaccharide precursors, and hence usually consist of repeating units (RU).[41,44]

The structural diversity of CPS of pathogenic bacteria is enormous (Figure 3.1A).[45,46] In *S. pneumoniae*, more than 90 different CPS repeating unit structures are known, and categorization into serotypes of this bacterium is carried out based on these structures.[45] The length of repeating units may range from mono- to octasaccharides.[27,45] Repeating units may be connected via phosphodiester linkages, for example in CPS from *Haemophilus influenzae* type b (Hib). Polysaccharides may be linear or branched and contain common or rare sugars. While human cell surface glycans are composed of just nine different monosaccharides (D-Glc, D-Gal, D-Man, L-Fuc, D-GlcNAc, D-GalNAc, D-Xyl, D-Sia and D-GlcA), the number of building blocks that make up the bacterial glycome easily exceeds one hundred.[47] Monosaccharides may be present in pyranose or furanose forms or as open chain polyalcohols. Variability is further increased by the presence of chemical modifications, such as acetate, pyruvate and glycerate groups.[48]

It is possible to raise a humoral immune response against CPS. The mammalian immune system is tailored to discriminate between host-derived ("self") and foreign ("non-self") structures. Rare monosaccharides, unusual linkages and modifications are likely to contribute to the non-self recognition of bacterial CPS by the human immune system.

Proteins on the cell surface of protozoan parasites, such as *Plasmodium* or *Trypanosoma*, can be highly variable and may not be suited to induce a protective immune response.[49,50] Thus, using cell surface glycan-based antigens is a promising approach to prevent infection by these pathogens.[51] Protozoan cell membranes are rich in glycosylphosphatidylinositol (GPI) anchors that are often linked to cell surface proteins. GPI anchors can activate components of the innate immune system, such as Toll-like receptors (TLR).[52] Protozoan GPIs have been shown to induce the generation of specific antibodies and may thus be the antigens of choice for vaccination against these parasites.[51,53] Indeed, a fully synthetic GPI provided efficient protection against *Plasmodium* infections when conjugated to a carrier protein.[54,55] Interestingly, immunization with GPI-anchored proteins derived from the multicellular trematode *Schistosoma mansoni* protected mice from this parasite.[56]

High variability of surface proteins is also seen in enveloped viruses, such as human immunodeficiency virus (HIV). On host-derived virus membranes, only those surface proteins that are needed for infection of and budding from a host cell are displayed. The high variability of HIV surface glycoproteins promotes

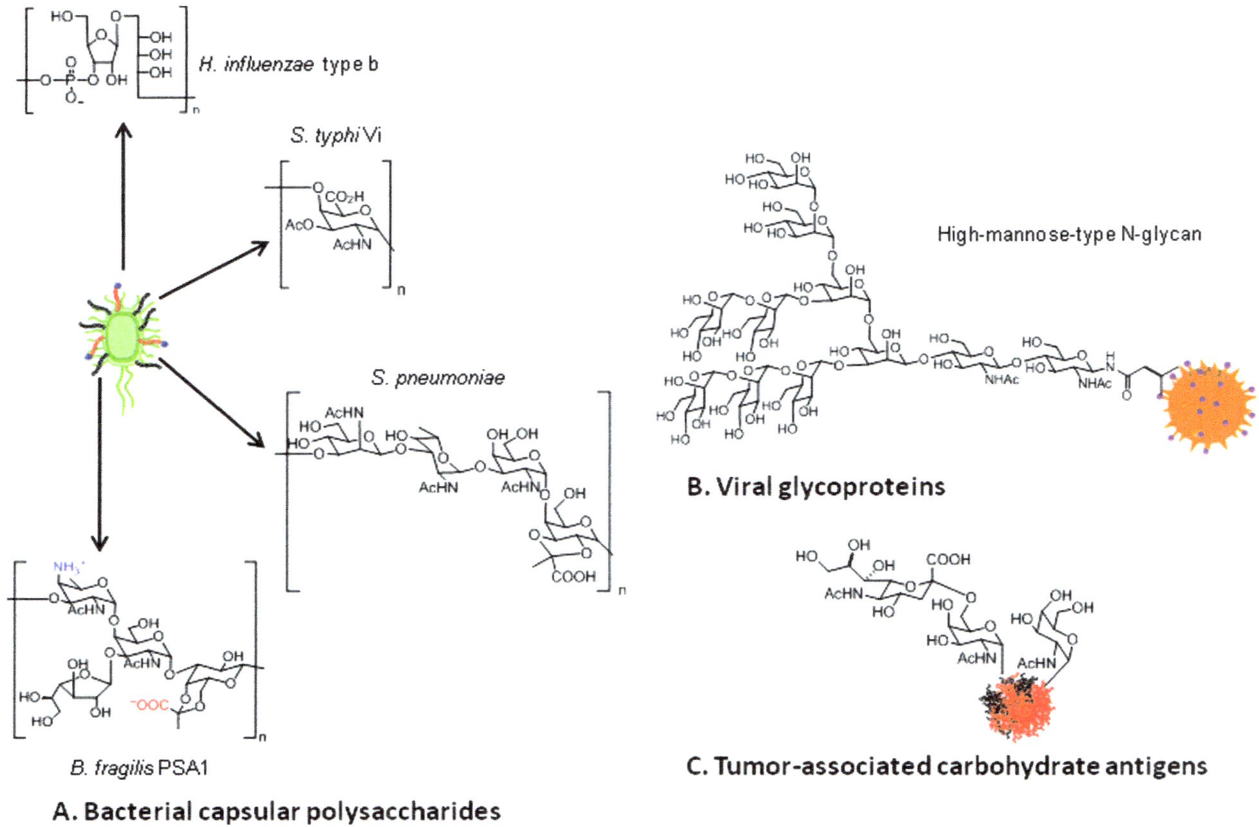

Figure 3.1 Examples of cell surface glycans from different organisms. (A) bacterial capsular polysaccharides. (B) high mannose N-glycans on HIV gp120. (C) truncated mucin-type O-glycans found on malignant cells.

immune evasion.[57] Consequently, there is no licensed HIV vaccine available yet. However, antibodies that bind to epitopes containing high mannose-type N-glycans on the surface glycoprotein gp120 (Figure 3.1B) can prevent the virus from infecting host cells and are called neutralizing antibodies.[58–61] Employing saccharides based on high-mannose N-glycans as antigens for HIV vaccination is the subject of current research.[62]

Many tumors exhibit low immunogenicity due to the expression of self-antigens, hence they are challenging to fight on a molecular level. In addition, tumors develop elaborate ways to evade the immune system.[63] Multiple principles of cancer vaccination have been explored and several vaccine candidates have entered clinical trials.[64] Still, no carbohydrate-based cancer vaccine is being marketed yet. Tumor-associated carbohydrate antigens (TACAs) are glycans found on the surface of cancer cells. Certain TACAs, such as truncated mucin-type O-glycan chains (Figure 3.1C), are rarely present on non-malignant cells.[65,66] Other TACAs, such as Lewis antigens, are found on normal cells, but are heavily overexpressed on tumor cells.[67,68] Cell surface glycan modification is found on cancers from many different tissues, and thus TACAs are important starting points for tumor vaccination.[66]

Despite the wide variety of cell surface glycans that are considered for vaccination, bacterial CPS are by far the best studied carbohydrate antigens. Therefore, the following paragraphs will mainly focus on vaccines based on bacterial glycans.

3.1.2 Immunology of Carbohydrate-based Vaccines

Antigens are categorized according to the type of immune response they invoke. Thymus-dependent (T_D) antigens are mainly proteins and can induce the formation of immunological memory, while thymus-independent (T_I) antigens are usually polysaccharide- or lipid-based and cannot elicit immunological memory.[69]

T_D antigens induce antibody production with the immediate help of $CD4^+$ T cells (T-helper or T_H cells). After internalization by antigen presenting cells (APC), such as B cells, pathogen-derived protein antigens are degraded in endosomal compartments.[38] The resulting peptide fragments are then loaded onto major histocompatibility complex (MHC) class II molecules and the MHC II-peptide complex is transported to the APC surface. The complex is recognized by the αβ-T cell receptor (TCR) of a specific T_H cell. Further APC-T cell interactions are established and cytokines, such as interleukin-4, are released by the T cell. Cell–cell interactions and cytokine release stimulate affinity maturation and class switch of immunoglobulins (Ig) in B cells to provide highly antigen-specific antibodies of the IgG type. B cells differentiate to plasma cells and produce large amounts of IgG that help the immune system to fight pathogens. In addition to plasma cells, antigen-specific memory B and T cells are generated, which are activated upon subsequent re-admission of the same T_D antigen. This immunological memory facilitates the build-up of an efficient immune response upon re-infection.[38]

T_I antigens can induce antibody production by B cells without direct help by T cells. They can be further subcategorized into two groups. T_I-1 antigens comprise a heterogeneous group of different B cell mitogens, such as LPS, and can activate even immature, neonatal B cells.[69] In contrast, T_I-2 antigens are high molecular weight polymers, such as CPS, that activate B cells by cross-linking their membrane-bound B cell receptors (BCR).[70]

T_I-2 antigens are characterized by their inability to induce antibody production in neonatal or otherwise immature B cells.[70] They cannot invoke $CD4^+$ T cell help by the classical MHC II-mediated pathway.[71] As a consequence, Ig class switch and affinity maturation are inefficient, leading to the secretion of low-affinity IgM and IgG antibodies.[70,72] It has been revealed that antibody generation against T_I-2 antigens depends on the presence of functional Bruton's tyrosine kinase, a signaling molecule important for B cell development.[72–75] Furthermore, cell types other than T cells, such as macrophages and natural killer cells, are needed to induce antibody production in response to T_I-2 antigens.[71,76,77]

Upon infection with pathogenic bacteria, a potent immune response can be generated due to the presence and interplay of T_I-1, T_I-2 and T_D antigens. A polysaccharide vaccine, however, consists exclusively of T_I-2 antigens and cannot efficiently stimulate neonatal B cells. Thus, in infants with an immature immune system, PSV fail to raise protective antibody levels to prevent infection.[78–80] There is no clear-cut age at which the immune system starts to produce antibodies against T_I-2 antigens. However, it is commonly believed that PSV are not effective in infants under the age of 18–24 months.[17,25,81,82] Children up to this age constitute a high risk group for *Haemophilus influenzae* type b infections. Consequently, the Hib PSV introduced in 1985 was withdrawn from the market in 1988 and replaced by CPS-protein conjugate vaccine formulations (see following text).[83] The pneumococcal polysaccharide vaccine PPV-23 (Pneumovax23®, Merck) is recommended for children older than 2 years.[84] PSV against *N. meningitidis* (Menomune®, Sanofi Pasteur) and *Salmonella enterica* sv. *Typhi* (Typhim Vi®, Sanofi Pasteur) are currently on the market, but not recommended for infants.[85–87]

The immune response to a polysaccharide can be enhanced significantly by conjugation to a carrier protein. Thereby, a T_I antigen is turned into a T_D antigen and the help of T_H cells is enforced. Widely used carrier proteins are inactivated bacterial toxoids, such as denatured diphtheria toxin (DT), tetanus toxin (TT), or the non-toxic DT mutant CRM197.[88,89] The glycan moiety of the polysaccharide-protein conjugate binds to a glycan-specific BCR. After internalization of the resulting complex, carrier-derived peptides are presented to T_H cells. Both cellular and cytokine-mediated signals then induce maturation and proliferation of the glycan-specific B cell. Consequently, antibodies are generated with high affinities towards CPS and immunological memory is induced even in children younger than 2 years.[17,29,32,36] Carbohydrate conjugate vaccines have almost completely replaced polysaccharide vaccines. CCV against Hib have been so successful that infectious diseases caused by this bacterium are nearly eradicated in vaccinated areas.[83] Pneumococcal conjugate vaccines (PCV) have been developed to cover an increasing number of serotypes, and current formulations are 10- (Synflorix®, GSK) and 13-valent (Prevenar13®, Pfizer).[90] PCV provide protection against serotypes that account for more than 70% of cases of invasive pneumococcal disease worldwide.[91] The impact of CCV has been reviewed elsewhere and novel glycoconjugate vaccines are in development against a variety of pathogens.[26,27,81]

3.1.3 Production and Manufacture of Glycoconjugate Vaccines

Carbohydrate conjugate vaccines are unique among therapeutic agents because they consist of multiple components from different origins (Figure 3.2A). Thereby, guidelines for quality control (QC) have to be met for each component prior to conjugation.[92–95]

Carrier proteins are isolated following either homologous or heterologous expression in bioreactors, using well-characterized bacterial strains.[93] Several chromatography steps are required to yield a solution of pure protein, which is sterile filtered and analyzed for homogeneity and structure.[96,97] Interestingly, if the non-toxic diphtheria toxin variant CRM197 is used as a carrier and produced in the same facility as wild type DT, toxicity studies must be performed to rule out cross-contamination of CRM197 with toxic DT.[96]

Most commercially available glycoconjugate vaccines contain saccharides isolated from cultured bacteria. Before harvesting CPS, the bacterial culture must be examined for contaminations.[92–94,96–98] Killing of bacteria can be performed either physically (e.g. by heat) or chemically (e.g. by phenol).[93,94] CPS are then precipitated by using suitable chemicals like cetyl trimethyl ammonium bromide (CTAB). Purification of CPS from other cellular components includes ultracentrifugation, gel permeation, enzyme treatment and ultrafiltration to give a polysaccharide preparation with low contents of nucleic acid and protein (usually $<3\%$ w/w).[92,94,96–98] Identity, monosaccharide content and purity are assessed using both wet chemical and analytical methods like NMR (e.g. ^{1}H, ^{13}C, ^{31}P), high performance liquid chromatography with multi-angle laser light scattering (HPSEC-MALLS) and high performance anion exchange chromatography with pulsed amperometric detection (HPAEC-PAD).[97,99–104] The existence of characteristic functional groups like *O*-acetates, phosphate groups and 6-deoxyhexoses is assessed using the same methods. Purified CPS are then subjected to mild depolymerization, for example by microfluidization.[93] The size distribution of the obtained fragments is analyzed using size exclusion chromatography.

To enable conjugation to the carrier protein, CPS fragments are chemically activated. Various conceptually different conjugation methods have profound impact on the physical properties of conjugates.[105] Single point attachment methods make use of the presence of one reactive functional group per CPS fragment, typically the anomeric carbon of the reducing end monosaccharide. Thereby, conjugation can be carried out via reductive amination between the anomeric carbon and a lysine side chain of the carrier.[106] Single point attachment methods yield non-cross-linked "neoglycoconjugates" (Figure 3.2A). In contrast, methods of multipoint attachment introduce multiple reactive groups per CPS fragment, leading to cross-linked carrier-glycan lattices (Figure 3.2B).[106–108] Multipoint attachment can be performed via carbodiimide coupling of uronic acids and cyanogen bromide activation of

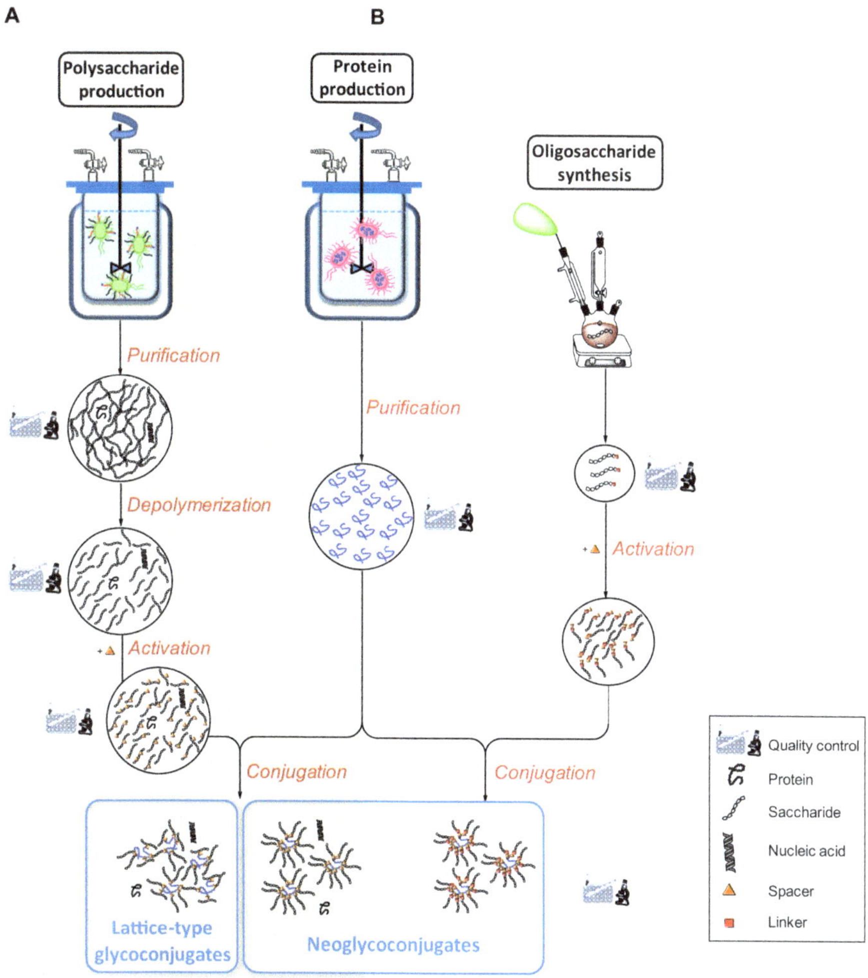

Figure 3.2 The production workflow of carbohydrate conjugate vaccines. (A) Manufacture of glycoconjugates based on isolated CPS. Both polysaccharide and carrier protein are purified after bacterial expression. Multiple steps of quality control are performed during isolation and processing of both moieties. Polysaccharides are degraded into fragments and activated. Depending on the conjugation method, either lattice-type glycoconjugates or neoglycoproteins are obtained. (B) Production of glycoconjugate vaccines based on synthetic oligosaccharides. Quality control is carried out by characterizing the final synthetic compound. Oligosaccharides are activated at a specific site, and conjugation results in neoglycoconjugates.

vicinal hydroxyl groups, for example.[109–112] Alternatively, periodate oxidation of saccharides generates aldehyde groups amenable for reductive amination.[110] A different conjugation method involves thioether formation. Thereby, polysaccharide fragments are thiolated and carrier proteins haloacylated (or vice versa) to ensure chemoselective coupling.[113,114] Methods used for the preparation of polysaccharide-protein conjugates have been reviewed elsewhere.[26,107,115]

The immunological implications of different conjugate structures have not been elucidated in detail. It has been proposed that lattice-type conjugates, although being highly immunogenic, display more artificial epitopes than neoglycoprotein-type conjugates.[108,116] As a consequence, the efficiency of lattice-type glycoconjugate vaccines may be hampered due to the induction of non-protective antibodies.

The choice of the conjugation method has to be matched to the functional groups available on the carrier protein. Toxins like DT and TT are usually inactivated by treatment with amine-reactive reagents like formaldehyde.[88,117] Inactivation reduces the availability of reactive lysine residues, and therefore toxoid carriers are often pre-activated and spacer molecules are incorporated during conjugation.[88,112,118,119]

After purification of the carbohydrate-protein conjugate, further characterization utilizes the methods described previously. When denatured DT or TT are used as carriers, non-toxicity needs to be confirmed and it has to be checked whether any reactive groups remain in the conjugate.[92] The saccharide-to-protein ratio must be reproducible for different vaccine lots. In PCV formulations, a ratio between 0.3 and 3.0 (w/w) is recommended by the World Health Organization.[96] Determining the amount of free, unconjugated glycan chains in a conjugate vaccine is another important step of QC because free CPS can lower vaccine efficacy considerably.[120] Finally, after formulating the conjugate with an adjuvant, the stability of the vaccine is assessed.[92–94,96] Further aspects of the production and quality control of CCV have been reviewed elsewhere.[95,105,107,119,121]

Despite the success of conjugate vaccines based on isolated polysaccharides, several challenges have to be met to enable the large-scale production of these vaccines: (i) Not every bacterial strain can be cultured easily. Even if a strain can be cultured, the production of the correct CPS in sufficient quantities must be optimized with respect to multiple variables.[37,122–125] (ii) Isolation of certain labile polysaccharides might pose a challenge for the manufacturer due to rapid decomposition.[126,127] Other polysaccharides may be very stable, and thus may be difficult to depolymerize.[128,129] (iii) Despite all the measures of QC, isolated polysaccharide preparations inevitably contain small amounts of impurities, such as the pneumococcal C-polysaccharide.[130,131] The implications of these contaminants on the immune response and possible side effects are not known yet. (iv) Shorter glycans, such as those found on tumor cells, are difficult to prepare on industrial scale using cell culture.

To address these shortcomings, alternative methods of glycan preparation have been developed. These include the enzymatic synthesis of oligosaccharides and the newly developed bacterial protein glycan coupling technology.[132–135] Most efforts have focused on synthetic oligosaccharide-based vaccines.

3.1.4 Opportunities and Challenges of Vaccines Based on Synthetic Oligosaccharides

Many of the quality control steps necessary when dealing with cell-derived preparations can be circumvented by chemical synthesis (Figure 3.2B). Since synthetic oligosaccharides are usually well-defined, they are comparably easy to characterize utilizing the tools of organic chemistry. The chemical assembly of oligosaccharides allows for the site-specific incorporation of a linker for conjugation. Thereby, the structural heterogeneity of the conjugate is drastically lowered, and a more defined antigen is produced.

Synthetic antigens facilitate the investigation of the mechanistic processes of immunization. It is not known whether our current understanding of the humoral immune response induced by glycoconjugate vaccines holds true, or if more sophisticated mechanisms are at play. Synthetic oligosaccharides are key to a better understanding of these mechanisms. In addition, synthetic haptens offer the opportunity of efficient rational vaccine design.

Carbohydrate chemistry has developed a plethora of efficient protecting group strategies, glycosylating agents and activation methods.[136] Synthetic routes can be established for nearly any oligosaccharide structure that can be imagined. However, synthetic carbohydrate chemistry has not yet become standard in the biology lab. When compared to the ease of preparing peptides and oligonucleotides, glycans are trailing far behind. Most oligosaccharide syntheses are still comparable to the total synthesis of a natural product.[137–141] Important details, such as modifications, linker length and conjugation chemistry have to be decided upon at the very beginning of the synthesis.

Over the past two decades, these challenges have been accepted by the glycoscience community, and strategies to facilitate the chemical assembly of glycans have been developed. Rare monosaccharides can be prepared via *de novo* synthesis from simple and inexpensive chiral precursors.[142–152] The concept of automated oligosaccharide synthesis has proven useful and is currently extended to meet the requirements of more elaborate targets.[33,34,153–171] Alternative approaches, such as programmable one-pot glycosylations, have also progressed.[172,173] These technologies are beginning to impact the development of carbohydrate-based vaccine candidates.

3.2 Chemical Immunology of Glycoconjugate Vaccines

Years of research have shaped our current understanding of the immunological mechanisms of carbohydrate conjugate vaccines. However, no consistent guidelines exist for the development and testing of new glycoconjugates. Different strategies are being employed regarding carrier proteins, conjugation chemistry, adjuvants and animal models. Care has to be taken when comparing experiments because they often differ in more than one variable. Nevertheless, general trends can be deduced on the effects of each component of a CCV on the immune response.

3.2.1 Effect of the Glycan on the Immune Response

Considering the structural heterogeneity of bacterial CPS, it is plausible that the immunogenic epitopes of these polysaccharides vary on a case-by-case basis. It is crucial to know the minimal determinant of a polysaccharide that can elicit a protective immune response to design a modern synthetic CCV.

A classical approach to assess the antigenicity of oligosaccharides involves immunization trials with the glycoconjugates of interest. The affinities of antisera from immunized animals to conjugated oligosaccharides and native CPS are determined.[174,175] Antisera or monoclonal antibodies are tested for their ability to bind pathogens and promote phagocytosis.[108,176] Ideally, immunized animals will be challenged with live pathogens to assess the potency of vaccination.[177]

More recent approaches of epitope mapping include the screening of antibodies from infected or immunized individuals for binding to synthetic oligosaccharides. Alternatively, oligosaccharides can be used to inhibit the interaction between antibodies and native polysaccharides.[178–180] Further novel methods of vaccine design are depicted in Section 3.3. Uncovering a potential antigenic epitope of a polysaccharide by methods other than immunization can significantly reduce the number of animal trials.

3.2.1.1 Saccharide Length

Insights into the minimal length of a glycan required to raise a protective immune response can facilitate oligosaccharide synthesis. Two opposing structural concepts come into play regarding the length of a saccharide. Sequential epitopes that comprise rare monosaccharides or unusual linkages may be immunodominant even in very small oligosaccharides. In contrast, polysaccharides may display conformational epitopes only above a certain chain length. Seminal experiments to assess the influence of saccharide length on immunogenicity have been carried out using protein conjugates of depolymerized dextran.[181] Thereby, polysaccharide fragments (1–10 kDa, corresponding to 6–60 monosaccharides) gave rise to higher immune responses than long polysaccharides (>10 kDa). This finding suggests that a certain degree of depolymerization is needed to provide functional glycan antigens.

Studies using both native and synthetic Hib CPS structures led to contradictory results regarding the optimal glycan chain length.[175,182–184] Initially, a CPS fragment comprising 20 RU (40 monosaccharides) was found to be more immunogenic than an 8 RU fragment (16 monosaccharides).[183] However, the same authors later attributed these findings to artifacts associated with the multipoint attachment conjugation strategy that was used.[116] When conjugated via single point attachment, the chain length of Hib CPS fragments was not a crucial determinant for immunogenicity.[116] Finally, it was found by using synthetic oligosaccharides that a hexasaccharide is the minimal structure required to invoke a Hib CPS-specific immune response.[175,184,185] A potent synthetic Hib glycoconjugate vaccine marketed in Cuba contains glycan chains with an average length of 8 RU (16 monosaccharides).[186,187] Thus, relatively short oligosaccharides seem to be sufficient to furnish effective Hib oligosaccharide-based vaccines, although a detailed size–antigenicity correlation study remains to be carried out.

Short glycans can be potent antigens in pneumococcal glycoconjugate vaccines. Synthetic oligosaccharides derived from CPS of *S. pneumoniae* serotypes 3,[177] 6B,[109] 14,[174,188] and others have been prepared and the respective conjugates tested in vaccination settings.[189] Haptens as small as disaccharides have been found to elicit specific antibody responses against the respective CPS. To date, a trisaccharide derived from serotype 6B CPS is the smallest structure that, as a glycoconjugate, confers cross-reactivity against the closely related serotype 6A.[109]

Conformational epitopes are crucial immunodeterminants in a number of bacterial CPS. For instance, conjugate vaccines containing Group B *Streptococcus* type III (GBS III) CPS require long polysaccharide structures to confer protection.[190] In fact, polysaccharide fragments with a size of 14 kDa (approx. 70 monosaccharides) were found to be optimal in TT conjugates against GBS III.[190,191] Fragments with a length of 8 RU (40 monosaccharides) were used to raise a specific immune response when conjugated to a carrier peptide.[192] In accordance with these findings, it was revealed by molecular dynamics calculations that the native polysaccharide adopts a helical structure at a molecular size above 5 RU (25 monosaccharides).[193] Thus, a fragment of this size is likely to be the smallest structure necessary in a conjugate vaccine against GBS III.

Oligosaccharides in CCV against certain *Shigella* strains need to be of intermediate chain length (8–16 monosaccharides corresponding to 2–3 RU) to raise a specific immune response.[194,195] It is currently unclear why shorter glycans are not antigenic, particularly when taking into account that human antisera have been shown to bind to oligosaccharides as short as one repeating unit.[195] Further work is required to elucidate the real protective epitope of these CPS.

If an oligosaccharide does not cover antigenic epitopes of the corresponding CPS, vaccination with the respective conjugate can still elicit robust immune responses. Thereby generated antibodies may not be cross-reactive against the natural antigen. High titers of non-specific antibodies are not desirable in vaccination.[190]

3.2.1.2 Epitope Charge

The majority of bacterial CPS are either uncharged or negatively charged. Anionic properties are usually caused by uronic acid residues or phosphodiester linkages. However, the repeating units of a class of bacterial CPS display both positive and negative charges (Figure 3.1A). These Zwitterionic Polysaccharides (ZPS) are found on certain commensal and symbiotic bacteria and exhibit unique immunomodulatory properties.[196–200] ZPS can be internalized by APC into endosomal compartments, where they are depolymerized in an oxidative fashion.[201,202] ZPS fragments are subsequently loaded onto MHC class II molecules and transported to the cell suface.[203] T cells recognize this ZPS–MHC II complex via their TCR and trigger an immune response.

It has been shown that ZPS induce the expansion of certain T cell subsets, thereby modulating the immune system.[199,201,204–207] Thus, ZPS are the first and only pure carbohydrate structures known to embark on the classical pathway of thymus-dependent antigens.

The immunomodulatory properties of ZPS have been exploited in vaccine design. Therein, zwitterionic charge motifs have been incorporated into polysaccharides to enhance their immunogenicity.[208–210]

It has been proposed that the ability of ZPS to act as T_D antigens is based on their unique secondary structure. In solution, ZPS adopt right-handed helices that are dependent on the presence of the charged residues.[211] To date, a number of ZPS fragments have been chemically synthesized.[140,212,213] These defined probes are free of contaminations with other cellular components. It will be interesting to evaluate the immunological effects of synthetic ZPS probes to gain a better understanding of the importance of charges on carbohydrate antigens.

3.2.1.3 Exposed Epitopes

The principle of conjugate vaccines is based upon the notion that BCR epitopes are exposed to the medium and therefore accessible to the immune system. Consequently, it may be speculated that the most exposed part of a hapten is most likely to interact with a BCR.[214] In addition, it is crucial that the immune system recognizes a glycan chain as "non-self" to trigger an efficient immune response. Thus, fashioning the most unusual moiety of an oligosaccharide, such as a rare bacterial sugar, at the most exposed position at the non-reducing end of the glycan chain is likely to increase immunogenicity. This strategy may be particularly important when the primary BCR epitope is non-conformational. In a recent study, the terminal Man residue of a human high mannose N-glycan-derived oligosaccharide was replaced by synthetic structural mimics.[62] A protein conjugate of one of these oligosaccharide congeners elicited a higher immune response against the native oligosaccharide than conjugates containing the native glycan itself, although no neutralizing antibodies against HIV were obtained.[62] Related structural studies are likely to have implications in future vaccine design (Section 3.3.3).

The reduction of self-like epitopes in conjugates of bacterial CPS can enhance immunogenicity against the pathogen. This concept has been demonstrated with GBS type V CPS fragments.[215] Glycans derived from this polysaccharide carry an α-D-Sia-(2→3)-D-Gal motif in each repeating unit, making them more self-like.[216] Chemically removing sialic acid residues from GBS V CPS fragments significantly enhanced the immunogenicity of the respective glycoconjugates in rhesus macaques.[215] These conjugates provided almost full protection of neonatal mice from a lethal challenge with GBS V after vaccination of the dams.

Synthetic oligosaccharides corresponding to the *O*-specific polysaccharide of *S. dysenteriae* serotype 1 LPS were prepared to evaluate the role of the non-reducing end on immunogenicity.[217] It was found that antigens containing terminal α-D-GlcNAc or α-D-Gal residues induce higher immune responses against the native polysaccharide than those carrying terminal α-L-Rha units. This finding is somewhat surprising because L-rhamnose is usually not found in mammals, and thus should be the least self-like structure present. It is conceivable that the linkages and/or spatial presentation of α-linked D-GlcNAc and D-Gal residues confer the non-self recognition of these oligosaccharides.

Candida albicans is a facultative pathogenic fungus causing infectious disease in immunocompromised people. *C. albicans* cell wall polysaccharides comprise both D-glucans and D-mannans.[218,219] Studies with synthetic β-(1→2)-D-mannan oligosaccharides revealed that an internal epitope is crucial to raise an anti-*Candida* immune response.[178] The authors speculate that the (1→2)-linkages in these glycans allow the exposure of a large part of each monosaccharide to the solvent, making internal residues resemble terminal monosaccharides.[178]

Capsular polysaccharides are usually built up from oligosaccharide building blocks *in vivo*. The sequences of these biological repeating units play an important role in vaccine design. Not taking into account degradation of CPS on the cell surface, the terminal monosaccharide of a repeating unit, together with branched sugars, is probably exposed to the medium in every polysaccharide chain. Thus, a synthetic oligosaccharide reflecting the biological repeating unit might be a more potent antigen than frameshifted glycans. Therefore, vaccine design should take into account the identity of the biological repeating unit by studying CPS biosynthesis.[220,221]

Branching points in saccharide chains can give rise to unique epitopes recognized by B cell receptors. A branched α-D-Glc residue was found to be crucial for immunogenicity of synthetic glycoconjugates against *Shigella flexneri* serotype 2a.[222] In addition, a branched glycan is the antigenic epitope of *S. pneumoniae* serotype 14 CPS. It was found using synthetic oligosaccharides that the sequence β-D-Gal(1→4)[β-D-Glc(1→6)]β-D-GlcNAc is required to raise an efficient antibacterial immune response in mice.[174]

However, incorporation of branching points can be detrimental for the antigenicity of a glycan. As part of an anti-*Candida* carbohydrate conjugate vaccine program, CRM197 conjugates of linear and branched synthetic β-D-glucan oligosaccharides were used to immunize mice.[223] The presence of a β-(1→6)-branch in the otherwise β-(1→3)-linked D-Glc chain prevented the generation of an antifungal immune response. It was speculated that the branching point covers epitopes crucial for recognition by the immune system, or that antibodies elicited against the branching point are non-protective.[224,225]

3.2.1.4 Monosaccharide Modifications

Covalent modifications can further increase the diversity of glycans beyond monosaccharide content and glycosidic linkage heterogeneity. The derivatization of hydroxyl groups confers structural variety, and comprises ester, ether or acetal linkages. These modifications have been reviewed in detail elsewhere.[46,226,227]

Modifications, especially in the form of *O*-acetates, are abundant in pneumococcal, meningococcal and salmonellal CPS.[45,228,229] Several pneumococcal serotypes, such as 9A and 9V or 18F, B and C, differ only in their CPS acetylation patterns.[45] It has been proposed that *O*-acetate groups in CPS promote immune evasion.[230,231] Variations of acetate number and positions can increase the structural heterogeneity of polysaccharides, hampering the ability to elicit a specific immune response.

In contrast, acetate groups are important for inducing a specific anti-CPS immune response in some cases.[232,233] For instance, immunogenicity of the Vi polysaccharide of *S. enterica* sv. *Typhi* is dependent on the presence of both *O*- and *N*-acetates. Vi is an α-(1→4)-linked D-GalNAcA homopolymer, where the monosaccharides are 3-*O*-acetylated to a large extent (Figure 3.1A). Complete removal of *O*-acetates reduces the antigenicity of Vi significantly.[229] The plant polysaccharide pectin differs from Vi only by the presence of a 2-hydroxyl group instead of an acetamide group. TT conjugates of fully *O*-acetylated pectin were used to raise potent anti-Vi immune responses in mice.[234] Thus, both acetyl groups on each monosaccharide in Vi seem to be important for immunogenicity irrespective of the nature of the heteroatom they are linked to.

The intestinal bacterium *Clostridium difficile* ribotype 27 is responsible for grave nosocomial infections. Two cell wall polysaccharides, PS-I and PS-II, are expressed by this pathogen and consist of repeating units connected to each other via phosphodiester linkages.[235] By using a combination of synthetic and isolated oligosaccharides conjugated to CRM197, the phosphate group was shown to be essential for the induction of a specific IgG response against native PS-II and whole bacteria.[236,237] However, our own studies have revealed that a synthetic, non-phosphorylated PS-II repeating unit is recognized by IgA molecules in the stool of infected patients.[180] Further studies will help to unravel the relevance of the phosphate group for immunogenicity.

O-glycans on highly dangerous *Bacillus anthracis* endospores display unusual glycan modifications. The non-reducing end monosaccharide, termed anthrose, contains a 3-hydroxy-3-methylbutamido group at the 4-position and an *O*-methyl ether at the 2-position.[238] A synthetic O-glycan-derived tetrasaccharide was conjugated to the carrier protein keyhole limpet hemocyanin (KLH) and used to generate monoclonal antibodies suitable as a diagnostic tool against anthrax spores.[141,176] Later studies revealed that only slight structural changes in the 4-*N*-acyl moiety are accepted in the induction of a specific immune response. Substitution with an acetamide leads to loss of specificity.[239–241] In contrast, the relevance of the 2-*O*-methyl group for immunogenicity seems to be negligible.[239,241] Recent work has uncovered that a disaccharide containing anthrose at the non-reducing end is the minimal antigenic epitope of *B. anthracis* O-glycans.[242] This finding highlights the importance of unusual monosaccharides at the exposed terminus of a saccharide hapten.

3.2.2 Effects of Conjugation on the Immune Response

3.2.2.1 Conjugation Chemistry

Working with synthetic oligosaccharides offers the advantage of knowing exactly what linkages are generated upon conjugation. Almost any desired functional group can be introduced using an orthogonal linker during conventional or automated synthesis.[35,154,243] Primary amines are the most widely used linker functional groups and are usually installed at the reducing end of the glycan through an aliphatic amino alcohol. Attention has to be paid when the target molecule itself harbors a monosaccharide with a free amino group. In this case, the linker should be orthogonal to free amines. Nevertheless, conjugates have been prepared with oligosaccharides containing two primary amino groups.[55,244] A myriad of different conjugation methods for synthetic oligosaccharides exist and have been extensively reviewed.[26,29,31,35,107,245–253]

A bifunctional spacer is often incorporated to activate the oligosaccharide (Figure 3.1B) and to increase the distance between glycan and carrier. In the case of an amino-functionalized oligosaccharide, typically an activated diacid is used, such as the bis-*N*-hydroxysuccinimide ester of adipic acid. The resulting mono-active ester is then attached to a lysine side chain of the carrier.

3.2.2.2 Immunogenicity of Linker/Spacer Constructs

Depending on the conjugation chemistry that is used, new epitopes may arise that can induce an immune response. Immunogenicity of these epitopes is of particular relevance in the case of weakly antigenic haptens. Thereby, immune responses against linker/spacer combinations may outcompete the desired anti-hapten immune responses.[254–256] In these cases, the use of alternative linker/spacer combinations and structurally heterogeneous linkers presents an alternative.[249,257]

Despite the success of azide-alkyne 1,3-dipolar cycloadditions that are commonly used in chemical biology,[258,259] these methods are not widely used for conjugation in vaccine design. Bulky hydrophobic functional groups have been associated with high immunogenicity.[254,255,260] However, detailed studies have not yet been conducted to confirm this finding.

3.2.2.3 Antigen Density

The amount of hapten that is linked to a carrier is referred to as the hapten density or loading. Although these terms are used for protein conjugates of both synthetic and isolated glycans, their meanings can differ. Reporting a mol/mol ratio of glycan and protein is meaningless for the multipoint attachment of long polysaccharides, and the density is given as a weight/weight ratio in these cases. In contrast, when handling conjugates generated by single point attachment, such as those containing synthetic oligosaccharides, the antigen density is reflected by the molar ratio of both components. It has been found that optimization of the loading can have an impact on repertoire, magnitude and kinetics of an antibody response.[261] Studies with conjugates comprising the model hapten dinitrophenol indicated that a very high hapten loading reduces immunogenicity.[262–264] More complex mechanisms seem to be at play for carbohydrate–protein conjugates, depending on the length of the glycan. While synthetic oligosaccharides of intermediate chain length (up to 16 monosaccharides) exhibited low immunogenicity at high loading,[194,265] this correlation was not observed with shorter oligosaccharides.[177,188] Thus, finding the optimal hapten density of a glycoconjugate may be essential to avoid unwanted side effects.

3.2.2.4 T Cell Epitopes

The classical dogma of thymus-dependence dictates that peptides are the determinants for invoking T cell help (Figure 3.3, pathway 1). Thereby, in case of a carbohydrate–protein conjugate, saccharide chains are only required for initial targeting to glycan-specific B cells. Carrier-derived peptides are presented by these B cells to $CD4^+$ T cells. These carrier-specific T cells finally induce maturation and proliferation of glycan-specific B cells.

Under this paradigm, the density and diversity of T cell epitopes in carrier proteins are important determinants for the immunogenicity of conjugate vaccines. By using bioinformatics and *in vitro* studies, major T cell epitopes have been mapped into carrier proteins, such as TT and CRM197.[266–268] Furthermore, peptides generated by intracellular processing of a commercial glycoconjugate vaccine have been identified.[269] A set of these peptides have been found to efficiently activate T cells after presentation by MHC II molecules. Derivatizing these T cell epitope peptides with oligosaccharides potentially abrogates binding to MHC II or TCR molecules and may result in decreased vaccine efficacy.[194,195] Interestingly, patterns of carrier processing and presentation have been found to depend on the serotype in pneumococcal conjugate vaccines.[269,270]

With the finding that other classes of compounds, such as ZPS and certain lipids,[271] can adopt the thymus-dependent immune response pathway, the dogma that proteins are the only T_D antigens had to be extended. Consequently, it may be asked whether the T cell epitopes contained in a glycoconjugate are exclusively non-glycosylated peptides. Several groups have reported the generation of carbohydrate-specific T cells.[192,272–275] These T cells recognize the saccharide moiety of glycopeptides presented by MHC I or II molecules in a TCR-dependent fashion (Figure 3.3, pathway 2). It was found recently that vaccination with a heterologous peptide conjugated to GBS III

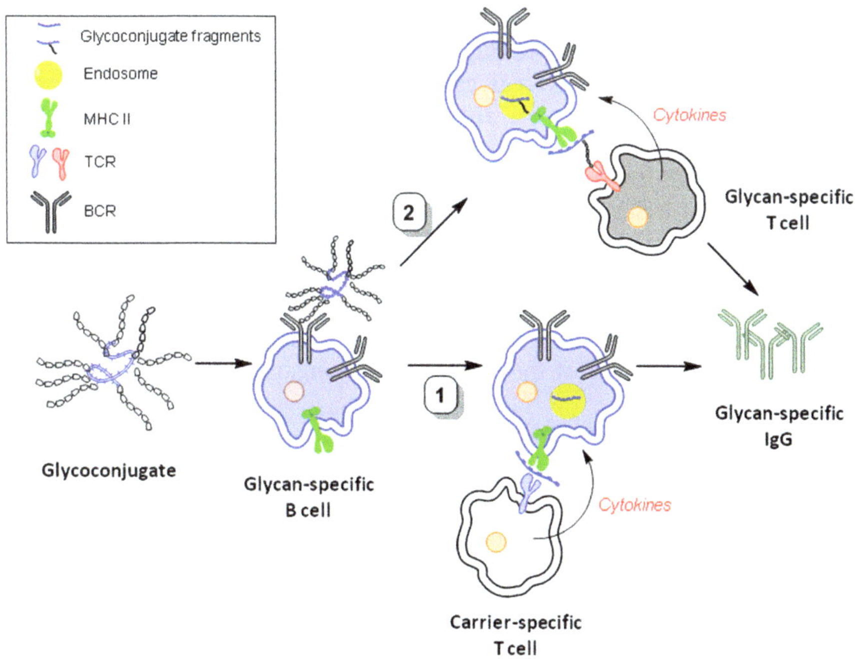

Figure 3.3 Processing and presentation of glycoconjugate vaccines. After uptake into carbohydrate-specific B cells, carbohydrate-protein conjugates are depolymerized by proteolysis. The resulting fragments are then loaded onto MHC class II molecules. Unmodified peptides are recognized by peptide-specific T cells (pathway 1), whereas glycopeptides can be recognized by carbohydrate-specific T cells (pathway 2). Both pathways lead to the secretion of glycan-specific antibodies of the IgG type.

CPS produces polysaccharide-specific $CD4^+$ T cells. Based on this finding, the glycan density on the peptide carrier was optimized to furnish a vaccine that is two orders of magnitude more efficient than conventional glycoconjugates.[192] If further studies reveal that glycopeptide presentation on MHC II generally enhances vaccine efficacy, this insight will influence the design of future conjugate vaccines.

Both mechanisms for T cell activation by glycoconjugates may hold true. Already in 1992, it was proposed that T cells may recognize the saccharide part of a glycopeptide depending on the position of the glycosylation site.[276] Further studies are required to unravel the precise mechanisms underlying T cell recruitment of glycoconjugates.

3.2.2.5 Carrier Priming and Carrier-Induced Epitope Suppression

Following the concept of thymus-dependence, administration of the carrier ("priming") before vaccinating with a glycoconjugate should result in an improved anti-glycan immune response. In this scenario, carrier-specific $CD4^+$ T cells are expanded during priming. Upon vaccination with the conjugate, glycan-specific B cells can present carrier peptides to the already activated set of T cells. This leads to an augmented immune response towards the conjugate. Indeed, carrier priming has been proven useful in Hib-TT,[277,278] but not in Hib-CPS conjugates with an outer membrane protein complex of *N. meningitidis* (Hib-OMPC).[279]

In contrast, pre-existing immunity to a carrier may also negatively influence the immune response to glycoconjugates. Carrier-induced epitope suppression (CIES) leads to lower titers of carbohydrate-specific antibodies due to the dominant carrier-directed immune response.[278] This phenomenon has been associated with the nature of the carrier, the conjugation chemistry used and the glycan attached.[278]

Inactivated toxins exhibit high intrinsic immunogenicity due to the abundance of B and T cell epitopes displayed.[88,278] Following priming, the number of carrier-specific immune cells may be so high that they completely override the immunogenicity of the hapten after vaccination.[278] An alternative is the use of non-denatured carrier proteins, such as CRM197, that are generally not associated with CIES.[278]

Only few proteins are approved for use as carriers in commercially available vaccines. These include DT, TT, CRM197, OMPC and Protein D from non-typeable *Haemophilus influenzae*, and safety profiles have been established for them.[90,92,96,98,280,281] However, special care has to be taken regarding the effects of pre-immunization with these carriers. DT and TT are contents of the childhood diphtheria-pertussis-tetanus (DPT) vaccine, and hence have been administered to many individuals. The efficiency of glycoconjugate vaccines based on these carriers might be hampered due to CIES.

3.3 Novel Developments in Vaccine Design

Since the introduction of carbohydrate conjugate vaccines, many creative approaches have been pursued to improve vaccine efficiency. Particularly, weakly immunogenic tumor antigens have been the subjects of extensive research. New concepts of antigen design, delivery and presentation have been scrutinized. Studies on carbohydrate-based vaccines have stimulated investigations into different areas of immunology, structural biology, chemical biology and materials science. Listed here are a few select examples to illustrate the creativity that is fueling the field.

3.3.1 Multicomponent Vaccines

Tumor-associated carbohydrate antigens are among the least immunogenic glycans used in vaccination studies. Short chain TACAs, such as the Tn or sTn antigens (Figure 3.1C), are highly abundant on the membrane protein Muc1 on the surface of malignant epithelial cells. TACAs do not invoke a strong non-self recognition by the human immune system.[282] Consequently, conventional

TACA–protein conjugates are unlikely to raise a potent anti-tumor immune response. Kunz and coworkers have studied conjugates of Muc1-derived glycopeptides to carrier proteins to put on display more defined tumor-derived BCR epitopes.[283–288] These glycopeptides–protein conjugates were used to elicit specific anti-tumor immune responses in mice.[284,286,288–290]

The immunological requirements on tumor vaccines differ conceptually from those against bacteria. While the production of saccharide-specific antibodies can induce immunity against most pathogens, an additional cellular immune response is required in cancer vaccination. Thus, a $CD4^+$ helper T cell response should be accompanied by activation of $CD8^+$ cytotoxic T cells. Therefore, antigens need to be loaded on MHC class I molecules after internalization, a process that is called cross-presentation and can be enhanced by TLR stimulation.[38,64,291] Several groups have combined Muc1 glycopeptides with TLR ligands and studied the effects on immunogenicity.[292–295] Synthetic lipopeptides containing the TLR2 agonist Pam_3CysSK_4 and Muc1 glycopeptides have been used to raise glycopeptide-specific immune responses in mice.[292] However, antibody titers were not as high as in the corresponding glycopeptide-TT conjugates, suggesting incomplete T cell recruitment. Indeed, incorporation of a T cell epitope peptide led to significantly higher IgG responses upon immunization than induced with the corresponding two-component antigen.[293] Very similar results have been observed by the Boons group, using a similar three-component vaccine in a liposomal formulation.[295,296] In addition to eliciting an efficient humoral immune response, recent results suggest that the tripartite antigen can also induce cytolytic activity by $CD8^+$ T cells and reduce tumor formation in mice.[294] Thus, introduction of additional immunomodulatory components into glycoconjugates offers the possibility to tailor the immune response according to the demands on an antigen.

3.3.2 Novel Carrier Platforms

Carrier proteins are widely used and well-established in carbohydrate-based vaccines. However, the side effects observed due to the high immunogenicity of commonly used carriers have prompted the search for alternatives. In addition, finding non-protein carrier platforms may alleviate the need of a cold chain for transportation and storage of vaccines. Not having to maintain a cold chain would tremendously lower the cost per dose of vaccine.[297,298]

Many new carrier strategies focus on a multivalent hapten presentation. Encouraging results in vaccination experiments have been obtained with oligosaccharide-bound virosomes, liposomes and gold nanoparticles (GNP).[299–301] A shortcoming of fully synthetic carriers, such as GNP, is that their fate inside an organism is often unclear. Thus, both *in vivo* toxicity and clearance pathways have to be accurately assessed for these materials.[302]

Recently, a novel carrier system was reported based on self-assembling peptide fibers.[303] Thereby, solid phase peptide synthesis was employed to fuse the well-known self-aggregating peptide Q11 to Muc1 glycopeptides. The resulting two-component peptides formed solid fibers with exposed Muc1 epitopes and were used to induce a tumor-specific antibody response.[303] No adjuvant was necessary in immunizations using these fibers, suggesting an adjuvant-like activity by the solid aggregates. The resulting immune response was indicative of a lack of direct T cell help. Thus, it may be interpreted that the high antibody titers are simply induced by the multivalent hapten display to B cells.[304]

The immunomodulatory properties of ZPS (see 3.2.1.2) have been exploited in vaccine design. *Bacteroides fragilis* PSA1, the best studied ZPS representative, has been used as a carrier for the Tn antigen to produce an all-carbohydrate cancer vaccine candidate.[305,306] The Tn–PSA1 conjugate was found to elicit a specific anti-Tn IgG response in mice without the need of an adjuvant.[305] Antisera from these mice bound to cancer cells slightly better than antisera raised against unmodified PSA1.[306] Although hapten presentation by ZPS requires further optimization, the concept of all-carbohydrate vaccines offers interesting prospects because they are likely to be biocompatible and may not require a cold chain.

3.3.3 Rational Vaccine Design

Finding the right antigen has always been a matter of trial and error and a process that is costly and time-consuming. Methods are needed to narrow down the number of potential antigens to those that are most likely to protect from disease.

The field of protein subunit vaccines has experienced a boost in the era of genomics, especially after the introduction of high-throughput genome sequencing technologies.[307,308] Thus, entire bacterial genomes can be analyzed and compared to identify cell surface structures that constitute potential antigens in a process called "reverse vaccinology".[307,309,310] However, it is hard to deduce the structure of cell surface glycans from the genome sequence.[309] Hence, reverse vaccinology is hardly applicable to carbohydrates.[220]

Alternative ways to find potential carbohydrate antigens involve chemical glycomics and structural vaccinology.[32,310–313] The key to success is the availability of antiserum or neutralizing antibodies against the target organism. Carbohydrate microarrays can then reveal an antigenic epitope based on the binding pattern of antiserum among a wide range of different glycans.[180,313–319] In contrast, knowing the structure of the antigen binding site of a neutralizing antibody can give information on a potential optimal epitope.[62]

An excellent example of structural vaccinology has been set in the search for a vaccine against the non-carbohydrate drug (+)-methamphetamine.[320] Molecular modeling was used to determine the lowest energy structures of the target molecule in water. Haptens were designed based on these calculations and the corresponding BSA conjugates used to immunize mice. The obtained antisera exhibited high antibody titers and affinities against (+)-methamphetamine. Hence, structural studies can provide essential information on the identity of potent antigens.

A view on the potential of rational glycoconjugate vaccine design has been provided by Bundle and coworkers. *C. albicans* cell wall mannans are composed of α-(1→6)-linked D-Man main chains carrying phosphodiester linkages and oligo-β-(1→2)-D-Man side chains.[321] Studies on several synthetic β-(1→2)-D-mannan oligosaccharides revealed that they adopt a helical secondary structure with the hydroxyl groups exposed to the medium.[179,322,323] Binding of the synthetic oligosaccharides to a neutralizing antibody was then assessed by inhibition experiments. Interestingly, di- and trisaccharides exhibited higher affinities to the antibody than longer oligosaccharides.[179] To further evaluate the nature of these interactions, oligosaccharide congeners were generated by either methylating or abstracting specific hydroxyl groups.[324,325] By using NMR spectroscopy, molecular modeling and inhibition studies, it was found that the primary epitope recognized by the neutralizing antibody is a disaccharide.[178,254] It was revealed that longer oligosaccharides need to undergo conformational changes to be accommodated in the binding pocket, a process that significantly reduces their affinities to the antibody. These results culminated in the design of a disaccharide conjugate vaccine that triggered a specific antibody response and reduced the loading of *C. albicans* in a challenge experiment in rabbits.[254] The structural studies described here were conducted to explain the results obtained from previous immunization experiments. Nevertheless, structure-based antigen design may gain importance in the development of carbohydrate-based vaccines, as it already has in the field of protein vaccines.[312]

3.4 Conclusions

Since the first experiments in the early 20th century, polysaccharide vaccines have emerged as an effective alternative to antibiotic treatment. Carbohydrate conjugate vaccines succeeded in protecting infants, the most important risk group for many infectious diseases, from pathogenic bacteria. The safety of glycoconjugate vaccine formulations is ensured by numerous checkpoints of quality control. Research has provided insights into the role of each component of carbohydrate conjugate vaccines in inducing a potent and protective immune response. However, many questions on the immunological mechanisms of carbohydrate-based vaccines are yet to be answered.

In the future, the design of carbohydrate-based vaccines will more than ever rely on new fundamental insights. Immunology must penetrate more deeply the mechanisms of vaccination to lay the groundwork for novel carriers and formulations. Particularly, the development of non-protein carriers is an important step towards structurally defined, adjuvant-free vaccines. Structural biochemistry and glycomics must provide insights into potential antigenic structures, thereby lowering the efforts of vaccine design. For instance, structure elucidation of the antigen binding site of a neutralizing antibody can precede the development of structurally optimized antigens that generate a highly potent humoral immune response. Oligosaccharide assembly must become efficient enough to provide large quantities of synthetic material for vaccination studies. It is exciting to anticipate the technical breakthroughs that will fuel vaccine development in the next decades.

Acknowledgements

We thank the Max Planck Society for generous financial support. This work was supported in part by a Kekulé doctoral fellowship by the Fonds der Chemischen Industrie (B.S.). We thank Dr Bernd Lepenies and Dr Jonathan Hudon (both Max Planck Institute of Colloids and Interfaces, Potsdam) for critically reviewing the manuscript, and Prof. Dr David Bundle (University of Alberta) for helpful discussions.

References

1. M. Heidelberger and O. T. Avery, *J Exp Med*, 1923, **38**, 73–79.
2. W. F. Goebel, *J Exp Med*, 1938, **68**, 469–484.
3. W. F. Goebel and R. D. Hotchkiss, *J Exp Med*, 1937, **66**, 191–205.
4. W. F. Goebel, *J Exp Med*, 1936, **64**, 29–38.
5. W. F. Goebel, O. T. Avery and F. H. Babers, *J Exp Med*, 1934, **60**, 599–617.
6. W. S. Tillett, O. T. Avery and W. F. Goebel, *J Exp Med*, 1929, **50**, 551–567.
7. W. F. Goebel and O. T. Avery, *J Exp Med*, 1929, **50**, 521–531.
8. O. T. Avery and W. F. Goebel, *J Exp Med*, 1929, **50**, 533–550.
9. K. Landsteiner, *Biochem Z*, 1921, **119**, 294–306.
10. M. Heidelberger and F. E. Kendall, *J Exp Med*, 1933, **57**, 373–379.
11. T. Francis and W. S. Tillett, *J Exp Med*, 1930, **52**, 573–585.
12. W. F. Goebel and O. T. Avery, *J Exp Med*, 1931, **54**, 431–436.
13. O. T. Avery and W. F. Goebel, *J Exp Med*, 1931, **54**, 437–447.
14. P. A. Rebers, S. Estrada-Parra and M. Heidelberger, *J Bacteriol*, 1963, **86**, 882–883.
15. A. E. Bridy-Pappas, M. B. Margolis, K. J. Center and D. J. Isaacman, *Pharmacotherapy*, 2005, **25**, 1193–1212.
16. N. Woodford and D. M. Livermore, *J Infect*, 2009, **59** (Suppl. 1), S4–16.
17. G. Ada and D. Isaacs, *Clin Microbiol Infect*, 2003, **9**, 79–85.
18. B. M. Gray and D. M. Musher, in *Pneumococcal Vaccines*, eds. G. R. Silber, K. P. Klugman and P. H. Mäkelä, ASM Press, Washington, DC, Editon edn., 2008, pp. 3–17.
19. P. H. Mäkelä and J. C. Butler, in *Pneumococcal Vaccines*, eds. G. R. Silber, K. P. Klugman and P. H. Mäkelä, ASM Press, Washington, DC, 2008, pp. 19–29.
20. R. S. Becker, *Springer Seminars in Immunopathology*, 1993, **15**, 217–226.
21. H. Käyhty, S. Lockhart and L. Schuerman, in *Pneumococcal Vaccines*, eds. G. R. Silber, K. P. Klugman and P. H. Mäkelä, ASM Press, Washington, DC, 2008, 227–243.

22. C. G. Whitney and M. R. Moore, in *Pneumococcal Vaccines*, eds. G. R. Silber, K. P. Klugman and P. H. Mäkelä, ASM Press, Washington, DC, Editon edn., 2008, 353–368.
23. P. Dro, *Genet Eng Biotechn N*, 2012, **32**, 38–38.
24. *Recommended Immunizations for Children*, http://www.cdc.gov/vaccines/parents/downloads/parent-ver-sch-0-6yrs.pdf, accessed December 2012.
25. J. F. Vliegenthart, *FEBS letters*, 2006, **580**, 2945–2950.
26. Y. L. Huang and C. Y. Wu, *Expert Rev Vaccines*, 2010, **9**, 1257–1274.
27. R. D. Astronomo and D. R. Burton, *Nat Rev Drug Discovery*, 2010, **9**, 308–324.
28. M. L. Hecht, P. Stallforth, D. V. Silva, A. Adibekian and P. H. Seeberger, *Curr Opin Chem Biol*, 2009, **13**, 354–359.
29. V. Pozsgay, *Curr Top Med Chem*, 2008, **8**, 126–140.
30. Z. W. Guo and Q. L. Wang, *Curr Opin Chem Biol*, 2009, **13**, 608–617.
31. D. R. Bundle, in *Vaccine Design: Innovative Approaches and Novel Strategies*, eds. R. Rappuoli and F. Bagnoli, Caister Academic Press, Norfolk, UK, 2011, pp. 69–107.
32. B. Lepenies and P. H. Seeberger, *Immunopharmacol Immunotoxicol*, 2010, **32**, 196–207.
33. L. Kröck and P. H. Seeberger, *Chimia*, 2008, **62**, 8–12.
34. P. H. Seeberger, *Abstr Pap Am Chem S*, 2003, **226**, U159–U159.
35. L. Morelli, L. Poletti and L. Lay, *Eur J Org Chem*, 2011, 5723–5777.
36. S. Lockhart, *Expert Rev Vaccines*, 2003, **2**, 633–648.
37. R. Rappuoli, S. Black and P. H. Lambert, *Lancet*, 2011, **378**, 360–368.
38. A. K. Abbas, A. H. Lichtman and S. Pillai, *Cellular and Molecular Immunology*, 7 edn., Elsevier Saunders, Philadelphia, 2012.
39. B. Pulendran and R. Ahmed, *Nat Immunol*, 2011, **12**, 509–517.
40. R. Arnon, in *Vaccine Design*, eds. R. Rappuoli and F. Bagnoli, Caister Academic Press, Norfolk, UK, 2011, pp. 1–19.
41. J. Yother, *Annu Rev Microbiol*, 2011, **65**, 563–581.
42. S. Merino and J. M. Tomás, in *Encyclopedia of Life Sciences (ELS)*, John Wiley & Sons, Chichester, UK, 2010.
43. T. Hallstrom and K. Riesbeck, *Trends Microbiol*, 2010, **18**, 258–265.
44. C. Whitfield, *Ann Rev Biochem*, 2006, **75**, 39–68.
45. J. P. Kamerling, in *Streptococcus pneumoniae*, A. Tomasz, Mary Ann Liebert, Inc., Larchmont, NY, 2000, pp. 81–114.
46. S. Herget, P. V. Toukach, R. Ranzinger, W. E. Hull, Y. A. Knirel and C. W. von der Lieth, *BMC Struct Biol*, 2008, **8**, 35.
47. N. K. Kochetkov, *Usp Khim*, 1996, **65**, 799–835.
48. I. W. Sutherland, *Int Dairy J*, 2001, **11**, 663–674.
49. A. Scherf, J. J. Lopez-Rubio and L. Riviere, *Annu Review Microbiol*, 2008, **62**, 445–470.
50. C. Stockdale, M. R. Swiderski, J. D. Barry and R. McCulloch, *PLoS Biol*, 2008, **6**, e185.
51. Y. H. Tsai, X. Liu and P. H. Seeberger, *Angew Chem Int Ed Engl*, 2012, **51**, 11438–11456.

52. F. Debierre-Grockiego and R. T. Schwarz, *Glycobiology*, 2010, **20**, 801–811.
53. R. S. Naik, O. H. Branch, A. S. Woods, M. Vijaykumar, D. J. Perkins, B. L. Nahlen, A. A. Lal, R. J. Cotter, C. E. Costello, C. F. Ockenhouse, E. A. Davidson and D. C. Gowda, *J Exp Med*, 2000, **192**, 1563–1576.
54. P. H. Seeberger, R. L. Soucy, Y. U. Kwon, D. A. Snyder and T. Kanemitsu, *Chem Commun (Camb)*, 2004, 1706–1707.
55. L. Schofield, M. C. Hewitt, K. Evans, M. A. Siomos and P. H. Seeberger, *Nature*, 2002, **418**, 785–789.
56. V. P. Martins, C. S. Pinheiro, B. C. Figueiredo, N. R. Assis, S. B. Morais, M. V. Caliari, V. Azevedo, W. Castro-Borges, R. A. Wilson and S. C. Oliveira, *Clin Dev Immunol*, 2012, 962538.
57. D. H. Barouch, *Nature*, 2008, **455**, 613–619.
58. C. N. Scanlan, R. Pantophlet, M. R. Wormald, E. Ollmann Saphire, R. Stanfield, I. A. Wilson, H. Katinger, R. A. Dwek, P. M. Rudd and D. R. Burton, *J Virol*, 2002, **76**, 7306–7321.
59. L. M. Walker, M. Huber, K. J. Doores, E. Falkowska, R. Pejchal, J. P. Julien, S. K. Wang, A. Ramos, P. Y. Chan-Hui, M. Moyle, J. L. Mitcham, P. W. Hammond, O. A. Olsen, P. Phung, S. Fling, C. H. Wong, S. Phogat, T. Wrin, M. D. Simek, G. P. I. Protocol, W. C. Koff, I. A. Wilson, D. R. Burton and P. Poignard, *Nature*, 2011, **477**, 466–470.
60. C. L. Lavine, S. Lao, D. C. Montefiori, B. F. Haynes, J. G. Sodroski, X. Yang and N. C. f. H. A. V. Immunology, *J Virol*, 2012, **86**, 2153–2164.
61. R. Pejchal, K. J. Doores, L. M. Walker, R. Khayat, P. S. Huang, S. K. Wang, R. L. Stanfield, J. P. Julien, A. Ramos, M. Crispin, R. Depetris, U. Katpally, A. Marozsan, A. Cupo, S. Maloveste, Y. Liu, R. McBride, Y. Ito, R. W. Sanders, C. Ogohara, J. C. Paulson, T. Feizi, C. N. Scanlan, C. H. Wong, J. P. Moore, W. C. Olson, A. B. Ward, P. Poignard, W. R. Schief, D. R. Burton and I. A. Wilson, *Science*, 2011, **334**, 1097–1103.
62. K. J. Doores, Z. Fulton, V. Hong, M. K. Patel, C. N. Scanlan, M. R. Wormald, M. G. Finn, D. R. Burton, I. A. Wilson and B. G. Davis, *Proc Natl Acad Sci USA*, 2010, **107**, 17107–17112.
63. M. Y. Mapara and M. Sykes, *J Clin Oncol*, 2004, **22**, 1136–1151.
64. G. Lizee, W. W. Overwijk, L. Radvanyi, J. Gao, P. Sharma and P. Hwu, *Ann Rev Med*, 2013, **64**, 25.1–25.20.
65. Y. Cao, P. Stosiek, G. F. Springer and U. Karsten, *Histochem Cell Biol*, 1996, **106**, 197–207.
66. Y. Xu, A. Sette, J. Sidney, S. J. Gendler and A. Franco, *Immunol Cell Biol*, 2005, **83**, 440–448.
67. A. Cazet, S. Julien, M. Bobowski, J. Burchell and P. Delannoy, *Breast Cancer Res*, 2010, **12**, 204.
68. T. Muramatsu, *Glycobiology*, 1993, **3**, 291–296.
69. K. E. Stein, *J Infect Dis*, 1992, **165** (Suppl. 1), S49–52.

70. J. J. Mond, A. Lees and C. M. Snapper, *Annu Rev Immunol*, 1995, **13**, 655–692.
71. P. Lane, *Clin Exp Immunol*, 1996, **105**, 10–11.
72. Q. Vos, A. Lees, Z. Q. Wu, C. M. Snapper and J. J. Mond, *Immunol Rev*, 2000, **176**, 154–170.
73. Y. Qiu and H. J. Kung, *Oncogene*, 2000, **19**, 5651–5661.
74. A. Q. Khan, G. Sen, S. Guo, O. N. Witte and C. M. Snapper, *Infect Immun*, 2006, **74**, 1419–1424.
75. A. Maas and R. W. Hendriks, *Dev Immunol*, 2001, **8**, 171–181.
76. C. M. Snapper and J. J. Mond, *J Immunol*, 1996, **157**, 2229–2233.
77. C. M. Snapper, H. Yamaguchi, M. A. Moorman and J. J. Mond, *J Immunol*, 1994, **152**, 4884–4892.
78. P. H. Mäkelä, M. Leinonen, J. Pukander and P. Karma, *Rev Infect Dis*, 1981, **3 Suppl**, S124–132.
79. R. M. Douglas, J. C. Paton, S. J. Duncan and D. J. Hansman, *J Infect Dis*, 1983, **148**, 131–137.
80. J. L. Sloyer, J. H. Ploussard and V. M. Howie, *Rev Infect Dis*, 1981, **3**, S119–S123.
81. A. Finn, *Br Med Bull*, 2004, **70**, 1–14.
82. V. Pozsgay, *Advances in carbohydrate chemistry and biochemistry*, 2000, **56**, 153–199.
83. *Hib Conjugate Vaccine Information*, http://www.cdc.gov/vaccines/pubs/pinkbook/downloads/hib.pdf, accessed October 31, 2012.
84. *PPV Product Information*, http:// www. cdc. gov/ vaccines/ pubs/ vis/ downloads/vis-ppv.pdf, Accessed October 31, 2012.
85. *Typhim Vi Product Information*, http://www.fda.gov/downloads/BiologicsBloodVaccines/Vaccines/ApprovedProducts/UCM142811.pdf, accessed October 31, 2012.
86. *Menomune Product Information*, http://www.fda.gov/downloads/BiologicsBloodVaccines/Vaccines/ApprovedProducts/UCM308370.pdf, accessed October 31, 2012.
87. A. L. Reingold, C. V. Broome, A. W. Hightower, G. W. Ajello, G. A. Bolan, C. Adamsbaum, E. E. Jones, C. Phillips, H. Tiendrebeogo and A. Yada, *Lancet*, 1985, **2**, 114–118.
88. J. M. Keith, in *Vaccine Design*, eds. R. Rappuoli and F. Bagnoli, Caister Academic Press, Norfolk, UK, 2011, pp. 109–138.
89. E. Malito, B. Bursulaya, C. Chen, P. Lo Surdo, M. Picchianti, E. Balducci, M. Biancucci, A. Brock, F. Berti, M. J. Bottomley, M. Nissum, P. Costantino, R. Rappuoli and G. Spraggon, *Proc Natl Acad Sci USA*, 2012, **109**, 5229–5234.
90. J. D. Croxtall and G. M. Keating, *Paediatr Drugs*, 2009, **11**, 349–357.
91. H. L. Johnson, M. Deloria-Knoll, O. S. Levine, S. K. Stoszek, L. Freimanis Hance, R. Reithinger, L. R. Muenz and K. L. O'Brien, *PLoS Med*, 2010, **7**.
92. *Recommendations for the production and control of Haemophilus influenzae type b conjugate vaccines*, World Health Organization, Geneva, 2000.

93. *Assessment Report for Synflorix*, European Medicines Agency, London, 2009.
94. *Recommendations to assure the quality, safety and efficacy of group A meningococcal conjugate vaccines*, World Health Organization, Geneva, 2006.
95. C. Jones, in *Carbohydrate-Based Vaccines*, R. Roy, American Chemical Society, Washington, DC, 2008, pp. 21–35.
96. *Recommendations for the production & control of pneumococcal conjugate vaccines*, World Health Organization, Geneva, 2003.
97. *Recommendations to assure the quality, safety and efficacy of pneumococcal conjugate vaccines*, World Health Organization, Geneva, 2009.
98. *Recommendations to Assure the Quality, Safety and Efficacy of DT-based Combined Vaccines*, World Health Organization, Geneva, 2012.
99. P. Talaga, S. Vialle and M. Moreau, *Vaccine*, 2002, **20**, 2474–2484.
100. C. Jones and F. Currie, *Biologicals*, 1991, **19**, 41–47.
101. J. S. Kim, E. R. Laskowich, R. G. Arumugham, R. E. Kaiser and G. J. MacMichael, *Anal Biochem*, 2005, **347**, 262–274.
102. L. Canaan-Haden, J. Cremata, J. Chang, Y. Valdes, F. Cardoso and V. V. Bencomo, *Vaccine*, 2006, **24**, S70–S71.
103. C. Vipond, B. Mulloy, P. Rigsby, K. Burkin, B. Bolgiano and MenC IS Working Group, *Biologicals*, 2012, **40**, 353–363.
104. C. M. Tsai, X. X. Gu and R. A. Byrd, *Vaccine*, 1994, **12**, 700–706.
105. C. Jones, *An Acad Bras Cienc*, 2005, **77**, 293–324.
106. F. Michon, C. Uitz, A. Sarkar, A. J. D'Ambra, M. Laude-Sharp, S. Moore and P. C. Fusco, *Clin Vaccine Immunol*, 2006, **13**, 936–943.
107. P. Costantino, R. Rappuoli and F. Berti, *Exp Opin Drug Discovery*, 2011, **6**, 1045–1066.
108. M. R. Wessels, L. C. Paoletti, H. K. Guttormsen, F. Michon, A. J. D'Ambra and D. L. Kasper, *Infect Immun*, 1998, **66**, 2186–2192.
109. W. T. M. Jansen, S. Hogenboom, M. J. L. Thijssen, J. P. Kamerling, J. F. G. Vliegenthart, J. Verhoef, H. Snippe and A. F. M. Verheul, *Infect Immun*, 2001, **69**, 787–793.
110. J. Poolman, C. Frasch, A. Nurkka, H. Kayhty, R. Biemans and L. Schuerman, *Clin Vaccine Immunol*, 2011, **18**, 327–336.
111. R. Schneerson, O. Barrera, A. Sutton and J. B. Robbins, *J Exp Med*, 1980, **152**, 361–376.
112. S. C. Szu, A. L. Stone, J. D. Robbins, R. Schneerson and J. B. Robbins, *J Exp Med*, 1987, **166**, 1510–1524.
113. A. F. Verheul, G. J. Boons, G. A. Van der Marel, J. H. Van Boom, H. J. Jennings, H. Snippe, J. Verhoef, P. Hoogerhout and J. T. Poolman, *Infect Immun*, 1991, **59**, 3566–3573.
114. A. Pawlowski, G. Källenius and S. B. Svenson, *Vaccine*, 1999, **17**, 1474–1483.
115. S. L. Demento, A. L. Siefert, A. Bandyopadhyay, F. A. Sharp and T. M. Fahmy, *Trends Biotechnol*, 2011, **29**, 294–306.
116. P. W. Anderson, M. E. Pichichero, E. C. Stein, S. Porcelli, R. F. Betts, D. M. Connuck, D. Korones, R. A. Insel, J. M. Zahradnik and R. Eby, *J Immunol*, 1989, **142**, 2464–2468.

117. H. Aggerbeck and I. Heron, *Biologicals*, 1992, **20**, 109–115.
118. D. F. Kelly, E. R. Moxon and A. J. Pollard, *Immunology*, 2004, **113**, 163–174.
119. A. Lees, V. Puvanesarajah and C. E. Frasch, in *Pneumococcal Vaccines*, eds. G. R. Silber, K. P. Klugman and P. H. Mäkelä, ASM Press, Washington, DC, 2008, pp. 163–174.
120. C. E. Frasch, *Vaccine*, 2009, **27**, 6468–6470.
121. N. Allison and H. S. Tranter, in *Vaccine Protocols*, eds. A. Robinson and M. J. Hudson, Humana Press, Inc., Totowa, NJ, 2003, pp. 391–407.
122. S. D. Jin, Y. M. Kim, H. K. Kang, S. J. Jung and D. Kim, *J Microbiol Biotechnol*, 2009, **19**, 1374–1378.
123. H. Jang, Y. K. Yoon, J. A. Kim, H. S. Kim, S. J. An, J. H. Seo, C. Cui and R. Carbis, *J Biotechnol*, 2008, **135**, 71–77.
124. V. M. Goncalves, T. C. Zangirolami, R. L. Giordano, I. Raw, M. M. Tanizaki and R. C. Giordano, *Appl Microbiol Biotechnol*, 2002, **59**, 713–717.
125. D. Cimini, O. F. Restaino, A. Catapano, M. De Rosa and C. Schiraldi, *Appl Microbiol Biotechnol*, 2010, **85**, 1779–1787.
126. A. W. Sturgess, K. Rush, R. J. Charbonneau, J. I. Lee, D. J. West, R. D. Sitrin and J. P. Hennessy, Jr., *Vaccine*, 1999, **17**, 1169–1178.
127. N. S. Pujar, N. F. Huang, C. L. Daniels, L. Dieter, M. G. Gayton and A. L. Lee, *Biopolymers*, 2004, **75**, 71–84.
128. *Requirements for Vi polysaccharide typhoid vaccine (Requirements for Biological Substances No. 48) Annex 1 (WHO Technical Report Series, No. 840)*, World Health Organization, Geneva, 1994.
129. C. L. Chaignat, J. D. Clemens, M. Favorov, B. Ivanoff, K. Klugmann, J. Kariuiki, M. Levine, D. DeRoeack, S. Szu and Z. Buhutta, *Background Paper on Vaccination against Typhoid Fever using New-Generation Vaccines – to be presented to WHO SAGE in November 2007*, World Health Organization, Geneva, 2007.
130. Q. Xu, C. Abeygunawardana, A. S. Ng, A. W. Sturgess, B. J. Harmon and J. P. Hennessey Jr., *Anal Biochem*, 2005, **336**, 262–272.
131. P. Talaga, L. Bellamy and M. Moreau, *Vaccine*, 2001, **19**, 2987–2994.
132. A. L. Sorensen, C. A. Reis, M. A. Tarp, U. Mandel, K. Ramachandran, V. Sankaranarayanan, T. Schwientek, R. Graham, J. Taylor-Papadimitriou, M. A. Hollingsworth, J. Burchell and H. Clausen, *Glycobiology*, 2006, **16**, 96–107.
133. R. Woodward, W. Yi, L. Li, G. Zhao, H. Eguchi, P. R. Sridhar, H. Guo, J. K. Song, E. Motari, L. Cai, P. Kelleher, X. Liu, W. Han, W. Zhang, Y. Ding, M. Li and P. G. Wang, *Nat Chem Biol*, 2010, **6**, 418–423.
134. V. S. Terra, D. C. Mills, L. E. Yates, S. Abouelhadid, J. Cuccui and B. W. Wren, *J Med Microbiol*, 2012, **61**, 919–926.
135. M. Wetter, M. Kowarik, M. Steffen, P. Carranza, G. Corradin and M. Wacker, *Glycoconjugate J ahead of print*, 2012.
136. D. E. Levy and P. Fügedi, eds., *The Organic Chemistry of Sugars*, CRC Press, Boca Raton, FL, 2006.

137. M. A. Walczak and S. J. Danishefsky, *J Am Chem Soc*, 2012, **134**, 16430–16433.
138. Y. Yang, C. E. Martin and P. H. Seeberger, *Chem Sci*, 2012, **3**, 896–899.
139. C. E. Martin, M. W. Weishaupt and P. H. Seeberger, *Chem Commun (Camb)*, 2011, **47**, 10260–10262.
140. R. Pragani and P. H. Seeberger, *J Am Chem Soc*, 2011, **133**, 102–107.
141. D. B. Werz and P. H. Seeberger, *Angew Chem Int Ed Engl*, 2005, **44**, 6315–6318.
142. R. Pragani, P. Stallforth and P. H. Seeberger, *Org Lett*, 2010, **12**, 1624–1627.
143. O. Calin, R. Pragani and P. H. Seeberger, *J Org Chem*, 2012, **77**, 870–877.
144. D. Leonori and P. H. Seeberger, *Org Lett*, 2012, **14**, 4954–4957.
145. P. Stallforth, S. Matthies, A. Adibekian, D. G. Gillingham, D. Hilvert and P. H. Seeberger, *Chem Commun (Camb)*, 2012, **48**, 11987–11989.
146. D. G. Gillingham, P. Stallforth, A. Adibekian, P. H. Seeberger and D. Hilvert, *Nat Chem*, 2010, **2**, 102–105.
147. A. Adibekian, M. S. Timmer, P. Stallforth, J. van Rijn, D. B. Werz and P. H. Seeberger, *Chem Commun (Camb)*, 2008, 3549–3551.
148. P. Stallforth, A. Adibekian and P. H. Seeberger, *Org Lett*, 2008, **10**, 1573–1576.
149. T. Ohara, A. Adibekian, D. Esposito, P. Stallforth and P. H. Seeberger, *Chem Commun (Camb)*, 2010, **46**, 4106–4108.
150. P. Bindschädler, A. Adibekian, D. Grunstein and P. H. Seeberger, *Carbohydr Res*, 2010, **345**, 948–955.
151. A. Adibekian, P. Bindschädler, M. S. Timmer, C. Noti, N. Schutzenmeister and P. H. Seeberger, *Chemistry*, 2007, **13**, 4510–4522.
152. M. S. Timmer, A. Adibekian and P. H. Seeberger, *Angew Chem Int Ed Engl*, 2005, **44**, 7605–7607.
153. P. H. Seeberger, *Chem Commun (Camb)*, 2003, 1115–1121.
154. M. T. Walvoort, H. van den Elst, O. J. Plante, L. Kröck, P. H. Seeberger, H. S. Overkleeft, G. A. van der Marel and J. D. Codee, *Angew Chem Int Ed Engl*, 2012, **51**, 4393–4396.
155. X. Liu, R. Wada, S. Boonyarattanakalin, B. Castagner and P. H. Seeberger, *Chem Commun (Camb)*, 2008, 3510–3512.
156. P. H. Seeberger, *Carbohydr Res*, 2008, **343**, 1889–1896.
157. J. D. Codee, L. Kröck, B. Castagner and P. H. Seeberger, *Chemistry*, 2008, **14**, 3987–3994.
158. P. H. Seeberger, *Chem Soc Rev*, 2008, **37**, 19–28.
159. D. B. Werz, B. Castagner and P. H. Seeberger, *J Am Chem Soc*, 2007, **129**, 2770–2771.
160. P. H. Seeberger and D. B. Werz, *Nat Rev Drug Discovery*, 2005, **4**, 751–763.
161. K. R. Love and P. H. Seeberger, *Angew Chem Int Ed Engl*, 2004, **43**, 602–605.
162. O. J. Plante, E. R. Palmacci and P. H. Seeberger, *Methods Enzymol*, 2003, **369**, 235–248.

163. O. J. Plante, E. R. Palmacci and P. H. Seeberger, *Adv Carbohydr Chem Biochem*, 2003, **58**, 35–54.
164. D. M. Ratner, E. R. Swanson and P. H. Seeberger, *Organic letters*, 2003, **5**, 4717–4720.
165. O. J. Plante and P. H. Seeberger, *Curr Opin Drug discovery Dev*, 2003, **6**, 521–525.
166. L. G. Melean, K. R. Love and P. H. Seeberger, *Carbohydr Res*, 2002, **337**, 1893–1916.
167. M. C. Hewitt, D. A. Snyder and P. H. Seeberger, *J Am Chem Soc*, 2002, **124**, 13434–13436.
168. M. C. Hewitt and P. H. Seeberger, *Org Lett*, 2001, **3**, 3699–3702.
169. O. J. Plante, E. R. Palmacci and P. H. Seeberger, *Science*, 2001, **291**, 1523–1527.
170. M. T. Walvoort, A. G. Volbeda, N. R. Reintjens, H. van den Elst, O. J. Plante, H. S. Overkleeft, G. A. van der Marel and J. D. Codee, *Org Lett*, 2012, 3776–3779.
171. W. F. Hogendorf, N. Meeuwenoord, H. S. Overkleeft, D. V. Filippov, D. Laverde, A. Kropec, J. Huebner, G. A. Van der Marel and J. D. Codee, *Chem Commun (Camb)*, 2011, **47**, 8961–8963.
172. C. Y. Wu and C. H. Wong, *Top Curr Chem*, 2011, **301**, 223–252.
173. J. C. Lee, W. A. Greenberg and C. H. Wong, *Nat Protoc*, 2006, **1**, 3143–3152.
174. D. Safari, H. A. T. Dekker, J. A. F. Joosten, D. Michalik, A. C. de Souza, R. Adamo, M. Lahmann, A. Sundgren, S. Oscarson, J. P. Kamerling and H. Snippe, *Infect Immun*, 2008, **76**, 4615–4623.
175. C. C. Peeters, D. Evenberg, P. Hoogerhout, H. Kayhty, L. Saarinen, C. A. van Boeckel, G. A. van der Marel, J. H. van Boom and J. T. Poolman, *Infect Immun*, 1992, **60**, 1826–1833.
176. M. Tamborrini, D. B. Werz, J. Frey, G. Pluschke and P. H. Seeberger, *Angew Chem Int Ed Engl*, 2006, **45**, 6581–6582.
177. B. Benaissa-Trouw, D. J. Lefeber, J. P. Kamerling, J. F. G. Vliegenthart, K. Kraaijeveld and H. Snippe, *Infect Immun*, 2001, **69**, 4698–4701.
178. M. A. Johnson, J. Cartmell, N. E. Weisser, R. J. Woods and D. R. Bundle, *J Biol Chem*, 2012, **287**, 18078–18090.
179. M. Nitz, C. C. Ling, A. Otter, J. E. Cutler and D. R. Bundle, *J Biol Chem*, 2002, **277**, 3440–3446.
180. M. A. Oberli, M. L. Hecht, P. Bindschädler, A. Adibekian, T. Adam and P. H. Seeberger, *Chem Biol*, 2011, **18**, 580–588.
181. I. Seppälä and O. Mäkelä, *J Immunol*, 1989, **143**, 1259–1264.
182. J. B. Robbins, R. Schneerson, P. Anderson and D. H. Smith, *J Am Med Assoc*, 1996, **276**, 1181–1185.
183. P. W. Anderson, M. E. Pichichero, R. A. Insel, R. Betts, R. Eby and D. H. Smith, *J Immunol*, 1986, **137**, 1181–1186.
184. P. Chong, N. Chan, A. Kandil, B. Tripet, O. James, Y. P. Yang, S. P. Shi and M. Klein, *Infect Immun*, 1997, **65**, 4918–4925.
185. P. Costantino, F. Norelli, A. Giannozzi, S. D'Ascenzi, A. Bartoloni, S. Kaur, D. Z. Tang, R. Seid, S. Viti, R. Paffetti, M. Bigio, C. Pennatini,

G. Averani, V. Guarnieri, E. Gallo, N. Ravenscroft, C. Lazzeroni, R. Rappuoli and C. Ceccarini, *Vaccine*, 1999, **17**, 1251–1263.
186. V. Verez-Bencomo, V. Fernandez-Santana, E. Hardy, M. E. Toledo, M. C. Rodriguez, L. Heynngnezz, A. Rodriguez, A. Baly, L. Herrera, M. Izquierdo, A. Villar, Y. Valdes, K. Cosme, M. L. Deler, M. Montane, E. Garcia, A. Ramos, A. Aguilar, E. Medina, G. Torano, I. Sosa, I. Hernandez, R. Martinez, A. Muzachio, A. Carmenates, L. Costa, F. Cardoso, C. Campa, M. Diaz and R. Roy, *Science*, 2004, **305**, 522–525.
187. V. Fernandez-Santana, F. Cardoso, A. Rodriguez, T. Carmenate, L. Pena, Y. Valdes, E. Hardy, F. Mawas, L. Heynngnezz, M. C. Rodriguez, I. Figueroa, J. Chang, M. E. Toledo, A. Musacchio, I. Hernandez, M. Izquierdo, K. Cosme, R. Roy and V. Verez-Bencomo, *Infect Immun*, 2004, **72**, 7115–7123.
188. F. Mawas, J. Niggemann, C. Jones, M. J. Corbel, J. P. Kamerling and J. F. G. Vliegenthart, *Infect Immun*, 2002, **70**, 5107–5114.
189. W. T. Jansen and H. Snippe, *Indian J Med Res*, 2004, **119 Suppl**, 7–12.
190. L. C. Paoletti, D. L. Kasper, F. Michon, J. DiFabio, H. J. Jennings, T. D. Tosteson and M. R. Wessels, *J Clin Invest*, 1992, **89**, 203–209.
191. L. C. Paoletti, D. L. Kasper, F. Michon, J. Difabio, K. Holme, H. J. Jennings and M. R. Wessels, *J Biol Chem*, 1990, **265**, 18278–18283.
192. F. Y. Avci, X. M. Li, M. Tsuji and D. L. Kasper, *Nat Med*, 2011, **17**, 1602–U1115.
193. J. Gonzalez-Outeirino, R. Kadirvelraj and R. J. Woods, *Carbohydr Res*, 2005, **340**, 1007–1018.
194. V. Pozsgay, C. Y. Chu, L. Pannell, J. Wolfe, J. B. Robbins and R. Schneerson, *P Natl Acad Sci USA*, 1999, **96**, 5194–5197.
195. A. Phalipon, M. Tanguy, C. Grandjean, C. Guerreiro, F. Bélot, D. Cohen, P. J. Sansonetti and L. A. Mulard, *J Immunol*, 2009, **182**, 2241–2247.
196. B. A. Cobb and D. L. Kasper, *Eur J Immunol*, 2005, **35**, 352–356.
197. S. K. Mazmanian and D. L. Kasper, *Nat Rev Immunol*, 2006, **6**, 849–858.
198. A. Tzianabos, J. Y. Wang and D. L. Kasper, *Carbohydr Res*, 2003, **338**, 2531–2538.
199. N. K. Surana and D. L. Kasper, *Immunol Rev*, 2012, **245**, 13–26.
200. S. K. Mazmanian, J. L. Round and D. L. Kasper, *Nature*, 2008, **453**, 620–625.
201. B. A. Cobb, O. Wang, A. O. Tzianabos and D. L. Kasper, *Cell*, 2004, **117**, 677–687.
202. C. D. Velez, C. J. Lewis, D. L. Kasper and B. A. Cobb, *Immunology*, 2009, **127**, 73–82.
203. T. L. Stephen, M. Fabri, L. Groneck, T. A. Rohn, H. Hafke, N. Robinson, J. Rietdorf, D. Schrama, J. C. Becker, G. Plum, M. Kronke, H. Kropshofer and W. M. Kalka-Moll, *PLoS Pathog*, 2007, **3**.
204. F. Stingele, B. Corthesy, N. Kusy, S. A. Porcelli, D. L. Kasper and A. O. Tzianabos, *J Immunol*, 2004, **172**, 1483–1490.
205. W. M. Kalka-Moll, A. O. Tzianabos, P. W. Bryant, M. Niemeyer, H. L. Ploegh and D. L. Kasper, *J of Immunol*, 2002, **169**, 6149–6153.

206. J. Mertens, M. Fabri, A. Zingarelli, T. Kubacki, S. Meemboor, L. Groneck, J. Seeger, M. Bessler, H. Hafke, M. Odenthal, J. G. Bieler, C. Kalka, J. P. Schneck, H. Kashkar and W. M. Kalka-Moll, *PLoS Pathog*, 2009, **5**.
207. L. Groneck, D. Schrama, M. Fabri, T. L. Stephen, F. Harms, S. Meemboor, H. Hafke, M. Bessler, J. C. Becker and W. M. Kalka-Moll, *Infect Immun*, 2009, **77**, 3705–3712.
208. S. Gallorini, F. Berti, G. Mancuso, R. Cozzi, M. Tortoli, G. Volpini, J. L. Telford, C. Beninati, D. Maione and A. Wack, *Proc Natl Acad Sci USA*, 2009, **106**, 17481–17486.
209. A. Wack and S. Gallorini, *Immunopharmacol Immunotoxicol*, 2008, **30**, 761–770.
210. S. Gallorini, F. Berti, P. Parente, R. Baronio, S. Aprea, U. D'Oro, M. Pizza, J. L. Telford and A. Wack, *J Immunol*, 2007, **179**, 8208–8215.
211. L. S. Kreisman, J. H. Friedman, A. Neaga and B. A. Cobb, *Glycobiology*, 2007, **17**, 46–55.
212. X. Y. Wu, L. N. Cui, T. Lipinski and D. R. Bundle, *Chem–Eur J*, 2010, **16**, 3476–3488.
213. A. E. Christina, L. J. van den Bos, H. S. Overkleeft, G. A. van der Marel and J. D. C. Codee, *J Org Chem*, 2011, **76**, 1692–1706.
214. C. Fernandez and E. Sverremark, *Cell immunol*, 1994, **153**, 67–78.
215. H.-K. Guttormsen, L. C. Paoletti, K. G. Mansfield, W. Jachymek, H. J. Jennings and D. L. Kasper, *Proc Natl Acad Sci USA*, 2008, **105**, 5903–5908.
216. A. F. Carlin, A. L. Lewis, A. Varki and V. Nizet, *J Bacteriol*, 2007, **189**, 1231–1237.
217. V. Pozsgay, J. Kubler-Kielb, R. Schneerson and J. B. Robbins, *Proc Natl Acad Sci USA*, 2007, **104**, 14478–14482.
218. R. Evron and J. A. Drewe, *Mycopathologia*, 1984, **84**, 141–149.
219. C. T. Bishop, F. Blank and P. E. Gardner, *Can J Chem*, 1960, **38**, 869–881.
220. S. D. Bentley, D. M. Aanensen, A. Mavroidi, D. Saunders, E. Rabbinowitsch, M. Collins, K. Donohoe, D. Harris, L. Murphy, M. A. Quail, G. Samuel, I. C. Skovsted, M. S. Kaltoft, B. Barrell, P. R. Reeves, J. Parkhill and B. G. Spratt, *PLoS Genet*, 2006, **2**, e31.
221. E. Garcia, D. Llull, R. Munoz, M. Mollerach and R. Lopez, *Res Microbiol*, 2000, **151**, 429–435.
222. A. Phalipon, C. Costachel, C. Grandjean, A. Thuizat, C. Guerreiro, M. Tanguy, F. Nato, B. Vulliez-Le Normand, F. Bélot, K. Wright, V. Marcel-Peyre, P. J. Sansonetti and L. A. Mulard, *J Immunol*, 2006, **176**, 1686–1694.
223. C. Bromuro, M. Romano, P. Chiani, F. Berti, M. Tontini, D. Proietti, E. Mori, A. Torosantucci, P. Costantino, R. Rappuoli and A. Cassone, *Vaccine*, 2010, **28**, 2615–2623.
224. A. Torosantucci, P. Chiani, C. Bromuro, F. De Bernardis, A. S. Palma, Y. Liu, G. Minogna, B. Maras, M. Colone, A. Stringaro, S. Zamboni, T. Feizi and A. Cassone, *PloS One*, 2009, **4**.

225. P. Chiani, C. Bromuro, A. Cassone and A. Torosantucci, *Vaccine*, 2009, **27**, 513–519.
226. S. M. Muthana, C. T. Campbell and J. C. Gildersleeve, *ACS Chem Biol*, 2012, **7**, 31–43.
227. E. Staudacher, *Biol Chem*, 2012, **393**, 675–685.
228. X. Lemercinier and C. Jones, *Carbohydr Res*, 1996, **296**, 83–96.
229. S. C. Szu, X. R. Li, A. L. Stone and J. B. Robbins, *Infect Immun*, 1991, **59**, 4555–4561.
230. T. B. McNeely, J. M. Staub, C. M. Rusk, M. J. Blum and J. J. Donnelly, *Infect Immun*, 1998, **66**, 3705–3710.
231. P. C. Fusco, E. K. Farley, C. H. Huang, S. Moore and F. Michon, *Clin Vaccine Immunol*, 2007, **14**, 577–584.
232. R. C. Guy, M. J. How and M. Stacey, *J Biol Chem*, 1967, **242**, 5106–5111.
233. D. A. Kennedy, J. G. Buchanan and J. Baddiley, *Biochem J*, 1969, **115**, 37–45.
234. S. C. Szu, S. Bystricky, M. Hinojosa-Ahumada, W. Egan and J. B. Robbins, *Infect Immun*, 1994, **62**, 5545–5549.
235. J. Ganeshapillai, E. Vinogradov, J. Rousseau, J. S. Weese and M. A. Monteiro, *Carbohydr Res*, 2008, **343**, 703–710.
236. E. Danieli, L. Lay, D. Proietti, F. Berti, P. Costantino and R. Adamo, *Org Lett*, 2011, **13**, 378–381.
237. R. Adamo, M. R. Romano, F. Berti, R. Leuzzi, M. Tontini, E. Danieli, E. Cappelletti, O. S. Cakici, E. Swennen, V. Pinto, B. Brogioni, D. Proietti, C. L. Galeotti, L. Lay, M. A. Monteiro, M. Scarselli and P. Costantino, *ACS Chem Biol*, 2012, **7**, 1420–1428.
238. J. M. Daubenspeck, H. Zeng, P. Chen, S. Dong, C. T. Steichen, N. R. Krishna, D. G. Pritchard and C. L. Turnbough, Jr., *J Biol Chem*, 2004, **279**, 30945–30953.
239. A. S. Mehta, E. Saile, W. Zhong, T. Buskas, R. Carlson, E. Kannenberg, Y. Reed, C. P. Quinn and G. J. Boons, *Chemistry*, 2006, **12**, 9136–9149.
240. E. Saile, G. J. Boons, T. Buskas, R. W. Carlson, E. L. Kannenberg, J. R. Barr, A. E. Boyer, M. Gallegos-Candela and C. P. Quinn, *Clin Vaccine Immunol*, 2011, **18**, 743–748.
241. J. Kubler-Kielb, E. Vinogradov, H. Hu, S. H. Leppla, J. B. Robbins and R. Schneerson, *Proc Natl Acad Sci USA*, 2008, **105**, 8709–8712.
242. O. Milhomme, S. M. Kohler, D. Ropartz, D. Lesur, S. Pilard, F. Djedaini-Pilard, W. Beyer and C. Grandjean, *Org Biomol Chem*, 2012, **10**, 8524–8532.
243. L. Kröck, D. Esposito, B. Castagner, C. C. Wang, P. Bindschädler and P. H. Seeberger, *Chem Sci*, 2012, **3**, 1617–1622.
244. K. F. Mo, X. Li, H. Li, L. Y. Low, C. P. Quinn and G. J. Boons, *J Am Chem Soc*, 2012, **134**, 15556–15562.
245. P. Farkas and S. Bystricky, *Chem Pap*, 2010, **64**, 683–695.
246. A. Dondoni and A. Marra, *Chem Soc Rev*, 2012, **41**, 573–586.
247. C. Grandjean, A. Boutonnier, C. Guerreiro, J. M. Fournier and L. A. Mulard, *J Org Chem*, 2005, **70**, 7123–7132.

248. S. Dziadek, S. Jacques and D. R. Bundle, *Chem-Eur J*, 2008, **14**, 5908–5917.
249. T. Lipinski, T. Luu, P. I. Kitov, A. Szpacenko and D. R. Bundle, *Glycoconjugate J*, 2011, **28**, 149–164.
250. J. Kubler-Kielb and V. Pozsgay, *J Org Chem*, 2005, **70**, 6987–6990.
251. P. B. van Seeventer, J. A. L. M. van Dorst, J. F. Siemerink, J. P. Kamerling and J. F. G. Vliegenthart, *Carbohydr Res*, 1997, **300**, 369–373.
252. D. Ponader, F. Wojcik, F. Beceren-Braun, J. Dernedde and L. Hartmann, *Biomacromolecules*, 2012, **13**, 1845–1852.
253. V. Pozsgay, N. E. Vieira and A. Yergey, *Org Lett*, 2002, **4**, 3191–3194.
254. D. R. Bundle, C. Nycholat, C. Costello, R. Rennie and T. Lipinski, *ACS Chem Biol*, 2012, **7**, 1754–1763.
255. T. Buskas, Y. Li and G.-J. Boons, *Chem–Eur J*, 2004, **10**, 3517–3524.
256. J. Ni, H. Song, Y. Wang, N. M. Stamatos and L.-X. Wang, *Bioconjugate Chem*, 2006, **17**, 493–500.
257. Q. Wang, S. A. Ekanayaka, J. Wu, J. Zhang and Z. Guo, *Bioconjugate Chem*, 2008, **19**, 2060–2067.
258. J. C. Jewett and C. R. Bertozzi, *Chem Soc Rev*, 2010, **39**, 1272–1279.
259. E. M. Sletten and C. R. Bertozzi, *Acc Chem Res*, 2011, **44**, 666–676.
260. J. M. Peeters, T. G. Hazendonk, E. C. Beuvery and G. I. Tesser, *J Immunol Meth*, 1989, **120**, 133–143.
261. Q. Li, L. G. Rodriguez, D. F. Farnsworth and J. C. Gildersleeve, *ChemBioChem*, 2010, **11**, 1686–1691.
262. C. Desaymard and J. G. Howard, *Eur J Immunol*, 1975, **5**, 541–545.
263. C. Desaymard, *Eur J Immunol*, 1977, **7**, 646–649.
264. M. Feldmann, J. G. Howard and C. Desaymard, *Transplant Rev*, 1975, **23**, 78–97.
265. A. Kabanova, I. Margarit, F. Berti, M. R. Romano, G. Grandi, G. Bensi, E. Chiarot, D. Proietti, E. Swennen, E. Cappelletti, P. Fontani, D. Casini, R. Adamo, V. Pinto, D. Skibinski, S. Capo, G. Buffi, M. Gallotta, W. J. Christ, A. Stewart Campbell, J. Pena, P. H. Seeberger, R. Rappuoli and P. Costantino, *Vaccine*, 2010, **29**, 104–114.
266. G. S. Bixler, Jr., R. Eby, K. M. Dermody, R. M. Woods, R. C. Seid and S. Pillai, *Adv Exp Med Biol*, 1989, **251**, 175–180.
267. J. C. Reece, H. M. Geysen and S. J. Rodda, *J Immunol*, 1993, **151**, 6175–6184.
268. B. M. Diethelm-Okita, R. Raju, D. K. Okita and B. M. Conti-Fine, *J Infect Dis*, 1997, **175**, 382–391.
269. E. G. Leonard, D. H. Canaday, C. V. Harding and J. R. Schreiber, *Infect Immun*, 2003, **71**, 4186–4189.
270. T. L. McCool, C. V. Harding, N. S. Greenspan and J. R. Schreiber, *Infect Immun*, 1999, **67**, 4862–4869.
271. J. Rossjohn, D. G. Pellicci, O. Patel, L. Gapin and D. I. Godfrey, *Nat Rev Immunol*, 2012, **12**, 845–857.
272. T. Dudler, F. Altmann, J. M. Carballido and K. Blaser, *Eur J Immunol*, 1995, **25**, 538–542.

273. J. A. Speir, U. M. Abdel-Motal, M. Jondal and I. A. Wilson, *Immunity*, 1999, **10**, 51–61.
274. M. B. Deck, P. Sjolin, E. R. Unanue and J. Kihlberg, *J Immunol*, 1999, **162**, 4740–4744.
275. S. Muthukkumar and K. E. Stein, *Vaccine*, 2004, **22**, 1290–1299.
276. G. Y. Ishioka, A. G. Lamont, D. Thomson, N. Bulbow, F. C. Gaeta, A. Sette and H. M. Grey, *J Immunol*, 1992, **148**, 2446–2451.
277. D. M. Granoff, S. J. Holmes, R. B. Beishe, M. T. Osterholm, J. E. Mchugh and E. L. Anderson, *J Am Med Assoc*, 1994, **272**, 1116–1121.
278. R. Dagan, J. Poolman and C. A. Siegrist, *Vaccine*, 2010, **28**, 5513–5523.
279. D. M. Granoff, M. H. Rathore, S. J. Holmes, P. D. Granoff and A. H. Lucas, *Vaccine*, 1993, **11**, S46–S51.
280. *Requirements for diphtheria, tetanus, pertussis and combined vaccines (Requirements for Biological Substances Nos. 8 and 10) Annex 2 (WHO Technical Report Series, No. 800)*, World Health Organization, Geneva, 1990.
281. *Recommendations for diphtheria, tetanus, pertussis and combined vaccines (Amendments 2003) Annex 5 (WHO Technical Report Series, No. 927)*, World Health Organization, Geneva, 2005.
282. T. Ju, V. I. Otto and R. D. Cummings, *Angew Chem Int Ed Engl*, 2011, **50**, 1770–1791.
283. U. Westerlind, A. Hobel, N. Gaidzik, E. Schmitt and H. Kunz, *Angew Chem Int Ed Engl*, 2008, **47**, 7551–7556.
284. A. Kaiser, N. Gaidzik, U. Westerlind, D. Kowalczyk, A. Hobel, E. Schmitt and H. Kunz, *Angew Chem Int Ed Engl*, 2009, **48**, 7551–7555.
285. U. Westerlind and H. Kunz, *Chimia*, 2011, **65**, 30–34.
286. S. Wittrock, T. Becker and H. Kunz, *Angew Chem Int Ed Engl*, 2007, **46**, 5226–5230.
287. S. Dziadek, A. Hobel, E. Schmitt and H. Kunz, *Angew Chem Int Ed Engl*, 2005, **44**, 7630–7635.
288. S. Dziadek, D. Kowalczyk and H. Kunz, *Angew Chem Int Ed Engl*, 2005, **44**, 7624–7630.
289. A. Hoffmann-Roder, A. Kaiser, S. Wagner, N. Gaidzik, D. Kowalczyk, U. Westerlind, B. Gerlitzki, E. Schmitt and H. Kunz, *Angew Chem Int Ed Engl*, 2010, **49**, 8498–8503.
290. N. Gaidzik, A. Kaiser, D. Kowalczyk, U. Westerlind, B. Gerlitzki, H. P. Sinn, E. Schmitt and H. Kunz, *Angew Chem Int Ed Engl*, 2011, **50**, 9977–9981.
291. A. M. McDonnell, B. W. Robinson and A. J. Currie, *Clin Dev Immunol*, 2010, **2010**, 539519.
292. A. Kaiser, N. Gaidzik, T. Becker, C. Menge, K. Groh, H. Cai, Y. M. Li, B. Gerlitzki, E. Schmitt and H. Kunz, *Angew Chem Int Ed Engl*, 2010, **49**, 3688–3692.
293. B. L. Wilkinson, S. Day, L. R. Malins, V. Apostolopoulos and R. J. Payne, *Angew Chem Int Ed Engl*, 2011, **50**, 1635–1639.

294. V. Lakshminarayanan, P. Thompson, M. A. Wolfert, T. Buskas, J. M. Bradley, L. B. Pathangey, C. S. Madsen, P. A. Cohen, S. J. Gendler and G. J. Boons, *Proc Natl Acad Sci USA*, 2012, **109**, 261–266.
295. S. Ingale, M. A. Wolfert, J. Gaekwad, T. Buskas and G. J. Boons, *Nat Chem Biol*, 2007, **3**, 663–667.
296. S. Ingale, M. A. Wolfert, T. Buskas and G. J. Boons, *Chembiochem*, 2009, **10**, 455–463.
297. L. Brenzel and A. Jones, *Estimating Cold Chain Requirements*, GAVI Alliance, The World Bank, 2010.
298. J. Milstein, M. Zaffran, Ü. Kartoglu and A. Galazka, *Temperature Sensitivity of Vaccines*, World Health Organization, Geneva, 2006.
299. F. Said Hassane, A. Phalipon, M. Tanguy, C. Guerreiro, F. Belot, B. Frisch, L. A. Mulard and F. Schuber, *Vaccine*, 2009, **27**, 5419–5426.
300. X. Liu, S. Siegrist, M. Amacker, R. Zurbriggen, G. Pluschke and P. H. Seeberger, *ACS Chem Biol*, 2006, **1**, 161–164.
301. D. Safari, M. Marradi, F. Chiodo, H. A. T. Dekker, Y. L. Shan, R. Adamo, S. Oscarson, G. T. Rijkers, M. Lahmann, J. P. Kamerling, S. Penades and H. Snippe, *Nanomedicine (London, UK)*, 2012, **7**, 1111–1112.
302. A. M. Alkilany and C. J. Murphy, *J Nanopart Res*, 2010, **12**, 2313–2333.
303. Z. H. Huang, L. Shi, J. W. Ma, Z. Y. Sun, H. Cai, Y. X. Chen, Y. F. Zhao and Y. M. Li, *J Am Chem Soc*, 2012, **134**, 8730–8733.
304. J. S. Rudra, Y. F. Tian, J. P. Jung and J. H. Collier, *Proc Natl Acad Sci USA*, 2010, **107**, 622–627.
305. R. A. De Silva, Q. L. Wang, T. Chidley, D. K. Appulage and P. R. Andreana, *J Am Chem Soc*, 2009, **131**, 9622–9623.
306. R. A. De Silva, D. K. Appulage, H. Pietraszkiewicz, K. R. Bobbitt, J. Media, J. Shaw, F. A. Valeriote and P. R. Andreana, *Cancer Immunol Immun*, 2012, **61**, 581–585.
307. R. Rappuoli, *Vaccine*, 2001, **19**, 2688–2691.
308. K. L. Seib, X. Zhao and R. Rappuoli, *Clin Microbiol Infect*, 2012, **18 Suppl 5**, 109–116.
309. A. Sette and R. Rappuoli, *Immunity*, 2010, **33**, 530–541.
310. D. Serruto and R. Rappuoli, *FEBS Lett*, 2006, **580**, 2985–2992.
311. D. C. Douek, P. D. Kwong and G. J. Nabel, *Cell*, 2006, **124**, 677–681.
312. P. R. Dormitzer, G. Grandi and R. Rappuoli, *Nat Rev Microbiol*, 2012, **10**, 807–813.
313. M. Tamborrini, X. Liu, J. P. Mugasa, Y. U. Kwon, F. Kamena, P. H. Seeberger and G. Pluschke, *Bioorg Med Chem*, 2010, **18**, 3747–3752.
314. C. D. Rillahan and J. C. Paulson, *Annu Rev Biochem*, 2011, **80**, 797–823.
315. M. A. Oberli, M. Tamborrini, Y. H. Tsai, D. B. Werz, T. Horlacher, A. Adibekian, D. Gauss, H. M. Moller, G. Pluschke and P. H. Seeberger, *J Am Chem Soc*, 2010, **132**, 10239–10241.
316. N. Azzouz, F. Kamena and P. H. Seeberger, *OMICS: J Integr Biol*, 2010, **14**, 445–454.
317. T. Horlacher and P. H. Seeberger, *Chem Soc Rev*, 2008, **37**, 1414–1422.

318. Y. Liu, A. S. Palma and T. Feizi, *Biol Chem*, 2009, **390**, 647–656.
319. S. Park, J. C. Gildersleeve, O. Blixt and I. Shin, *Chem Soc Rev*, 2012.
320. A. Y. Moreno, A. V. Mayorov and K. D. Janda, *J Am Chem Soc*, 2011, **133**, 6587–6595.
321. H. Kobayashi, N. Shibata, M. Nakada, S. Chaki, K. Mizugami, Y. Ohkubo and S. Suzuki, *Arch Biochem Biophys*, 1990, **278**, 195–204.
322. M. Nitz and D. R. Bundle, *J Org Chem*, 2001, **66**, 8411–8423.
323. M. Nitz, B. W. Purse and D. R. Bundle, *Org Lett*, 2000, **2**, 2939–2942.
324. C. M. Nycholat and D. R. Bundle, *Carbohydr Res*, 2009, **344**, 1397–1411.
325. C. Costello and D. R. Bundle, *Carbohydr Res*, 2012, **357**, 7–15.

CHAPTER 4

Generation and Maturation of Therapeutic Antibodies via In Vitro Somatic Hypermutation

DAVID J. KING

AnaptysBio Inc., 10421 Pacific Center Court, San Diego, CA 92121, USA
Email: dking@anaptysbio.com

4.1 Antibodies as Therapeutic Agents

Antibodies are a central component of the adaptive immune system, which has evolved to be able to produce antibodies with high specificity and affinity to a multitude of target antigens. Such high specificity and binding affinity for a particular target has led to intense interest in harnessing antibodies as targeted therapeutics for a number of different diseases, such that antibody therapeutics are now a well established class of pharmaceuticals, and one of the major growth drivers for the pharmaceutical industry.

Antibodies are attractive for a number of reasons, and can be used in a number of different therapeutic modes. In their natural role in the immune system, antibodies are generated against invading pathogens, or other foreign antigens, to clear them and direct a number of immune effector functions through their interaction with Fc receptors and the complement system. Therapeutic antibodies can make use of these mechanisms and others through antigen binding via the Fab regions as well as interaction with effector mechanisms. Potential therapeutic modes therefore include those mediated solely through antigen binding, such as blocking of receptor–ligand

RSC Drug Discovery Series No. 36
Biotherapeutics: Recent Developments using Chemical and Molecular Biology
Edited by Lyn H. Jones and Andrew J. McKnight

Published by the Royal Society of Chemistry, www.rsc.org

interactions, agonist activity on a receptor to trigger a downstream signaling pathway, direct cell killing via induction of apoptosis, inhibition of enzymatic activity, or antigen internalization and down-regulation. And those mediated through more complex immune pathways, such as antibody-dependent cellular cytotoxicity (ADCC), complement dependent cytotoxicity (CDC) or the induction of phagocytosis. In addition, antibodies are also being exploited as targeting agents, for example to deliver small molecule cytotoxics to tumors as antibody–drug conjugates, or radioisotope conjugates. Such diverse uses require optimization of antibodies in several different ways, and the ability to identify suitable antibodies and tailor them for their intended application through antibody engineering has become crucial to meeting the evermore stringent requirements for a viable therapeutic antibody.

In addition, to meeting the requirements for specificity, affinity and functional properties, therapeutic antibodies also have to be efficiently produced and suitable for pharmaceutical development. Therefore biophysical properties of the antibody are also important. High stability to enable a long shelf-life, high solubility, especially for formulation for subcutaneous delivery, resistance to unwanted post-translational modifications and aggregation, are all important properties, as well as the ability for the antibody to be efficiently expressed in a manufacturing cell line. Advances in antibody technology are also crucial to meeting these requirements.

4.2 Approaches to the Generation of Therapeutic Antibodies

Since the approval of the first antibody therapeutic, the murine monoclonal antibody OKT3, and subsequent development of many recombinant antibodies, there has been continuous evolution in antibody technology for the design and development of improved antibody-based pharmaceutical agents for therapy of an increasing number of diseases. Approaches to the generation of therapeutic antibodies can be broadly categorized as those from antibody diversity generated *in vivo,* and those derived by *in vitro* selection from recombinant antibody libraries. Therapeutic antibodies approved by the FDA have come from both of these sources, although the vast majority approved to date have come from *in vivo* immunization based approaches (Figure 4.1).

It was recognized early on that murine monoclonal antibodies were not ideal for therapeutic use due to the host anti-mouse immunoglobulin response generated, which led to rapid clearance of injected antibody on subsequent injections, and the risk of toxicity. For example, OKT3 led to an anti-mouse immune response in the majority of patients dosed with the antibody.[1] Therefore, murine monoclonal antibodies were used as a starting point to generate recombinant antibodies where much of the sequence was replaced by human antibody sequence, resulting in antibodies that are better tolerated by the human immune system. Chimeric antibodies in which the entire variable domains from the mouse mAb are grafted onto human constant regions were

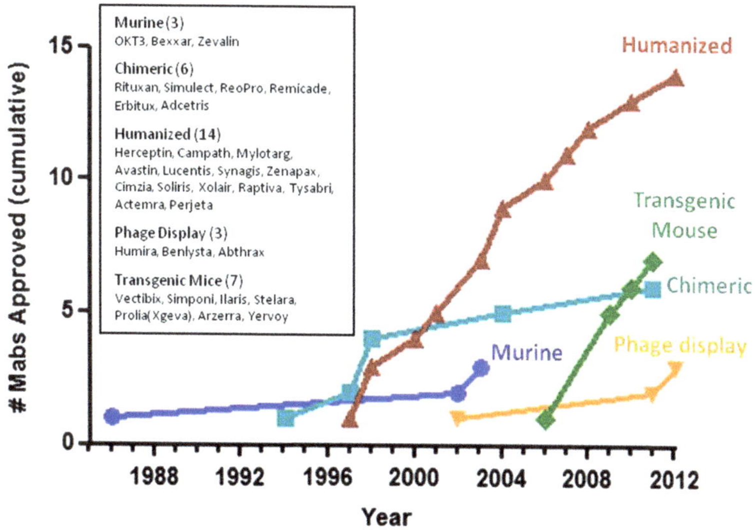

Figure 4.1 FDA-approved therapeutic antibodies by technology. Therapeutic antibodies are plotted by the year of first approval (cumulative). Inset box lists trade names of the antibodies. Note the antibody in the products Prolia and Xgeva, denosumab, is the same. Similarly the antibody in Herceptin and the recently approved product Kadcyla is the same, trastuzumab, which in Kadcyla is conjugated to the cytotoxic drug DM1.

developed first, and subsequently, humanized antibodies in which only the CDR loops responsible for antigen binding are grafted onto a human antibody sequence were developed.[2] Chimeric antibodies were largely replaced by humanized antibodies as the technology was developed, except in a few instances. For example, the antibody–drug conjugate Adcetris™ approved by the FDA in 2011 was developed as a chimeric antibody conjugate.[3] Humanized antibodies have become the largest class of FDA-approved therapeutic antibodies and continue to be successfully developed. Nevertheless, there are a number of disadvantages to this approach to therapeutic antibody generation. Some antigens can be problematic to raise antibodies by immunization of mice or other animal species, especially if the antigen is highly conserved between human and mouse, or poorly immunogenic. It is difficult to control the output of hybridoma technology and the mAbs isolated may be dominated by antibodies by one, or few, immunodominant epitopes. In addition, this is a relatively time-consuming approach that requires generation and characterization of a panel of mouse mAbs, followed by humanization, which may lead to altered properties compared to the original monoclonals.

Subsequently, sources of antibody diversity have been expanded to enable the development of fully human antibodies. This has been achieved through the use of techniques such as phage display of antibody libraries, immunization of transgenic mice in which murine antibody genes have been replaced with human antibody genes, and more recently recovery of antibodies directly from human B cells.

4.2.1 Recovery of Antibodies from *In Vivo* Sources

Transgenic mice, in which mouse IgG production has been disabled and human IgG genes introduced, have been made by several groups,[4–6] enabling human antibodies to be produced directly using standard mouse immunization and hybridoma technology. The first transgenic mice produced included only a portion of the immunoglobulin locus with partial coverage of the immunoglobulin variable region genes, but later generations of mice were able to increase this to increase the diversity and robustness of immune response.[7] Such mice have proven successful in generation of human antibodies and have led to a rapidly increasing number of FDA-approved therapeutic agents (Figure 4.1). Subsequently, transgenic mice have been developed with human variable regions, but retaining the mouse constant regions to enable improved interaction with the mouse immune system.[8] Such mice further improve immune responses, and the chimeric antibodies produced with human V regions and mouse constant regions can readily be converted to fully human antibodies. This approach has also been extended to transgenic rats with similar results.[9] As an *in vivo* immunization-based technique, many of the disadvantages of using transgenic animals are similar to those of traditional murine mAb generation described, such as limited epitope diversity, difficulty in raising antibodies to highly conserved antigens and creating mouse/human cross-reactive antibodies. In addition, generation of very high affinity antibodies may be difficult. Typical affinities of mAbs generated from these approaches are in the low nM KD range, with occasional sub-nM antibodies generated.[10] This is consistent with the predicted affinity ceiling for antibodies generated through an *in vivo* B cell response. It has been proposed that there is rarely enough selective pressure *in vivo* to drive affinity maturation of antibodies to very high affinities because antibody KD values in the 10^{-9}–10^{-10} range are usually sufficient to clear the relevant antigen.[11,12] Higher affinity antibodies can be observed, but are rare.[13] Similar limits have been observed in hypervaccinated humans. In a study with tetanus toxoid, hypervaccination resulted in antibody affinity maturation peaking in the low nM range.[14]

Recently, there has been much interest in the direct isolation of antibodies from human B cells. Antigen-specific antibodies have been recovered from human B cells by a number of approaches, but until recently the technology for recovering such antibodies using techniques such as generation of human hybridomas or EBV transformation of B cells, was unreliable and only sporadically successful. However, several advances in the ability to prolong the survival of human antibody-producing cells and to recover immunoglobulin sequences directly from them, have reinvigorated the field.[15–18] A major limitation however, is that the human donor needs to have raised a high-affinity antibody response to the antigen in question. This essentially limits the field to the isolation of antibodies to infectious disease antigens, where the donor has been exposed to the antigen in question, although it may also be possible to isolate antibodies to some cancer antigens where a patient may have raised an immune response to a particular tumor-associated antigen.

Activated B cells differentiate into memory B cells and antibody secreting cells, including the short-lived plasmablasts and long-lived plasma cells. Antibody secreting cells have typically been difficult to immortalize and keep in prolonged *in vitro* culture, although recently conditions have been established to allow prolonged survival and screening from plasmablasts.[15,16] Memory B cells have also been cultured and differentiated into antibody-secreting cells *in vitro*, allowing screening of the secreted antibodies.[17] Improved methodology is also available for immortalization of memory B cells, for example, using activation through Toll-like receptor 9 (TLR9) agonists, especially CpG DNA, alongside EBV transformation,[18] which has enabled generation of mAbs to a number of infectious disease agents.[19] An alternative method has been to immortalize memory B cells by transduction with a retroviral vector enabling expression of the anti-apoptopic factors bcl-x and bcl-6.[20] On culture with appropriate cytokines, the cells can be differentiated into antibody secreting cells with retention of antibody on the cell surface, to allow antigen-specific antibodies to be identified using both selection of antigen-specific clones by flow cytometry and screening of the cell supernatants for the properties of the secreted antibodies.

One of the major advances in the field has been the ability to recover antibody heavy and light chain sequences from single B cells by real-time polymerase chain reaction (RT-PCR).[21] This allows antibodies to be recovered from rare B cells, providing the cells display antibody on the cell surface and can be isolated by flow cytometry with fluorescently labeled antigen.[22,23] Antibody sequences can then be recovered from the cells and re-expressed to characterize the antibodies further.

This approach has also been applied to the isolation of antibody sequences from the short-lived plasmablasts. Plasmablasts are found in the circulation in high concentrations for a short period 5–7 days following vaccination, and have been harvested for direct recovery of antibody sequences. The assumption being that most of the sequences recovered will encode antibodies specific to the vaccinating antigen.[24]

Recently, high-throughput sequencing techniques have also been applied to the isolation of antibody sequences from B cells. High-throughput sequencing has advanced to the point where sequencing of entire repertoires of antibody sequences may be possible. Bone marrow plasma cells produce the majority of circulating antibodies, and have not proven possible to immortalize. Sequencing of antibody repertoires from bone marrow plasma cells of immunized mice using the Roche/454 methodology, revealed a highly polarized repertoire in the bone marrow plasma cells with some sequences represented at 1–10% of the total.[25] Recovering antigen-specific antibodies is still problematic however because during the process the information of heavy and light pairings is lost. Nevertheless, it proved possible to recover a handful of specific clones to three different antigens by simply pairing the most abundant heavy and light chain sequences.[25] This approach has subsequently been refined and extended through the combination of high throughput sequencing of the B cell repertoire with a proteomics approach to the isolation of high affinity antibodies from the circulation.[26] In this methodology, high affinity antibodies to the target of interest are purified from the serum by antigen affinity chromatography and analyzed by tandem mass spectrometry. Peptide sequences identified can then be correlated with DNA sequences from the B cell repertoire to identify complete V-region sequences for both heavy and light chains. On re-expression of the antibodies the specific activity of the purified antibodies could be recapitulated.[26]

High throughput sequencing of antibodies from human B cells has also been investigated, and is an exciting approach to isolation of antibodies from immunized individuals. Although repertoires of sequences for both heavy and light chain variable regions can be obtained, as in the mouse system described previously, retaining the information of which heavy chain is paired with which light chain is critical to recovery of functional antibodies. One methodology to achieve this has been described recently in which single B cells ($>5 \times 10^4$ per experiment) are plated into high density microwell plates and lysed, prior to recovery of mRNA and RT-PCR of paired heavy and light chains from each well in a single product. The linked transcripts can then be pooled and subject to high-throughput sequencing.[27] This methodology allowed sequencing of isolated memory B cells, or plasmablasts, and allowed the recovery of a panel of high affinity antibodies to tetanus toxoid from sequencing of plasmablasts after immunization.

4.2.2 Recovery of Antibodies from *In Vitro* Sources

Phage display was the first *in vitro* technique developed for the display of human antibody libraries, and has been widely used. This technique was enabled through the work of George Smith, who demonstrated that peptide sequences fused to the gene III protein of the *Escherichia coli* bacteriophage M13 could result in display of the peptide on the surface of the phage.[28] Since that time, display techniques have been developed in a number of alternate systems, including display on ribosomes, bacteria, yeast and mammalian cells. All of the techniques rely on the generation of a library of antibody sequences, the ability to display the library in a form that can be efficiently screened, and the ability to recover the encoding DNA sequences from the displayed phenotype. *In vitro* selection systems share a number of common attributes, such as the ability to control the selection conditions and rapid access to antibody sequences, and overcome many of the limitations imposed *in vivo*, such as antigen homology and epitope dominance. In addition, the use of toxic antigens is possible in *in vitro* selections, as are antigens that are poorly immunogenic in animals.

Phage display was rapidly developed from the early peptide display to the display of antibody variable domains,[29] and the ability to generate large antibody libraries and display them on phage was demonstrated.[30–32] A diverse library of antibody genes is a prerequisite for successful *in vitro* antibody discovery, and these can be generated from a number of different sources. These include recovery of antibody genes from human B cells by PCR,[29] synthetic libraries generated for example with synthetic framework sequences and randomized CDRs,[30] and combinations of genes from B cells with synthetic diversity.[33] Large naïve libraries can be constructed, and have been shown to be able to be used to identify antibodies against a wide range of antigens. Indeed a key finding from early experiments with phage display was that antibodies could be isolated to human (self) proteins from naïve libraries, even though the donors would be expected to be tolerized to the antigens in question.[31] The random reassortment of heavy and light chains in a combinatorial library may contribute to this ability. Library sizes in phage are relatively large, achieving diversity of 10^{10}–10^{11} members.[34] This practical limit is due to the practical limit of phage particle density of approx. 10^{13} per ml and *E. coli* transformation efficiency.

Although several phage surface proteins have been investigated for protein display, the pIII protein has been most widely used for a number of reasons, including its low copy number of approximately five per phage and its tolerance to the incorporation of antibody fragments. There are two generally used formats for phage display. In the first of these, the pIII-antibody fusion is incorporated directly into the phage genome and results in five copies of the antibody fragment per phage particle. Although this works well for library selections, the multivalent display results in high-avidity binding and makes it difficult to resolve high affinity binders from low affinity. The second version uses a pIII-antibody fragment fusion encoded on a separate plasmid, a phagemid, which is incorporated into newly synthesized phage on infection with helper phage. This results in lower copy number, with many phage displaying a single antibody fragment, and therefore reduces the problems associated with high-avidity binding masking high affinity.

Phage display has the advantages of being simple to operate, able to use large antibody libraries from a number of sources, and well established. However, due to the size limitation for efficient expression, protein folding in *E. coli*, and phage assembly, phage display is restricted to antibody fragments such as the V-domain containing, scFv and Fab fragments. This leads to a number of disadvantages as an antibody discovery methodology. Screening of antibody fragments is possible for binding, but assessment of functional activity often requires reformatting the isolated variable domains as an IgG and expression in mammalian cells. Therefore direct screening for functional activity can be problematic. In addition, phage libraries are expressed in *E. coli*, and this can lead to a number of issues. Libraries may be dominated by those antibody fragments that express well in *E. coli* as pIII fusions, which are not necessarily those that are the best behaved soluble antibody fragments.[35] Reconstitution as IgGs may be even more problematic and in some cases may result in antibodies that are expressed at low level, which may have poor solubility with potentially variable protein folding properties.[36,37] This has led to much interest in the design of improved antibody libraries for phage display that attempts to improve mammalian cell expression properties.[38]

Although relatively large, phage presented libraries rarely yield high affinity antibodies directly. This means that further optimization is required prior to therapeutic utility. This has been achieved through a number of affinity maturation methods that are described in more detail in the following text. An alternative approach to the isolation of high affinity antibodies has been to use phage display of libraries constructed from immunized donors, combining harvesting a human immune response with display of the B cell repertoire on phage as antibody fragments. As with other human B cell directed approaches this is useful for isolating antibodies to infectious disease antigens, and it has been successfully used in a number of cases, such as antibodies to influenza virus,[39] SARS virus[40] and anthrax toxin.[41] This approach is difficult however, due to the low frequency of antigen specific B cells that can be found in peripheral blood, and the limitations in immunizing humans.

A number of alternative microbial systems have also been investigated for antibody display, including the bacterial and yeast cell surface. These systems have one big advantage, which is the ability to use flow cytometry (FACS)-based techniques for isolation of cells displaying the best antibodies. This allows simultaneous determination of cell surface expression level and antigen binding, which avoids selection of only the best expressing antibodies, and allows much finer resolution of selected antibodies than can be achieved by standard phage panning techniques.[42]

Bacterial cell surface display has not been widely adopted, despite being investigated in both *E. coli* and gram positive bacteria such as *Staphylococcus carnosus*.[43–45] Transformation of *E. coli* is highly efficient, and therefore it should be well suited to the creation of large libraries, but selection may be limited by the display system. The most studied system uses anchoring to the cytoplasmic membrane via fusion to a fragment of lipoprotein A.[43] Access for screening requires permeabilization of the outer membrane, and removal of the peptidoglycan layer. The resulting *E. coli* spheroplasts can be screened by FACs with labeled antigen, but are not viable for growth, such that selected populations must be analyzed by gene recovery and re-expression. The system has also been tested for periplasmic display of full-length aglycosylated IgG, and was able to recover antibodies to *Bacillus anthracis* from a library constructed from an immunized mouse.[44] However, the system may be more suited to affinity maturation of antibodies, where the fine resolution of FACS selection is particularly important, and recent studies have combined phage screening with affinity maturation on the bacterial cell surface.[46] Similarly, presentation on the surface of *S. carnosus* requires presentation in a complex environment of peptidoglycan, and again more recent studies have used this in combination with phage display and suggested that the technique is more useful for maturation of antibodies from another source.[47]

Bacterial display suffers from many of the same limitations as phage display with regard to expression biases and the potential to isolate antibodies well-behaved in a prokaryotic expression system, but not necessarily in mammalian cells. Yeast has the advantage of being closer to mammalian cells with regard to protein folding and secretory pathways. Yeast display has been accomplished using fusion to the Aga2p mating adhesion receptor,[48] or by a secretion capture method in which *in vivo* biotinylated IgG is captured on the avidinated surface of the parent cell.[49] Library sizes available in yeast are smaller than those which can be generated in the phage system, and therefore, as in phage display, discovery of very high affinity antibodies is not commonly observed. A non-immune library of approx. 10^9 scFv displayed on yeast via the aga2p system was able to yield antibodies to a range of different antigens, with affinities similar to those obtained via phage display techniques.[50] Like the bacterial system, yeast display takes advantage of the fine resolution offered by FACS, allowing fine resolution of antibodies of different affinities.[42] A direct comparison of phage display with yeast display has been carried out using the same library, an immune library of approx. 10^7 scFv members, with selection by the same antigen.[51] In this case yeast display was able to outperform phage, finding all of the antibodies isolated from phage and twice as many novel ones, likely reflecting improved folding, expression and presentation at the yeast cell surface. Affinity maturation using yeast display has been successfully used in a number of cases. Random mutagenesis of yeast displayed antibodies has resulted in affinity improvement into the femtomolar range, illustrating that the limit on affinity maturation reached *in vivo* is not an intrinsic limit of antibody structure.[52]

One interesting study highlights the danger of *in vitro* selections carried out using antibody fragments such as scFv fragments. In selection of antibodies to calmodulin from a yeast display library, clones were obtained with affinity and specificity for the target protein, but on conversion to IgG the binding ability of these library hits was lost.[53] In this case it may have been due to the hydrophobic surface of the VH : VL interface, normally buried in an intact IgG, being involved in binding. Yeast libraries have also been constructed using Fab fragments,[54] and through extensive strain engineering, yeast strains have been developed that are capable of expression of aglycosylated IgG, which may also be possible through the secretion-capture methodology.[49] Selection in a format as close as possible to native IgG is preferred for efficient selections.

4.3 Mammalian Cell Display

The essential link between genotype and phenotype for an *in vitro* display system may also be satisfied by the use of mammalian cells. Mammalian cells are able to express, fold, assemble, post-translationally modify and secrete antibodies as authentic, fully functional IgG. In addition, they are well suited to FACS and offer the ability to co-select antibodies for multiple properties, including expression level in mammalian cells and other biophysical properties important for development.

A number of different cell types have been used for cell surface display, and frequently used laboratory cell lines such as HEK293, CHO and COS cells are particularly useful due to their ability to grow rapidly, their competence for efficient transfection, and robustness, especially through FACs cell sorting procedures. In addition, a number of B cell lines have been utilized for introduction of antibody libraries,[55] or for evolution of endogenous immunoglobulins.[56] The chicken cell line DT40, which has the advantage of a very fast growth rate, has also been used for display of its native immunoglobulin, which can be diversified through gene conversion.[57,58] Viral delivery systems can extend the number of usable cell lines, for example, a Sindbis virus-based vector has been used for presentation of an antibody library produced from human antigen specific B cells on BHK cells.[59]

Display at the surface is usually accomplished simply by fusion of the heavy chain of an IgG to a heterologous transmembrane domain. Few studies have been reported comparing transmembrane domains, although those from CD8 and CD28 outperformed the transmembrane region from HIV gp41 for presentation of a single chain Fv–Fc fusion on HEK293 cells.[60] Fusion to a GPI anchor sequence has been used to display IgG on HEK293 cells, and may have the advantage that antibodies can be readily released by treatment with phospholipase, PI-PLC.[61] The PDGFR transmembrane region has been used, probably due to its ready availability in commercially available pDisplay vectors, as have those from thrombomodulin and CD4.[62–64]

The size of combinatorial libraries that can be effectively screened in mammalian cell-based systems is similar to those that are practical in yeast display. These typically cover combinatorial diversity of approx. 10^9. As discussed above, these are smaller than those that can be screened by phage display, although in the case of mammalian cells, there are no issues with bias in microbial expression, and all members of the library are usually expressed and functional. As with yeast display, direct isolation of high affinity antibodies from naïve libraries is unlikely, and the majority of published studies have used libraries biased toward a particular antigen, such as those prepared from immunized donors[59,62,65] or by diversification of a single antibody.[63] More recently however, strategies have been devised and implemented for generation of high affinity antibodies from naïve human libraries. This is achieved by recovery of lower affinity antibodies from naïve human germline libraries, designed to mimic the natural mechanism of antibody generation from B lymphocytes, followed by *in vitro* somatic hypermutation, and is discussed in more detail in the following text.[66]

The first example of display of recombinant antibodies on a mammalian cell surface used retroviral vectors in packaging cell lines during the development of a retroviral display system.[67] This was subsequently developed for display on B cells.[55] Since then a number of viral vectors have been explored, including those based on Adenovirus[68] and Vaccinia.[69] A novel system has been developed to harness display on both lentivirus and lentiviral transduced human cells using self-inactivating lentiviral vectors.[60] Display of scFv–Fc in this system was achieved, allowing FACS based enrichment with labeled antigen. The importance of authentic post-translational modification was revealed as antibodies in this system were shown to be able to be sulfated on specific CDR tyrosine residues, a requirement for the antibody specificity being examined in this study.

Both transient expression systems and stable cell lines have been used for mammalian cell display. Transient systems have the advantage of being faster, but permit only a single cycle of selection before gene recovery and re-transfection. In addition it is difficult to control gene copy number during transient expression experiments, and usually library transfection results in many different antibodies being displayed on each cell. Approaches to control this include diluting the expression vector with control DNA to reduce the number of genes per cell.[64] Nevertheless, transient display has been successful for isolation from small libraries. In one example, a library of approx. 10^4 variants was constructed by randomizing a short sequence in an anti-CD22 scFv, and transient display on HEK293 cells was able to isolate a variant with improved affinity after a single selection round.[63] A larger library of approx. 10^6 diversity, constructed from PBMC of an immunized human donor, has been used to isolate antibodies to hepatitis B surface antigen over three rounds of selection using transient display of IgG on COS cells.[62] Antibody genes were recovered between each round of selection and re-transfected for subsequent rounds. Recovery and re-transfection of the antibody genes is relatively straightforward, but to maintain heavy and light chain pairings the genes must be linked in some way. In this case heavy and light chain were encoded on a single expression vector, but generation of more complex libraries in this format with high diversity in both heavy and light chains may be technically challenging. Other means of linking the genes, such as creation of single chain Fv or single-chain Fab fragments, have been explored[59,63,64] that solve the pairing issue but lose the advantages of screening with authentic functional IgG. A new approach that may prove particularly useful in this area is to isolate single cells, link individual heavy and light chains by PCR, re-pool and use deep sequencing techniques to investigate selections.[27] The presence of multiple heavy and light chains per cell may complicate the analysis, but such techniques may allow rapid transient selections to be performed.

Stable expression in mammalian cell lines is well established for high-level antibody production, and such stable expression is ideal for multiple rounds of selection to enrich desired antibody expressing cells to the majority of the population. However, most stable expression systems developed in mammalian cells result in integration of the introduced DNA into the host cell genome, with selection through an antibiotic, or other, selectable marker introduced at the same time. Integration of the DNA into an active transcriptional site is a rare event, resulting in the majority of the cells being killed, while the few drug resistant cells, which are also expressing the antibody, grow. This works well for expression of a single antibody, but expression of large libraries is very inefficient. In addition, such techniques often result in the integration of multiple copies of the antibody genes, which, with libraries, would result in several different antibodies being present on the surface of each cell, making identification of the best antibodies challenging. There are a number of expression systems that have now been developed for high-level expression from a single copy, using co-integration of a chromosomal opening element[70] or by directed integration into an active transcriptional site. One of these, the commercially available *Flp-In*™ system, has been explored for mammalian cell display.[71] This system uses homologous recombination through introduction of a *Flp* recombination (FRT) site into the host cell genome, with subsequent integration of an expression cassette containing the antibody genes via *Flp* recombinase mediated recombination at the single FRT site. The selectable marker in the antibody expression vector, has no promoter or start codon and can only be expressed when integrated downstream of the promoter and in-frame with the ATG codon of the FRT site, resulting in the selection of single copy integrants. The use of this system with large libraries remains challenging, for the reasons highlighted previously, but it has been used for the display of small libraries. Screening of a library derived from human donor PBMCs of an individual immunized with hepatitis B surface antigen resulted in the isolation of three unique antibodies.[65] In this study transfection of the library of approx. 10^5 diversity resulted in approx. 40% of the selected cells expressing antibody on the cell surface, with approx. 1% positive for antigen binding, allowing two rounds of FACS selection to enrich antigen binding clones. The same system was also used to enrich affinity-matured variants of an anti-OX40L antibody from a library of approx. 10^5 variants.

Larger libraries have now been enabled by use of an alternate system based on episomally maintained vectors for antibody expression. Such vectors are self-replicating and able to be stably maintained in the cell without integration into the host cell genome. Epstein-Barr virus (EBV)-based vectors fit this description, and constructs containing the EBV oriP origin of replication support maintenance in cells expressing EBV nuclear antigen (EBNA-1).[72] Vectors containing these elements are maintained at low copy number and are well suited to combinatorial gene expression.[73] Antibody expression can be maintained for months at low copy number using such an episomal vector system without integration into the genome.[61,66]

Transfection of an antibody library derived from immunized chicken spleen cells, with diversity of approx. 10^5 made up of heavy and light chains cloned into a single episomal vector, resulted in approx 90% of the cells expressing IgG on the surface of HEK293 cells.[61] After a bead-based negative selection to remove non-specific binders, 3–4 rounds of FACS selection were used to isolate a panel of antibodies cross-reactive to human and mouse IL-12.

Episomal vectors have the advantage of being able to be rapidly selected, with most of the transfected cells being able to survive selection and present antibody on the cell surface. This has enabled large naïve libraries to be designed and used for antibody isolation in a wide range of different targets. In this case independent episomal vectors are used for IgG heavy and light chain expression, each independently selected with a different antibiotic resistance marker, which facilitates combinatorial experiments.[66] Library construction is also simplified in this format with separate heavy and light chain libraries constructed for co-transfection. Combinatorial libraries of 10^9 members have been shown to be efficiently presented using this methodology. A naïve library presented in this way has allowed selection of specific antibodies for a large number of target antigens, as exemplified by a panel of antibodies to βNGF.[66] The library used for these experiments, termed the ABEL library, was constructed from human germline V segments linked to rearranged heavy and light chain (D)J diversity recovered from human donor PBMC, resulting in a repertoire of rearranged V(D)J sequences. This library is intended to mimic the naïve antibody repertoire expressed by B cells *in vivo*, which can be successfully affinity matured to recognize a constellation of different antigens. As such the ABEL library is expected to yield low to moderate affinity antibodies, with fully human germline sequences, that can be subsequently matured to high affinity using *in vitro* somatic hypermutation, as discussed next.

4.4 Somatic Hypermutation (SHM)

The immune system has developed to be able to generate effective immunity against a wide variety of invading pathogens. As part of this, antibody generation against a plethora of different antigens is possible despite a limited number of germline antibody genes. Diversity in the antibody binding site is therefore introduced at different stages. Initial diversity in the variable region is produced entirely within the CDR3 regions of the antibody through recombination of the VDJ genes for the heavy chain and VJ genes for the light chain, during B cell development in the bone marrow. This is an antigen independent process and includes not just recombination of the different gene segments but also nucleotide addition and excision at the junctional sites mediated by terminal deoxynucleotidyl transferase (TdT) and exonucleases, respectively. This leads to diversity in CDR3 sequences and in length, and permits the weak recognition of a wide range of different antigens. The second stage of antibody diversification takes place following antigenic stimulation of B cells in the periphery. Known as somatic hypermutation (SHM), this process introduces mutations into the whole of the antibody variable region, allowing the evolution of antibodies to high binding affinity and specificity to a particular antigen. SHM is initiated by the action of the enzyme activation-induced cytidine deaminase (AID), acting directly on antibody V-region DNA. AID operates on single-stranded DNA, both the transcribed and opposite strand, exposed during transcription of the antibody V-region DNA, resulting in the deamination of cytidine to uridine.[74,75] The resulting U/G mismatches are subject to one of several pathways of DNA repair, including base excision and mismatch repair, with the error-prone repair leading to the mutations that are termed somatic hypermutation (Figure 4.2).

SHM occurs in the immunoglobulin variable region at a rate of approx. 10^{-3} mutations/bp/generation, about a million-fold higher frequency than the background rate of mutation.[76] AID-initiated mutations occur at all four nucleotides with roughly equivalent frequencies despite the direct action of AID being restricted to deamination of C. The AID generated U/G mismatch is a substrate for uracil DNA glycosylase (UNG) or mismatch repair, or alternatively, it may be replicated over resulting in a transition mutation (Figure 4.2).

SHM is targeted to antibody variable regions, yet the exact targeting mechanisms remain a subject of active research. Over 20 different proteins have been suggested to be involved as co-factors important for AID targeting in B cells.[75] It is clear that AID activity is preferentially directed to specific hotspot sequences within the V-region sequence, and the motif WRCH

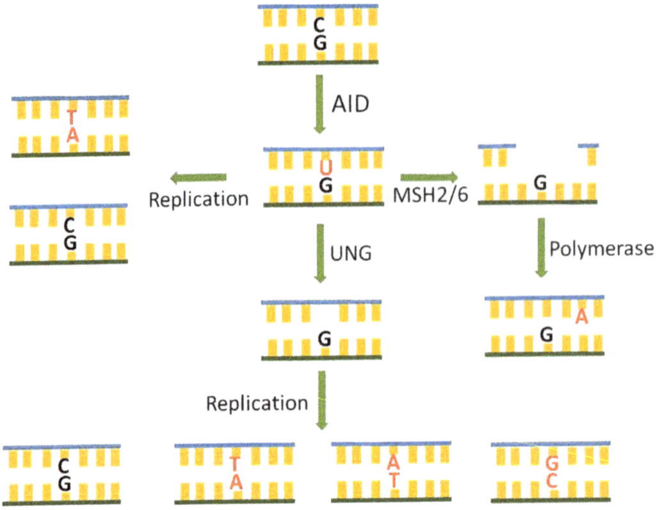

Figure 4.2 Mechanism of somatic hypermutation initiated by AID. AID results in deamination of cytidine to uridine and the resulting U/G mismatches undergo a number of possible routes of DNA repair as indicated.

(AT/GA/C/TAC) has been reported to be optimal for AID activity, likely due to a combination of the sequence preference of AID and those sequences preferentially recognized by the enzymes of DNA repair.[77] AID introduces mutations into the antibody V-region at a far higher rate than other genes in the B cell[78] and antibody genes have evolved to use AID-mediated SHM efficiently, with hotspots arranged to promote changes likely to alter antibody binding characteristics. For example, amino acid replacements are more likely in the antigen-binding CDR loops than in the V-region framework because AID action at hotspots in the framework more frequently leads to silent mutations.[66,79]

Although the majority of SHM mutations are point mutations, AID-initiated SHM can also lead to insertions and deletions in the antibody variable region, especially in the CDR loops.[80,81] Estimates range up to 6.5% of human antibodies containing SHM-associated insertions and deletions.[82] A number of mechanisms have been proposed for the creation of these, and insertions appear to arise from local sequence duplication combined with point mutation events.[83,84] The location of SHM-associated insertions and deletions has been examined through deep sequence analysis of donor B cells, revealing approx. 76% of them are found within CDRs, with another cluster found at the important framework 3 region which corresponds to "CDR4" in T cell receptors.[85] The additional diversity afforded by insertions and deletions may assist the generation of high affinity antibody responses, and a number of antibodies have been characterized where they are important for antigen binding. For example, a 3 amino acid insertion in the CDRH2 region of an antibody to influenza virus was responsible for a 35-fold improvement in binding affinity,[85] whereas a 2 amino acid deletion in an anti-HIV antibody had at least a 5-fold effect on affinity.[86]

4.5 Combining Mammalian Cell Display with *In Vitro* SHM

The natural process of antibody generation in B cells combines mammalian cell display with SHM and is a model for improved selection of antibodies *in vitro*. Combining these two processes allows the benefits of mammalian cell display outlined previously to be utilized, and overcomes limitations in library sizes through the *in situ* generation of antibody diversity by SHM. In addition, this diversity can be concentrated in antigen-specific clones, allowing more relevant antibody sequences to be generated and sampled. Expression of AID is sufficient to drive SHM in both B and non-B cells *in vitro*.[87] The remainder of the biochemical pathways leading to SHM consist of DNA-repair, which is widely conserved in mammalian cells and therefore a wide range of cell types can support SHM upon introduction of exogenous AID. These include robust laboratory cell lines such as CHO[87] and HEK293.[66] The concept of *in vitro* maturation of antibodies using SHM was initially demonstrated by the evolution of the endogenous IgM of Ramos cells to recognize streptavidin with an apparent affinity of 11nM, as well as the endogenous chicken immunoglobulin of DT40 cells to recognize protein A and IgG.[56]

B cells have been investigated further, especially the chicken cell line DT40 due to its rapid growth rate. Chicken IgM antibodies have been derived that recognize human IgG or streptavidin.[57] In chicken and other avian cell lines, affinity maturation of antibodies is driven through the process of gene conversion as well as SHM, both of which are initiated by the action of AID. Both of these activities were important in this study, and this system has been modified to generate chicken:human chimeric antibodies with human constant regions and chicken V-regions to a number of additional antigens.[88]

Although B cells might seem like an obvious choice for antibody display and *in vitro* SHM, they may not be the most practical system. Mammalian B cell lines grow relatively slowly and are difficult to transfect at high efficiency. The use of cell types such as CHO and HEK293, which are also used for antibody manufacturing, offers the advantage of co-selecting antibodies for high-level expression that are well behaved in these cell types. Both CHO and HEK293 are competent for SHM with mutation rates essentially identical to those reported in B cells.[66,87] Other cell types have also been investigated. For example, the human non-small cell lung carcinoma line H1299 was used to affinity mature an scFv to TNFα via SHM, with higher affinity variants identified after two rounds of FACS sorting.[89]

The SHM–XEL™ system is a fully integrated technology for antibody discovery and maturation that has been developed by coupling mammalian cell display of antibody germline libraries with *in vitro* SHM.[66,90] This system is designed to mimic many of the features of the human adaptive immune response as shown in Figure 4.3.

Human antibody libraries, termed ABEL libraries, are designed to mimic the primary gene rearrangements seen in naïve B cells. This is achieved through the

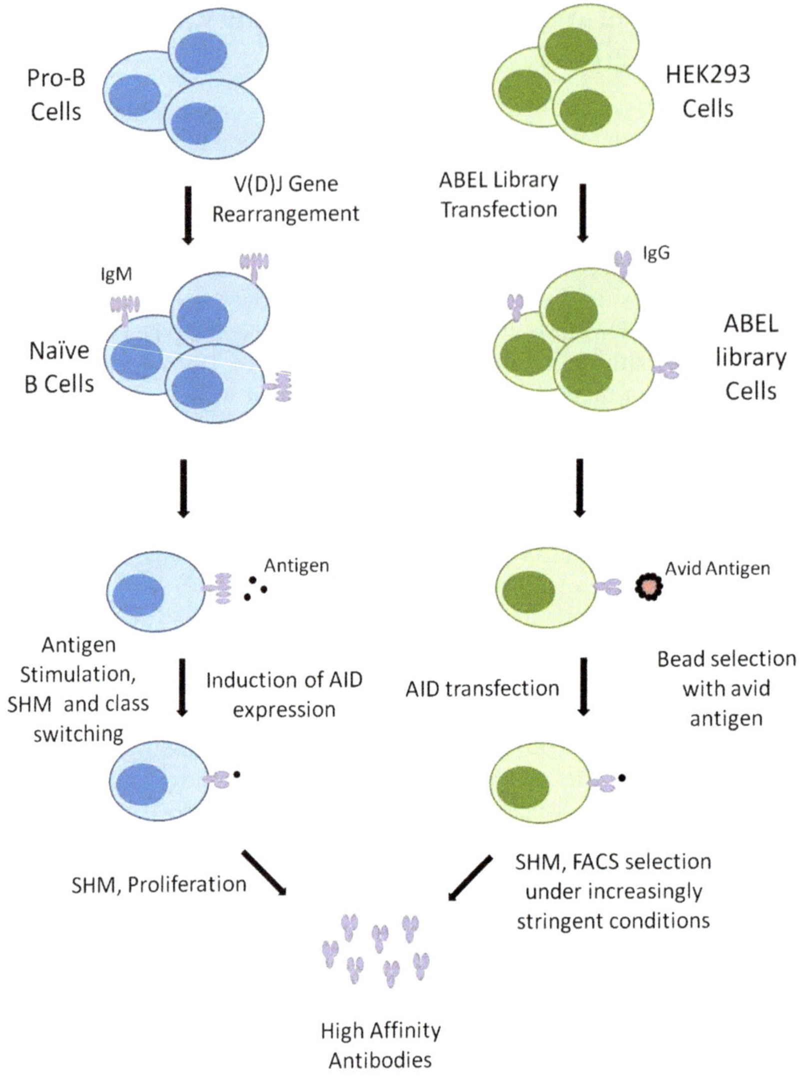

Figure 4.3 Comparison of antibody generation by B cells *in vivo* with the SHM–XEL™ system using the ABEL library in HEK293 cells. In the natural process of antibody generation pro-B cells undergo V(D)J gene rearrangements to present antibody as IgM on the surface of B cells. In the presence of antigen, cells presenting antibodies with weak affinity to the antigen are stimulated to proliferate and express AID, resulting in class switching and affinity maturation of antibodies to the target antigen. The SHM–XEL™ system uses, for example, HEK293 cells transfected with a library of human IgG, designed to mimic the naïve rearranged germline sequences in a B cell. Antigen selection is carried out under avid conditions, resulting in the isolation of cells presenting antigen-specific antibodies. These cells are then transfected with AID resulting in affinity maturation to the antigen.

use of synthetic human germline V gene segments fused to the rearranged (D)J regions isolated by PCR from PBMC of a panel of human donors.[66,91] The resulting libraries contain the donor-derived diversity in the antibody CDR3 regions, which includes both length and sequence diversity as expected. ABEL libraries have been constructed and displayed on the surface of HEK293 cells with diversity of 10^8–10^9 expressed sequences.

As discussed previously, cell surface display is accomplished simply by fusion of the IgG heavy chain to a type I transmembrane sequence, allowing the linkage between a selectable phenotype and genotype. However, in order to select ideal antibodies for development and to take full advantage of screening fully functional IgG, it is also desirable to test soluble antibody for important characteristics, such as activity in cell-based functional assays and biophysical properties including stability and solubility. This has been achieved by constructing a heavy chain expression construct with splice donor and acceptor sites surrounding the transmembrane sequence,[92] resulting in two alternatively spliced heavy chain products being produced, one resulting in cell surface display and the other resulting in secretion of IgG with the same V-region sequences. In addition, the ratio of cell-surface to secreted IgG may be readily manipulated by adjusting the sequences flanking the splice donor site.[92]

Interrogation of libraries such as the ABEL library with diversity solely in the CDR3 region has a number of advantages. A library constructed with synthetic diversity only in the heavy chain CDR3, which was interrogated by phage display, was able to identify hits to all of the seven targets screened, and CDR3 lengths in the selected antibodies trended toward the same length distribution as natural antibodies.[93] Screening of the ABEL library, which comprises natural CDR3 diversity in both heavy and light chain CDR3 sequences, has resulted in hits to each of many different antigens used.[91] Although hits are usually relatively low affinity, the remainder of the sequence is human germline, with no "passenger" mutations present, allowing antibodies to be derived with minimal sequence variation from human germline, which are likely to be well tolerated by the human immune system. Antibodies evolved using *in vitro* SHM are selected to only incorporate sequence changes that impact important properties of the antibody.

Screening of ABEL libraries and subsequent maturation is accomplished under conditions of increasing stringency. Libraries may be initially screened using high avidity forms of antigen, such as soluble antigen coated on magnetic beads, or by binding to cells displaying high density target antigen. Such highly avid systems facilitate the isolation of the low affinity germline antibodies. Isolated cells displaying low affinity antibodies are then transfected with AID and subjected to AID-induced SHM with selection of improved variants by iterative rounds of FACS sorting with lower avidity forms of antigen at diminishing concentrations to drive maturation. Conditions may be manipulated to drive maturation specifically in on-rates or off-rates, for example, by shorter contact times with antigen or extended washing conditions respectively.[91] In addition, unlabelled antigen or other molecules may be added as competitors to drive evolution to the required end points. In one example, an antibody to human βNGF was derived from screening of an ABEL library using mammalian cell display on HEK293 cells, and SHM *in vitro* resulted in evolution of the antibody to very high affinity (K_D 25pM), with functional activity simultaneously selected for using the soluble form of the antibody.[66]

The advent of deep sequencing technology is also making an impact on the practice of *in vitro* SHM. Deep sequencing of cell populations, in which AID has been actively generating SHM mutations, can be carried out before and after FACS sorting, allowing rare mutations to be identified at an early stage. Individual mutations have widely different effects on the affinity of antibodies; some improving mutations may result in only a 1.5–2-fold improvement, whereas other single events can result in 10-fold improvements or more. Deep sequencing enables multiple SHM events to be identified and recombined before multiple rounds of FACS sorting, which might result in displacement of some helpful mutations by those that have a more profound effect.[66,91]

One advantage of FACS-based selection techniques is that cell sorting may be performed in multiple colors, allowing multiple parameter selections. Simultaneous evolution in antibody affinity and expression level is routinely carried out by staining cells with both labeled antigen and anti-IgG reagents. In addition, biophysical properties can be improved. For example, antibody engineering of a stable variable region scaffold, and *in vitro* SHM, has been employed to improve antibody thermal stability from a melting temperature of 68 °C to over 90 °C, with a simultaneous improvement in binding affinity of 100-fold.[94]

Modulation of antibody specificity may also be achieved. Unwanted specificities are readily screened for by inclusion of unlabeled protein to block the unwanted specificity. Selections with two different antigens, such as murine and human orthologs of an antigen, with antigens labeled with compatible fluorescent groups, can be used to derive murine/human cross-reactive antibodies. Antibodies with rearranged germline sequences, such as those present in the ABEL library, have been reported to have binding sites with a high degree of structural plasticity and may evolve along multiple pathways.[95,96] SHM may drive evolution of polyreactive antibodies toward specificity to a single antigen, but by controlled *in vitro* SHM, it may also be possible to drive evolution toward multiple specificity. A number of artificially created antibodies have been generated with specificity for two non-homologous antigens,[97,98] and it may be possible to derive similarly dual reactive antibodies directly by *in vitro* SHM.

Maturation via SHM may be also be used to derive humanized antibodies with minimal murine derived sequence content. During the traditional CDR-grafting process, in which murine derived CDR sequences are grafted onto a human antibody variable region framework, significant loss of antigen binding affinity may occur. This may be regained by back-mutation of some framework amino acids to the original murine sequence, but at the cost of reducing the overall human derived sequence content. An alternative method is to mature a CDR-grafted antibody via *in vitro* SHM, where the original binding affinity may be regained and surpassed. This has been extended to a method using transfer of only the murine mAb CDR3 to an otherwise human germline sequence antibody. Sufficient binding specificity may be retained to allow maturation to high affinity but with significantly less murine derived antibody sequence. This has been exemplified by grafting the CDR3 from a murine mAb recognizing βNGF into a non-homologous human germline sequence antibody framework.[99] Following *in vitro* SHM, a humanized antibody was derived with a low pM dissociation constant that was considerably more potent than the parental antibody.

This methodology has now been extended to obviate the requirement to make murine hybridomas, with direct derivation of humanized antibodies.[89] In this study CDR3 sequences were isolated from the spleen and lymph nodes of mice immunized with the human complement protein C5. Following the cloning of these sequences fused with human germline heavy chain V-genes to form intact humanized heavy chains, they were paired with a human light chain library, displayed on the surface of HEK293 cells, and co-transfected with AID. Numerous clones were isolated by FACS, and SHM resulted in the rapid generation and selection of high affinity, functional antibodies.

The ability to reproduce SHM *in vitro* coupled with mammalian cell display is a powerful tool to study antibody maturation pathways. It is now becoming apparent that SHM can influence not only binding affinity and specificity, but also other important properties of the antibody such as expression level and stability. In a study of SHM events within the antibody 9F3F3, a subset of mutations were sufficient to confer the high binding affinity to the antibody, yet the remaining mutations selected in the antibody were crucial for stability of the molecule.[100] Similar results were observed with the antibody OKT3, and in both cases many of these stabilizing mutations were required for optimal organization of the interface between the heavy and light chain variable regions. Many of these mutations may have previously been thought to be neutral, but this data lends support to the interplay of SHM in optimization of all properties of the antibody molecule to enable it to be an effective constituent of the immune system.

References

1. T. J. Schroeder, S. Hariharan and M. R. First, *Transplant Proc*, 1993, **25**, 77–80.
2. J. C. Almagro and J. Fransson, *Front Biosci*, 2008, **13**, 1619–1633.
3. P. D. Senter and E. L. Sievers, *Nat Biotechnol*, 2012, **30**, 631–637.
4. M. Bruggeman, H. M. Caskey, C. Teale, H. Waldmann, G. T. Williams, M. A. Surani and M. S. Neuberger, *Proc Natl Acad Sci USA*, 1989, **86**, 6709–6713.
5. N. Lonberg, L. D. Taylor, F. A. Harding, M. Trounstine, K. M. Higgins, S. R. Schramm, C. C. Kuo, R. Mashayekh, K. Wymore, J. G. McCabe, D. Munoz-O'Regan, S. L. O'Donell, E. S. G. Lapachet, T. Bengoechea, D. M. Fishwild, C. E. Carmack, R. M. Kay and D. Huszar, *Nature*, 1994, **368**, 856–859.

6. L. L. Green, M. C. Hardy, C. E. Maynard-Currie, H. Tsuda, D. K. Louie, M. J. Mendez, H. Abderrahim, M. Noguchi, D. H. Smith, Y. Zeng, N. E. David, H. Sasai, D. Garza, D. G. Brenner, J. F. Hales, R. P. McGuiness, D. J. Capon, S. Klapholz and A. Jakobovits, *Nat Genet*, 1994, **7**, 13–21.
7. N. Lonberg, *Handb Exp Pharmacol*, 2008, **181**, 69–97.
8. T. M. Dechiara, W. T. Poueymirou, W. Auerbach, D. Frendewey, G. D. Yancopoulos and D. M. Valenzuela, *Methods Mol Biol*, 2009, **530**, 311–324.
9. M. J. Osborn, B. Ma, S. Avis, A. Binnie, J. Dilley, X. Yang, K. Lindquist, S. Menoret, A. L. Iscache, L. H. Ouisse, A. Rajpal, I. Anegon, M. S. Neuberger, R. Buelow and M. Bruggemann, *J Immunol*, 2013, **190**, 1481–1490.
10. C. Tkacyzk, L. Hua, R. Varkey, Y. Shi, L. Dettinger, R. Woods, A. Barnes, R. S. MacGill, S. Wilson, P. Chowdhury, C. K. Stover and B. R. Sellman, *Clin Vacc Immunol*, 2012, **19**, 377–385.
11. F. D. Batista and M. S. Neuberger, *Immunity*, 1998, **8**, 751–759.
12. J. Foote and H. N. Eisen, *Proc Natl Acad Sci (USA)*, 1995, **92**, 1254–1256.
13. P. Rathanaswami, S. Roalstad, L. Roskos, Q. J. Su, S. Lackie and J. Babcock, *Biochem Biophys Res Commun*, **334**, 1004–1013.
14. T. R. Poulsen, A. Jensen, J. S. Haurum and P. S. Anderson, *J Immunol*, 2011, **187**, 4229–4235.
15. A. Jin, T. Ozawa, K. Tajiri, T. Obata, S. Kondo, K. Kinoshita, S. Kadowaki, K. Takahashi, T. Sugiyama, H. Kishi and A. Muraguchi, *Nat Med*, 2009, **15**, 1088–1092.
16. D. Corti, J. Voss, S. J. Gamblin, G. Codoni, A. Macagno, D. Jarrossay, S. G. Vachieri, D. Pinna, A. Minola, F. Vanzetta, C. Silacci, B. M. Fernandez-Rodriguez, G. Agatic, S. Bianchi, I. Giacchetto-Sasselli, L. Calder, F. Sallusto, P. Collins, L. F. Haire, N. Temperton, J. P. Langedijk, J. J. Skehel and A. Lanzavecchia, *Science*, 2011, **333**, 850–856.
17. L. M. Walker, S. K. Phogat, P. Y. Chan-Hui, D. Wagner, P. Phung, J. L. Goss, T. Wrin, M. D. Simek, S. Fling, J. L. Mitcham, J. K. Lehrman, F. H. Priddy, O. A. Olsen, S. M. Frey, P. W. Hammond, Protocol G Principal Investigators, S. Kaminsky, T. Zamb, M. Moyle, W. C. Koff, P. Poignard and D. R. Burton, *Science*, 2009, **326**, 285–289.
18. E. Traggiai, S. Becker, K. Subbarao, L. Kolesnikova, Y. Uematsu, M. R. Gismondo, B. R. Murphy, R. Rappuoli and A. Lanzavecchia, *Nat Med*, 2004, **10**, 871–875.
19. J. Fraussen, K. Vrolix, P. Martinez-Martinez, M. Losen, E. Meulemans, M. H. De Baets, P. Stinissen and V. Somers, *J Autoimmun*, 2010, **35**, 130–134.
20. M. J. Kwakkenbos, S. A. Diehl, E. Yasuda, A. Q. Bakker, C. M. van Geelen, M. V. Lukens, G. M. van Bleek, M. N. Widjojoatmodjo, W. M. Bogers, H. Mei, A. Radbruch, F. A. Scheeren, H. Spits and T. Beaumont, *Nat Med*, 2010, **16**, 123–128.

21. T. Tiller, E. Meffre, S. Yurasov, M. Tsuiji, M. C. Nussenzweig and H. Wardemann, *J Immunol Methods*, 2008, **329**, 112–124.
22. J. F. Scheid, H. Mouquet, N. Feldhahn, B. D. Walker, F. Pereyra, E. Cutrell, M. S. Seaman, J. R. Mascola, R. T. Wyatt, H. Wardemann and M. C. Nussenzweig, *J Immunol Methods*, 2009, **343**, 65–67.
23. R. Di Niro, L. Mesin, M. Raki, N. Y. Zheng, F. Lund-Johansen, K. E. Lundin, A. Charpilienne, D. Poncet, P. C. Wilson and L. M. Sollid, *J Immunol*, 2010, **185**, 5377–5383.
24. J. Wrammert, K. Smith, J. Miller, W. A. Langley, K. Kokko, C. Larsen, N. Y. Zheng, I. Mays, L. Garman, C. Helms, J. James, G. M. Air, J. D. Capra, R. Ahmed and P. C. Wilson, *Nature*, 2008, **453**, 667–671.
25. S. T. Reddy, X. Ge, A. E. Miklos, R. A. Hughes, S. H. Kang, K. H. Hoi, C. Chrysostomou, S. P. Hunicke-Smith, B. L. Iverson, P. W. Tucker, A. D. Ellington and G. Georgiou, *Nat Biotechnol*, 2010, **28**, 965–969.
26. W. C. Cheung, S. A. Beausoleil, X. Zhang, S. Sato, S. M. Schieferl, J. S. Wieler, J. G. Beaudet, R. K. Ramenani, L. Popova, M. J. Comb, J. Rush and R. D. Polakiewicz, *Nat Biotechnol*, 2012, **30**, 447–452.
27. B. J. DeKosky, G. C. Ippolito, R. P. Deschner, J. J. Lavinder, Y. Wine, B. M. Rawlings, N. Varadarajan, C. Giesecke, T. Dörner, S. F. Andrews, P. C Wilson, S. P. Hunicke-Smith, C. G. Willson, A. D. Ellington and G. Georgiou, *Nat Biotechnol*, 2013, **31**, 166–169.
28. G. P. Smith, *Science*, 1985, **228**, 1315–1317.
29. J. McCafferty, A. D. Griffiths, G. Winter and D. J. Chiswell, *Nature*, 1990, **348**, 552–554.
30. C. F. Barbas, J. D. Bain, D. M. Hoekstra and R. A. Lerner, *Proc Natl Acad Sci USA*, 1992, **89**, 4457–4461.
31. J. D. Marks, H. R. Hoogenboom, T. P. Bonnert, J. McCafferty, A. D. Griffiths and G. Winter, *J Mol Biol*, 1991, **222**, 581–597.
32. A. D. Griffiths, S. C. Williams, O. Hartley, I. M. Tomlinson, P. Waterhouse, W. L. Crosby, R. E. Kontermann, P. T. Jones, N. M. Low, T. J. Allison, T. D. Prospero, H. R. Hoogenboom, A. Nissim, J. P. L. Cox, J. L. Harrison, M. Zaccolo, E. Gherardi and G. Winter, *EMBO J*, 1994, **13**, 3245–3260.
33. R. M. Hoet, E. H. Cohen, R. B. Kent, K. Rookey, S. Schoonbroodt, S. Hogan, L. Rem, N. Frans, M. Daukandt, H. Pieters, R. van Hegelsom, N. C. Neer, H. G. Nastri, I. J. Rondon, J. A. Leeds, S. E. Hufton, L. Huang, I. Kashin, M. Devlin, G. Kuang, M. Steukers, M. Viswanathan, A. E. Nixon, D. J. Sexton, H. R. Hoogenboom and R. C. Ladner, *Nat Biotechnol*, 2005, **23**, 344–348.
34. C. Lloyd, D. Lowe, B. Edwards, F. Welsh, T. Dilks, C. Hardman and T. Vaughan, *Protein Eng Des Sel*, **22**, 159–168.
35. E. Pavoni, G. Monteriù, M. Cianfriglia and O. Minenkova, *Gene*, 2007, **391**, 120–129.
36. L. Persic, M. Righi, A. Roberts, H. R. Hoogenboom, A. Cattaneo and A. Bradbury, *Gene*, 1997, **187**, 1–8.

37. M. C. Vendel, M. Favis, W. B. Snyder, F. Huang, A. D. Capili, J. Dong, S. M. Glaser, B. R. Miller and S. J. Demarest, *Arch Biochem Biophys*, 2012, **526**, 188–193.
38. J. Prassler, S. Thiel, C. Pracht, A. Polzer, S. Peters, M. Bauer, S. Nörenberg, Y. Stark, J. Kölln, A. Popp, S. Urlinger and M. Enzelberger, *J Mol Biol*, 2011, **413**, 261–278.
39. M. Throsby, E. van den Brink, M. Jongeneelen, L. L. Poon, P. Alard, L. Cornelissen, A. Bakker, F. Cox, E. van Deventer, Y. Guan, J. Cinatl, J. ter Meulen, I. Lasters, R. Carsetti, M. Peiris, J. de Kruif and J. Goudsmit, *PLoS One*, 2008, **3**, e3942.
40. J. ter Meulen, E. N. van den Brink, L. L. Poon, W. E. Marissen, C. S. Leung, F. Cox, C. Y. Cheung, A. Q. Bakker, J. A. Bogaards, E. van Deventer, W. Preiser, H. W. Doerr, V. T. Chow, J. de Kruif, J. S. Peiris and J. Goudsmit, *PLoS Med*, 2006, **3**, e237.
41. M. A. Wild, H. Xin, T. Maruyama, M. J. Nolan, P. M. Calveley, J. D. Malone, M. R. Wallace and K. S. Bowdish, *Nat Biotechnol*, 2003, **21**, 1305–1306.
42. S. A. Gai and K. D. Wittrup, *Curr Opin Struct Biol*, 2007, **17**, 467–473.
43. B. R. Harvey, G. Georgiou, A. Hayhurst, K. J. Jeong, B. L. Iverson and G. K. Rogers, *Proc Natl Acad Sci USA*, 2004, **101**, 9193–9198.
44. Y. Mazor, T. Van Blarcom, R. Mabry, B. L. Iverson and G. Georgiou, *Nat Biotechnol*, 2007, **25**, 563–565.
45. J. Lofblom, *Biotechnol J*, 2011, **6**, 1115–1129.
46. Y. Mazor, T. Van Blarcom, S. Carroll and G. Georgiou, *FEBS J*, 2010, **277**, 2291–2303.
47. N. Kronqvist, M. Malm, L. Göstring, E. Gunneriusson, M. Nilsson, I. Höidén Guthenberg, L. Gedda, F. Y. Frejd, S. Ståhl and J. Löfblom, *Protein Eng Des Sel*, 2011, **24**, 385–396.
48. E. T. Boder and K. D. Wittrup, *Nat Biotechnol*, 1997, **15**, 553–557.
49. J. A. Rakestraw, D. Aird, P. M. Aha, B. M. Baynes and D. Lipovsek, *Protein Eng Des Sel*, 2011, **24**, 525–530.
50. M. J. Feldhaus, R. W. Siegel, L. K. Opresko, J. R. Coleman, J. M. Feldhaus, Y. A. Yeung, J. R. Cochran, P. Heinzelman, D. Colby, J. Swers, C. Graff, H. S. Wiley and K. D. Wittrup, *Nat Biotechnol*, 2003, **21**, 163–170.
51. D. R. Bowley, A. F. Labrijn, M. B. Zwick and D. R. Burton, *Protein Eng Des Sel*, 2007, **20**, 81–90.
52. E. T. Boder, K. S. Midelfort and K. D. Wittrup, *Proc Natl Acad Sci USA*, 2000, **97**, 10701–10705.
53. J. M. Weaver-Feldhaus, K. D. Miller, M. J. Feldhaus and R. W. Siegel, *Protein Eng Des Sel*, 2005, **18**, 527–536.
54. L. Blaise, A. Wehnert, M. P. Steukers, T. van den Beucken, H. R. Hoogenboom and S. E. Hufton, *Gene*, 2004, **342**, 211–218.
55. U. Grawunder and J. Stitz, Identification of antigen or ligand specific binding proteins International patent application WO2009/109368 A1 September 11, 2009.

56. S. J. Cumbers, G. T. Williams, S. L. Davies, R. L. Grenfell, S. Takeda, F. D. Batista, J. E. Sale and M. S. Neuberger, *Nat Biotechnol*, 2002, **20**, 1129–1134.
57. H. Seo, M. Masuoka, H. Murofushi, S. Takeda, T. Shibata and K. Ohta, *Nat Biotechnol*, 2005, **23**, 731–735.
58. M. Kajita, T. Okazawa, M. Ikeda, K. Todo, M. Magari, N. Kanayama and H. Ohmori, *J Biosci Bioeng*, 2010, **110**, 351–358.
59. R. R. Beerli, M. Bauer, R. B. Buser, M. Gwerder, S. Muntwiler, P. Maurer, P. Saudan and M. F. Bachmann, *Proc Natl Acad Sci USA*, 2008, **105**, 14336–14341.
60. R. Taube, Q. Zhu, C. Xu, F. Diaz-Griffero, J. Sui, E. Kamau, M. Dwyer, D. Aird and W. A. Marasco, *PLoS One*, 2008, **3**, e3181.
61. Y. Akamatsu, K. Pakabunto, Z. Xu, Y. Zhang and N. Tsurushita, *J Immunol Methods*, 2007, **327**, 40–52.
62. K. Higuchi, T. Araki, O. Matsuzaki, A. Sato, K. Kanno, N. Kitaguchi and H. Ito, *J Immunol Methods*, 1997, **202**, 193–204.
63. M. Ho, S. Nagata and I. Pastan, *Proc Natl Acad Sci USA*, 2006, **103**, 9637–9642.
64. A. L. Gurney, A. L. L. Lazetic and C. J. Bond, US Patent Application US2011/0287979 A1, November 24, 2011.
65. C. Z. Li, Z. K. Liang, Z. R. Chen, H. B. Lou, Y. Zhou, Z. H. Zhang, F. Yu, S. Liu, Y. Zhou, S. Wu, W. Zheng, W. Tan, S. Jiang and C. Zhou, *Cell Mol Immunol*, 2012, **9**, 184–190.
66. P. M. Bowers, R. A. Horlick, T. Y. Neben, R. M. Toobian, G. L. Tomlinson, J. L. Dalton, H. A. Jones, A. Chen, L. Altobell 3rd, X. Zhang, J. L. Macomber, I. P. Krapf, B. F. Wu, A. McConnell, B. Chau, T. Holland, A. D. Berkebile, S. S. Neben, W. J. Boyle and D. J. King, *Proc Natl Acad Sci USA*, 2011, **108**, 20455–20460.
67. S. J. Russell, R. E. Hawkins and G. Winter, *Nucleic Acids Res*, 1993, **21**, 1081–1085.
68. G. Gao and H. Wu, Cell display of antibody libraries US patent 7,790,655 September 7, 2010.
69. M. Zauderer and E. S. Smith, US patent application US2008/0167193 A1, July 10, 2008.
70. T. Benton, T. Chen, M. McEntee, B. Fox, D. King, R. Crombie, T. C. Thomas and C. Bebbington, *Cytotechnology*, 2002, **38**, 43–46.
71. C. Zhou, F. W. Jacobsen, L. Cai, Q. Chen and W. D. Shen, *MAbs*, 2010, **2**, 508–518.
72. S. E. Lindner and B. Sugden, *Plasmid*, 2007, **58**, 1–12.
73. R. A. Horlick, A. E. Schilling, P Samama, R. N. Swanson, V. D. Fitzpatrick, A. K. Robbins and B. Damaj, *Gene*, 2000, **243**, 187–194.
74. J. M. Di Noia and M. S. Neuberger, *Annu Rev Biochem*, 2007, **76**, 1–22.
75. M. Larijani and A. Martin, *Semin Immunol*, 2012, **24**, 255–263.
76. A. Martin and M. D. Scharff, *Nat Rev Immunol*, 2002, **2**, 605–614.
77. I. B. Rogozin and M. Diaz, *J Immunol*, 2004, **172**, 3382–3884.

78. M. Liu, J. L. Duke, D. J. Richter, C. G. Vinuesa, C. C. Goodnow, S. H. Kleinstein and D. G. Schatz, *Nature*, 2008, **451**, 841–845.
79. N. Y. Zheng, K. Wilson, M. Jared and P. C. Wilson, *J Exp Med*, 2005, **201**, 1467–1478.
80. P. C. Wilson, O. de Bouteiller, Y. J. Liu, K. Potter, J. Banchereau, J. D. Capra and V. Pascual, *J. Exp. Med.*, 1998, **18**, 59–70.
81. D. C. Reason and J. Zhou, *Biol. Direct*, 2006, **1**, 24.
82. R. M. de Wildt, W. J. van Venrooij, G. Winter, R. M. Hoet and I. M. Tomlinson, *J Mol Biol*, 1999, **294**, 701–710.
83. J. E. Sale and M. S. Neuberger, *Immunity*, 1998, **9**, 859–869.
84. J. C. Krause, D. C. Ekiert, T. M. Tumpey, P. B. Smith, I. A. Wilson and J. E. Crowe Jr, *MBio*, 2011, **2**, e00345.
85. B. S. Briney, J. R. Willis and J. E. Crowe Jr, *Genes Immun*, 2012, **13**, 523–529.
86. T. Zhou, I. Georgiev, X. Wu, Z. Y. Yang, K. Dai, A. Finzi, Y. D. Kwon, J. F. Scheid, W. Shi, L. Xu, Y. Yang, J. Zhu, M. C. Nussenzweig, J. Sodroski, L. Shapiro, G. J. Nabel, J. R. Mascola and P. D. Kwong, *Science*, 2010, **329**, 811–817.
87. A. Martin and M. D. Scharff, *Proc Natl Acad Sci USA*, 2002, **99**, 12304–12308.
88. W. Lin, K. Kurosawa, A. Murayama, E. Kagaya and K. Ohta, *Nucleic Acids Res*, 2011, **39**, e14.
89. S. Chen, J. Qiu, C. Chen, C. Liu, Y. Liu, L. An, J. Jia, J. Tang, L. Wu and H. Hang, *Protein Cell*, 2012, **3**, 460–469.
90. A. D. McConnell, M. Do, T. Y. Neben, V. Spasojevic, J. MacLaren, A. P. Chen, L. Altobell III, J. L. Macomber, A. D. Berkebile, R. A. Horlick, P. M. Bowers and D. J. King, *PLOS One*, 2012, **7**, e49458.
91. P. M. Bowers, R. A. Horlick, M. R. Kehry, T. Y. Neben, G. L. Tomlinson, L. Altobell, X. Zhang, J. L. Macomber, I. P. Krapf, B. F. Wu, A. D. McConnell, B. Chau, A. D. Berkebile, E. Hare, P. Verdino and D. J. King, *Methods*, 2013, In press.
92. R. A. Horlick, J. L. Macomber, P. M. Bowers, T. Y. Neben, G. Tomlinson, I. P. Krapf, H. A. Jones, J. Dalton and D. J. King, *J. Biol. Chem.*, doi:10.174/jbc.M113.452482
93. C. M. Mahon, M. A. Lambert, J. Glanville, J. M. Wade, B. J. Fennell, M. R. Krebs, D. Armellino, S. Yang, X. Liu, C. M. O'Sullivan, B. Autin, K. Oficjalska, L. Bloom, J. Paulsen, D. Gill, M. Damelin, O. Cunningham and W. J. Finlay, *J Mol Biol*, 2013, **425**, 1712–1730.
94. A. D. McConnell, V. Spasojevich, J. L. Macomber, I. P. Krapf, A. Chen, J. C. Sheffer, A. Berkebile, R. A. Horlick, S. Neben, D. J. King and P. M. Bowers, *Protein Eng DesSel*, 2013, **26**, 151–164.
95. J. Yin, A. E. Beuscher 4th, S. E. Andryski, R. C. Stevens and P. G. Schultz, *J Mol Biol*, 2003, **330**, 651–656.
96. G. J. Wedemayer, P. A. Patten, L. H. Wang, P. G. Schultz and R. C. Stevens, *Science*, 1997, **276**, 1665–1669.

97. J. Bostrom, S. F. Yu, D. Kan, B. A. Appleton, C. V. Lee, K. Billeci, W. Man, F. Peale, S. Ross, C. Wiesmann and G. Fuh, *Science*, 2009, **323**, 1610–1614.
98. G. Schaefer, L. Haber, L. M. Crocker, S. Shia, L. Shao, D. Dowbenko, K. Totpal, A. Wong, C. V. Lee, S. Stawicki, R. Clark, C. Fields, G. D. Lewis Phillips, R. A. Prell, D. M. Danilenko, Y. Franke, J. P. Stephan, J. Hwang, Y. Wu, J. Bostrom, M. X. Sliwkowski, G. Fuh and C. Eigenbrot, *Cancer Cell*, 2011, **20**, 472–486.
99. P. M. Bowers, T. Y. Neben, G. L. Tomlinson, J. L. Dalton, L. Altobell, X. Zhang, J. L. Macomber, B. F. Wu, R. M. Toobian, A. D. McConnell, P. Verdino, B. Chau, R. A. Horlick and D. J. King, *J Biol Chem*, 2013, **288**, 7688–7696.
100. F. Wang, S. Sen, Y. Zhang, I. Ahmad, X. Zhu, I. A. Wislon, V. V. Smider, T. J. Magliery and P. G. Schulz, *Proc Natl Acad Sci USA*, 2013, **110**, 4261–4266.

CHAPTER 5

Synthetic Protein Biologics

BENJAMIN G. DAVIS

Department of Chemistry, University of Oxford, Chemistry Research Laboratory, Mansfield Road, Oxford, OX1 3TA, UK
Email: Ben.Davis@chem.ox.ac.uk

5.1 Introduction and Strategy

5.1.1 Synthetic Biology as a Parallel to Synthetic Organic Chemistry

The development of synthetic biology has many facets. One intriguing aspect allows parallels to be drawn with the development of synthetic organic chemistry[1] in which the scope and ambition of synthetic biology represents a key part of the future development of biology in the same way that synthetic chemistry enabled organic chemistry and the development of the small molecule drug industry (Figure 5.1). For example, work at the end of the 19th and early 20th century led to the vital isolation and characterisation of natural products – what might be termed a discovery phase. Elegant total syntheses and dedicated degradation studies over the first part of the 20th century led to identification and characterisation of natural products as well as some early understanding of their function, and through this ultimately giving an understanding of both function and structure. In time this led to the development of useful structure-activity relationships (SARs), logical redesign and the discovery of 'hits and leads' within our traditional view of small molecule drug development. This has undoubtedly led to some remarkable interventions in biological systems. The consequent application of small molecules to manipulate

RSC Drug Discovery Series No. 36
Biotherapeutics: Recent Developments using Chemical and Molecular Biology
Edited by Lyn H. Jones and Andrew J. McKnight

Published by the Royal Society of Chemistry, www.rsc.org

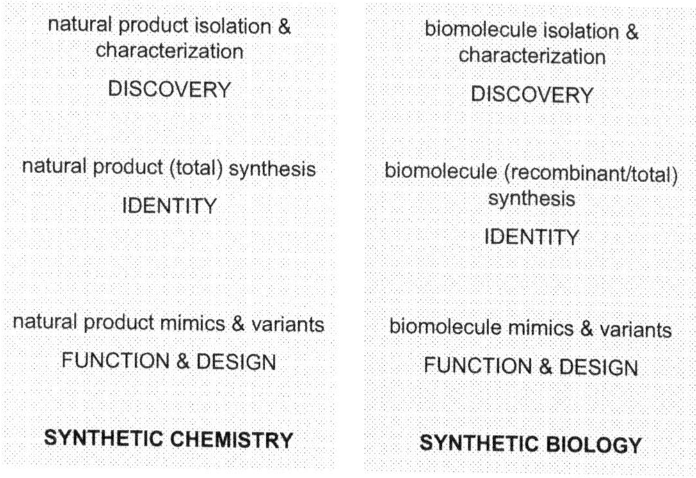

Figure 5.1 The parallels that may be drawn between the past development of Synthetic Chemistry and the current development of Synthetic Biology.

biological systems should, no doubt, therefore be taken as one of the great hallmarks of the 20th century.

By the same token, the 21st century could well be defined by a wave of innovation fuelled by isolation and characterisation of larger structures (well beyond those of traditional guides[2]), and primarily by larger biomolecules typically thought to be beyond the reach of direct chemical synthesis. Such larger molecules have already played a role in therapy, typically in the form of proteins (the 'workhorses' of biology). Isolated, biologically-derived examples have led to some excellent detailed knowledge of their roles: of their ability to interact with other molecules; their ability to catalyse the formation of bonds and ultimately their ability to control, in a discrete way, key aspects of functional biology in whole systems. Sequencing technologies for both proteins and nucleic acids coupled with more recent developments in the field of oligosaccharide analysis have allowed us to identify with reasonable precision the structural components of these biomolecules, although some aspects (global roles of post-translational modification and wide-ranging configuration analysis in oligosaccharides with absolute certainty) still elude.

In the latter half of the 20th century the ability to be able to design experiments that give individual indications of protein function had become a hallmark of biochemistry. So much so, in fact, that although only a small fraction of the proteins in sequenced genomes have been expressed and genuinely characterised, we are often sufficiently confident to assign putative annotations to many of the proteins in a given genome. This growing confidence in functional assignment, based on approaches driven by innovative developments in bioinformatics, has been coupled with increasingly more innovative and rapid structural biology. Together, these have perhaps created a cultural and technological environment in which function can be taken somewhat for granted (this overconfidence is also worthy of a note of caution – if cyclical thinking is to be avoided, true wide-ranging function determination is still an enormous gap to be overcome).

Biomolecules, and particularly proteins, offer another aspect to the chemist. They are remarkable spatially addressable scaffolds with often great biases in conformation. They may sample and often operate in only a very tiny fraction of the possible structural (and therefore possible functional) states of such scaffolds. Even simple permutation calculations based on variants of natural amino acids in relatively short peptides quickly reveal possible combinations well beyond those that have ever been created by nature. These observations of enormous putative diversity have struck a chord with many researchers, leading to exciting discoveries surrounding the use of proteins as scaffolds to create proteins that have not existed (sometimes referred to as 'unborn' or 'unknown' proteins). These studies have explored, for example, the creation and use of proteins as scaffolds in nanotechnology and simple use of peptidic variation to explore putative secondary structural variations.

In the current absence of any accurate method for *de novo* prediction of complete protein structure or function from sequence, then synthetic biology has particular potential to create useful new molecules based on the structural and functional paradigms of *existing* biomolecules. This provides a wonderful 'stomping ground' for discovery. In the context of Figure 5.1 (and through parallels with organic chemistry), isolation and characterisation, for example of proteins, has led to their recombinant, and even now total, synthesis through both molecular biological and chemical methods. In recent years this has led to the construction of a host of biomolecule mimics and variants in which the introduction of unnatural/unusual alterations, linkages and prosthetics has created biological function. When logically extended this has the potential to expand the limited molecular pallet that nature has thus far employed (for proteins typically the 20 well-known proteinogenic amino acids) by using the power of organic chemistry to access almost any structural motif. Given that almost a third of Investigational New Drug (IND) applications are now for protein therapeutics, the ability to create structural and functional variants are probably a dominant theme that can be led by synthetic biology.

5.1.2 Synthetic Biology as the Driver for Chemical Medicine

In this way we can imagine the development of not only science, but also associated technology that combines the power of biology with chemistry, allowing access to spectacularly exciting scaffolds with potential unlimited molecular diversity to create synthetic variants of biomolecules that could use some or all of their evolved bias to interact with biological targets. Expressed in another way (and set in the language of small molecule drug discovery) we can view the structural motifs in, for example, protein chemistry as being 'privileged' scaffolds that have been 'redesigned' and evolved to interact with biological systems. These provide, as it were, molecular keys for unlocking the functional complexity of biological systems, potentially providing methods for reporting on, interacting with, manipulating and ultimately fine-tuning and modulating that function in a way that is compatible with diagnosis and therapy. This suggests a form of what might be called chemical medicine that is distinctly different from traditional small molecule medicinal chemistry. Such chemical medicine is likely to be based much more on an informed approach to understanding fundamental biological science and interfacing the molecules or direct relevance to those processes. These will therefore, almost by definition, have a higher probability of being based on existing biomolecules. Given the enormous strides that have been made in the required underpinning technologies such as mass spectrometry, protein chemistry, structural biology, and molecular imaging, the stage is now perfectly set for future decades of molecular interrogation of whole-organism biology using synthetic biomolecular variants. This will create something that could be called a *true* systems biology, understanding the biology of a system from the inside out (without the limitations of some form of abstracted *in vitro* or *in silico* modelling of that system, as is now sometimes practised).

5.1.3 Current Strategic Small-Mindedness and Opportunities Beyond It

Current approaches to traditional biologics developed in the latter part of the 20th century have, understandably, focused on well-defined and well-understood processes for making proteins. These include, typically, methods of fermentation or production that take advantage of well-known expression hosts and their ability to express or over-express proteins in high levels, in addition to well-established methods for isolating and purifying those proteins. This can lead to their isolation in a form that may reflect, in some cases, not only the gene sequence of the intended protein, but (by virtue of post-translational modification) they may also bear the hallmark of the organism that has produced it. Whilst this represents a pragmatic method for creating well-known biopharmaceuticals such as erythropoietin or insulin, there are associated problems. Post-translational modifications arising from such expression systems have in some cases dominated aspects of their production.[3]

The convergence of accelerating technological methodology for *genuine* characterisation of proteins, advances in molecular biology and new methods for *in vivo* examination of molecule function has presented (and is presenting) timely opportunities to create synthetic proteins that might ultimately end up in the market as biopharmaceuticals: synthetic biologics. Whilst the broad definition of synthetic biologics could (and should) stretch to, for example, antisense or synthetic siRNA, in this review I will focus on one area of synthetic biology using illustrative strategies – that of synthetic proteins and their development towards pharmaceutical ends (synthetic protein biologics). This reflects the pre-eminence, at least until recently, of proteins as the main form of traditional biopharmaceuticals or biologics. Current approaches in the pharmaceutical industry for producing biologics have been dominated by large-scale (e.g. fermenter or roller-bottle facilities) pragmatic solutions to allow the use and control of certain known cell lines/types. The ability to chemically synthesise or chemically manipulate existing protein structures in an alternative way is likely to necessitate new forms of Good Manufacturing Practice (GMP) and corresponding GMP facility. Moreover, new strategies for the development of the underpinning science to generate methodology robust enough for translation to GMP will be another key exciting aspect.

This review will not discuss smaller peptidic constructs. These provide an interesting conceptual midpoint between what may be considered to be small molecules and those that may be considered to be biologics. Often peptides of 50 residues or less, whilst challenging, can be recognised as genuine possible targets in chemical synthesis and have been produced and therapeutically developed in a number of examples over the last 10–15 years. However, chemically-created larger targets (typically >50 amino acids) have, on the whole, not yet broken through and it is the exciting development in these new protein biologics that this review will focus on.

Of course, smaller peptide variants can illustrate some of the strategic approaches that may be applied ultimately to larger molecular scaffolds, especially in peptides with a high degree of inherent secondary structural bias – these may themselves be viewed as potential components for larger synthetic proteins. One related key aspect is the large amount of excellent background developmental work that has analysed the theory of protein folding on one hand and on the other attempts to classify the structure and function of proteins into various analytical frameworks. The lynchpin studies behind the CATH structural database[4–6] and the work of those who have demonstrated that enzymatic function may be maintained in protein scaffolds that are reduced to homologues containing only eight different amino acid types[7] are just two such examples. These methods of analysis of function have been usefully supported by methods to create focused relevant function in *de novo* proteins.[8]

5.1.4 Current Limitations Provide Further Opportunities

In addition to such analyses, there are the limitations to the notion of protein-based therapy that are currently imposed by their perceived, typical mode of action. The use of monoclonal antibodies to probe and interact with biology, in many ways, can actually be considered to be quite abiotic. The use of adaptive immunity to create protein probes for use *in vivo* to recognise targets is certainly not a mimic of the way that most protein/protein interfaces are constructed – exploring the scope of protein/protein interface formation by using relevant scaffolds and using, for example, relevant functional or prosthetic groups might provide more (and arguably more relevant) information.

It has been estimated that only 10% of all human proteins are displayed in a sufficiently accessible extracellular form for ready targeting by molecules.[9] Yet synthetic biology has the ability to create modified variants that can combine the function of a protein that typically acts extracellularly with transport properties of one that might be taken up through active processes such as receptor-mediated endocytosis or displayed in the cytosol with a relevant and suitable tag for extracellular transport or nuclear localisation. In this way combining the relevant pharmacophores and functional motifs of a protein might even become a simple case of domain combination or creation of a suitably chemically (or molecular biologically) fused protein variant.

Select examples of the use of protein scaffolds with a focus on potential pharmaceutical utility have been reviewed elegantly by others.[10,11] The approach adopted by several is to reason that many years of evolution have created elegant solutions to protein interaction – by utilising aspects of what we have learned from these solutions we can create our own useful variants. It is often the case that the entire protein scaffold is necessary. Several examples have shown that the removal or abstraction of small peptides, often even those with important secondary structure, from their protein scaffold results in a corresponding inactive peptide structure; the clear implication is that the entire protein provides the essential scaffold that stabilises the needed conformation of the component that is removed. This has, in turn, led to the development of interesting strategies for stabilising conformers in much smaller peptidic units.[12]

The aim of this review is to take select, strategically relevant examples from the development of synthetic proteins as alternative illustrations of the central strategy to their assembly. These have been chosen with an eye to a pragmatic approach that could eventually be translated into a relevant pharmaceutical application. Each different example study has been chosen to illustrate different facets of potential outcomes to a burgeoning synthetic biologics industry. Consequently, most of the examples chosen in this chapter have been those that are *en route* to clinical development, including up to phase three trials, with a view to clinical application.

5.2 Synthetic Vaccines

Since the 1920s[13] it has been possible to attach carbohydrates to proteins using chemical methods to allow the production of synthetic conjugates with the potential for generating immunogenic response. This has led to an extensive application of this methodology, typically using isolated polysaccharide fractions from pathogen cell walls and their attachment to a relatively limited number of immunogenic proteins, to be used in vaccination. In this way, a relatively small subset of chemistries has allowed the creation of many examples of well accepted, but nonetheless relatively chemically heterogeneous products.

In considering the role for potential application of more precise chemical synthesis, an early example that reached advanced clinical stages is the putative cancer vaccine *Theratope* (developed by Biomira). This conjugate contains the relatively small carbohydrate sialyl-T_N present in the mucin antigen, which is associated with breast cancer, perirectal cancer and ovarian cancer, and was attached in Theratope to the highly immunogenic keyhole limpet hemocyanin (KLH) carrier protein. Variations of attachment method (using reductive amination and/or amide chemistry) allowed exploration of the potential role of copy number and linker variation[14] Theratope's phase three trials illustrated several aspects of the potential strategy that might emerge for synthetic biologics. Whilst performing relatively poorly in the overall patient population, sub-populations receiving particular hormone therapy showed beneficial effects. Recent developments of the vaccine have included optimisation of adjuvants needed during vaccination and continued synthetic variation in an attempt to generate a fully efficacious, broad-population protection for breast cancer.[15,16] Moreover, similar conjugate vaccines have entered phase three trials for related uses, e.g. GMK, a GM2 ganglioside–KLH conjugate, is being used for melanoma treatment and this shows particular promise in patients who have had tumours removed through surgery.

A shining example of scientific collaboration between Cuba and Canada, (in the Bencomo and Roy groups), led to the construction of a fully synthetic conjugate vaccine antigen based on the polysaccharide of *Haemophilus influenzae* B (HIB) and its use to create a vaccine when attached to a carrier protein.[17] In pioneering development, this conjugate containing sythetic antigen was evaluated in clinical trials and showed long-term protective antibody titres that outstripped those of licensed products based on extracted (rather than synthetic) polysaccharide. It provided one of the very first demonstrations in a large-scale clinical context that vaccines containing such synthetic and complex carbohydrates would be putative agents in prophylaxis against bacterial infections. Construction of an oligomeric polyribosyl ribitol-phosphate appended with a primary amine allowed its conjugation as a maleimide to a tetanus toxoid platform. Phase one trials with 40 volunteers (largely recruited from the research groups that had conducted the work in synthesis and conjugate preparation) was followed by a phase one clinical trial with over 100 4–5-year-old children. A subsequent phase two trial in over 1000 children demonstrated large cohort efficacy. More detailed clinical trials in 2-month-old infants, followed by over 1000 phase two infant candidates led to the large-scale adoption of the vaccine, which is now protecting many thousands of children, not only in Cuba but beyond. It is interesting to note that early investment decades ago by the Cuban Government in farsighted vaccine technology led to rapid translation of key chemical methodology. In this way, this advanced strategy was applied against a serious pathogen in only a few years and gave widespread protection to a population that would not have been able to afford or obtain access to higher cost (and indeed less efficacious) variants available in the developed world. Many millions of doses of this vaccine have now been administered.

These examples illustrate not only the power but also the frustrations (and the ways of circumventing them) of potential synthetic biologics. The typically poor B-cell response to some target epitopes such as sugars (sometimes leading to a 'self' recognition process) and therefore poor elicitation of long-term immunological memory means that certain biochemically logical targets are often difficult epitopes and immunogens when used alone as vaccines or biologics. The required control of antibody responses is therefore typically accessed through the creation of some form of synthetic conjugate. The developments of the last 15 years (as highlighted by the examples given previously) have shown that synthetically produced antigens can allow greater definition of these critical portions of such conjugates, and therefore a striking potential for increased precision and affinity. Many have pointed out, however, that the potential for heterogeneity remains in many of these constructs in either the nature of the initial protein carrier (although this may be controlled by expression variation or purification), and more importantly in the subsequent conjugation chemistry. It is often the irony that very precisely defined carbohydrates are then 'sprayed' onto protein carriers with a variety of different methods that typically do not control the site of attachment and lead to a statistical distribution that may be difficult to control or reproduce, and within which certain components may generate quite different immunogenic responses. Some have also argued, conversely, that it is perhaps this diversity in structure that might lead to a more polyclonal and therefore more robust broad immunological memory. The ability to construct precise synthetic protein vaccines is therefore a potential future opportunity for synthetic biology, which might allow the structure activity relationships of these synthetic vaccines and synthetic biologics to be dissected. This will, in turn, identify the key influence of parameters such as copy number, linking method and identity of a given carrier in addition to the antigen identity itself (which has been somewhat the primary focus of the last 20 years). These opportunities are yet to be tested fully. Exciting examples might be developed in the future in which the site and the spacing of antigens within a given carrier protein may have dramatic influences upon outcome and efficacy.

5.3 Pegylation Technology and the Pegylation of Proteins

The use of protein pegylation (the attachment of polyethyleneglycol, PEG) has long dominated the methods used for modification of proteins with artificial functional groups. Its rapid and widespread adoption in pharmaceutical technology provides some interesting and intriguing lessons for developmental aspects of protein design.[18]

Although it has partly led to a more general interest in the development of methods for attaching other polymers (not just PEG), PEG still dominates and notably little emphasis has been placed on potential site-specific modification.[19] Several chemical pegylation methods have been employed for decades now and they are usefully representative of the often-limited strategic adventure currently employed in the whole field. Typically, these reactions involve activation of a PEG polymer with an aldehyde functionality for reaction through reductive amination or the use of an activated hydroxyl group, typically as a tresyl- or tosyl-PEG sulfonate ester. Other variants include the use of activated esters or carbonates and these have been well-reviewed elsewhere.[18,20] Most of these therefore rely on reactions with amino groups in proteins, the N-terminus and eta-amino groups of lysines primarily, and suffer from the aspects of conjugation heterogeneity discussed previously for synthetic conjugate vaccines. Indeed aspects of heterogeneity are complicated further still in pegylation methodology by the inherent heterogeneity (present through polymer polydispersity) of PEG itself.[21] The resulting statistical distribution of site attachment and copy number in the protein, as well as the statistical distribution based on polymer length, creates diverse pegylated products for which characterisation can only be limited. Nonetheless, despite this inability to create a pure product, pegylated biologics are widely authorised for use. Strategies to control both polymer-heterogeneity[21] and, through site-specific modification, protein-heterogeneity[19] could both be combined in future applications to allow perhaps more precise assessment of pegylation in proteins. It has been argued, for example, that pegylation represents a functional and effective mimic of protein glycosylation, as demonstrated in certain key biomolecules. A number of pegylated proteins are now employed as marketed pharmaceuticals primarily due to the beneficial effects of PEG upon clearance (reducing sequestration and prolonging serum half life[22]). Particularly successful examples have included modification of a mutant variant of tumour necrosis factor alpha in which pegylation did not compromise bioactivity.[23]

Given the early acceptance of such technology, it is somewhat ironic that the display of variance in the resulting synthetic proteins has been dominated by this simple modification, PEG. This one-track application perhaps highlights a tension in current technology development strategies between trying to create innovative and disruptive technologies that create new markets on one hand whilst at the same time attempting to generate sufficient income from existing markets on the other (adventure *versus* conservatism). Notably, a number of exciting synthetic protein technologies have often been used simply to enable the attachment of PEG (see following text) – perhaps in trying to strike some sort of happy balance between these two opposing strategic demands. Certain examples illustrate this well. Although not all have yet reached (and might not reach) clinical trials, these might achieve/could have achieved much if given sufficient strategic and logistical latitude. One primary example is the example of the development by Gryphon Therapeutics of the Synthetic Erythropoietin Protein (SEP).[24] This was spectacularly assembled using native chemical ligation (NCL) methodologies based on the use of thioester-mediated peptide assembly. Sites typically glycosylated in erythropoeitin (EPO) were instead decorated in SEP through site-specific incorporation of branched PEG variants. SEP briefly entered phase one clinical trials in 2003. The NCL linear-assembly methodology used to create SEP has been used widely to generate synthetic proteins in academia,[25] and one could imagine using such an approach in many more ways. Its use here to create a synthetic, clinically-applicable protein variant (albeit here with the relatively dull incorporation of PEG as a modification) therefore demonstrates an important strategic principle.

Strategic principle has also been demonstrated through the use of enzymes to modify proteins. Elegant methodology for site-selective attachment of certain sugars onto proteins using glycosyltransferases (and which could have diverse applications in terms of protein glycosylation) was ultimately applied again to pegylation. This method, termed 'glycoPEG', involved the sequential modification of protein sites specifically, initially with enzymatic transfer of N-acetyl-galactosamine at specific Ser or Thr residues followed by enzymatic transfer of a pegylated variant of sialic acid using an alpha-2,6-sialyltransferase.[26] This methodology has been developed by Neose and applied in a collaboration between Neose and Biogenerix to create 'glyco-pegylated' versions of GCSF (Granulocyte colony-stimulating factor). This entered phase two clinical trials in 2008. Enrolling 200 patients in a randomised double-blind-controlled format, direct comparison sucessfully was made with another pegylated GCSF variant, Neulasta.

Pegylated human growth hormones from AMBRX Technologies that make use of unnatural amino acids entered clinical trials in 2007. Amber codon reassignment methodology, based upon methods for expanding the genetic code in *Escherichia coli*,[27] allows the installation of a chemical functional group to which PEG can be attached. The resulting candidate protein, ARX201, utilises para-acetyl-phenylalanine incorporated at position 134. Reaction with aminoxy-terminated PEG allows oxime formation with the protein in a manner that does not affect other amino acids in the protein.

One highly relevant and intriguing aspect of site-specific pegylation methodology has been its use to identify potential limitations of non-specific modification methods and so it provides an important general lesson. It is particularly the case that blocking of recognition domains in antibodies leads to much lower binding levels; site-specific alterations and attachments to engineered antibody fragments has proved successful in solving aspects of this issue. For example, the use of more specific cysteine-to-maleimide conjugation has allowed the creation of Fab' variants, which retain activity, in contexts where random pegylation of the same Fab' leads to loss of 50% of activity.[28]

5.4 Synthetic Biologics with Payloads

Another key intriguing aspect of synthetic proteins is the ability to construct protein-to-drug conjugates or hybrids in which the drug can achieve potentially enhanced therapeutic function by, for example, benefiting from potential circulatory and targeting benefits that are not present for the non-conjugated variant. A primary example is in the use of antibodies coupled to drugs to target the potent cytotoxcity of those drugs to given organs. This use of antibodies to target drugs has in fact been around since the late 1950s, when methotrexate was coupled to antibodies derived from human serum and used in a mouse model.[29] Various attachment groups have been used for the linkage chemistry, including amide formation and the use of thiol-based linkages. Many of the chemistries have been well established for decades and rely again on non-site-selective attachment to existing amino acids within antibodies. Site of incorporation can, however, have an important influence. With many methods, ε-amino groups of lysines dominate and, as for pegylation, modification of lysine residues can affect function dramatically – alteration of those found in antibody variable regions can affect the binding ability of the resultant modified antibodies. This lack of control can also lead to different levels of drug loading. Introduction of additional functional groups can increase the loading of lysines; disulfides of existing cysteines within antibodies have also been altered, as well as some alterations made through carbohydrate modification.[30]

Some drugs have even been directly attached through inherent functional groups such as Asp/Glu carboxyls. In all of these cases, some loose evidence suggests that even these more direct types of conjugates lead to the release of drugs ultimately by degradative hydrolysis, although additional subsequent studies suggest strongly that linkages may be retained in many systems, thereby creating cytotoxin–peptide fragments that then display activity and that, in certain cases, there remains the need for some sort of fragmentable linker.[31]

Many of these approaches have focused on carrier strategies for the delivery of drugs, although it could be more widely extended. An obvious limitation of the antibody carrier is the relatively low drug load (or ratio of drug-to-antibody) in any of these systems, although the attachment to an antibody can bring strong additional benefits such as reduced toxicity. Limitations therefore can suround the number of antigens on a given target cell and the concomitant requirements for toxin delivery to kill any target cell. This places a strong requirement on the cytotoxicity of the drug since typical antibody affinities are relatively low and their binding equilibria are established *in vivo* only slowly. One early example, Mylotarg, was developed by both Wyeth and Celltech/UCB. This work used the toxin calicheamicin conjugated to IgG4 humanized antibodies that recognise the CD33 antigen found on myeloid leukaemia cells.[32] Many positive signs were seen with these types of targeted constructs, for example, when administered to patients who were strongly CD33 positive. In a first-recurrence, single-agent treatment trial, 26% remission rates were observed with acceptable safety profiles. These early positive results for the use of antibody conjugates have now been followed by many other examples and there has been a recent resurgence in their use. For example, a number of trials are currently being conducted by Seattle Genetics using, for example, anti-CD30 antibodies against Hodgkin's lymphoma[33] as well as anti-CD33 antibodies.[34] All of these conjugates contain a linkage system, designed to be cleaved *in vivo,* which is attached to the monoclonal antibody via a cysteine-to-maleimide unit. One example linked this to the toxin auristatin via carbonates or esters.[35]

5.5 Future Prospects and Conclusions

These foregoing examples have been viewed through the self-imposed filter of 'synthetic proteins' that have reached significantly advanced clinical stages or the market (and choosing those that have required significant chemical input). They illustrate certain points well:

(i) The current forerunners and primary examples display relatively little scientific novelty – examples of chemically conjugated vaccines and antibody-targeted drug conjugates have been around since the 1920s and 1950s, respectively. The current clinical examples in these fields have made only minor advances;
(ii) The underpinning technologies of many of the molecules that enter trials are often moderated in the candidates that emerge. Thus, whilst we can

now cite translational examples of chemical biology, e.g. unnatural amino acids incorporated via Ambrx's ReCode™; the creation of synthetic eryrthopoeitins via thioester-mediated native ligation;[24] or the use of biocatalytic functionalization in Neose's GlycoPEG™,[26] they bear the hallmarks of a tension between the need to disrupt existing technologies whilst being able to show sufficient access to existing markets to provide vital lifeblood income. All three examples used striking chemical technologies to ultimately attach something as functionally bland as PEG. In this way, it is an ironic curiosity of the industry that the early insight,[22] which created a much-needed technological rallying point for the potential of what could be done with modified proteins in therapy, has now led to an apparent reluctance to take the next braver steps.

The result is a translational model that, to my mind, has bred a dullness and a lack of endeavour in what should (and will) be the start of an exciting new industry. Perhaps this is the case at the start of many such 'step changes' in approach. It is true that existing conservatism is now being tested: new concepts as to how one can create flexible facilities (for example, for lead development or GMP chemistry on proteins) and how we should characterise and regulate such molecules are being developed. How should the associated intellectual property for such molecules be protected? How and when will the method or composition of matter approaches interface with synthetic biologics and biogenerics? The notion of protein Markush[36] groupings is a fearsome and unwieldy prospect for synthetic biologics, although some still try[37] despite this. These issues threaten to (and should) change the way that biologics are considered more generally. Within a relatively short space of time, it may become as unthinkable that we grow, extract and sell biologics as the heterogeneous mixtures of isoforms/modiforms/glycoforms that are currently used, as it is to use crude extracts of willow bark as our source of salicylic acid. In essence, science has moved on and indeed moved past current industrial approaches to biologics.

These, now largely academically-driven, changes will lead to a new approach in biologics that perhaps sits more comfortably with the analytical molecular precision of the field of chemistry than with traditional biology. This is being dramatically enabled by the convergence of technologies that, for example, make rigorous protein analysis much more feasible. Yet this burgeoning area will not be populated by chemical chauvinists since its strategies and methods will be focused instead on those of biomolecules and dissecting fundamental biological and medicinal goals. There are some mouth-watering prospects: genuine free-ranging structure activity relationships (uncurtailed by the typical 20 proteinogenic amino acids) to unpick function; hybrids and chimaeras beyond those built on the traditional 'linear' backbone afforded by biological expression.

In this chapter and section, I have resisted citing up-and-coming examples of synthetic protein creation that have not yet reached translation. There are many shining examples and a brief concluding summary would not do any of them comprehensive justice; however, three examples are illustrative of strategies that could be envisaged.

Firstly, work illustrating the functional efficacy of artificial proteins (even biocatalysts[7]) that pick from a restricted subset of amino acids have shown that in the design of synthetic proteins, variants could be created in which sequence-structural permutations could be minimised (to make iteration and SARs more manageable) in some protein regions whilst maintaining utility. This would, in turn, allow focused alteration at smaller, perhaps more relevant, regions. In this way, a partitioning between a structural 'scaffold' with strong conformational bias (based on often uniquely addressable structural information when compared to small molecules) and a putative biologic 'pharmacophore domain' could be envisaged.

Secondly, even *de novo* design based on simple building blocks might provide aspects of such (and other) partitioned roles. Advances in the use of so-called 'foldamers' based on systematically variable artificial amino acid building blocks with greater or different conformational biases have allowed the *de novo* creation of folds/domains that have been structurally interrogated[38] or that have putative biological function.[39]

Thirdly, free-ranging manipulation of such scaffolds provides wonderful future opportunities and targets. 'Mutagenesis' has often come to be a shorthand way to describe the powerful approach of altering the DNA that codes for a protein product (i.e. site-directed gene mutagenesis, possibly better considered to be a *pre-expression* mutagenesis). Reassignment of stop codons in various cell types in an effective manner has given us a hint of the power that might be achieved with reinterpreting/redefining the message with *co-expression* mutagenesis. For example, the creation of parallel triplet/quadruplet reading systems[40] that might work without interference (and that can be used to insert more than one, different unnatural amino acid) multiplies the potential enormously.

Finally, *post-expression* mutagenesis (directly at the level of the protein) might prove strikingly powerful. Nature has already given us a clear glimpse of the potential for functional fine-tuning of proteins through enzymatic post-translational modification. By learning this lesson,[41] combined with the development of chemistries that allow wide-ranging and multi-site alterations,[42] we can imagine almost limitless possibilities. This would perhaps lead us beyond traditional appendage/conjugation chemistries to those where we might reprogramme from amino acid to amino acid directly without any need for additional genetic encoding.[43] Indeed, long before the advent of widespread recombinant molecular biology, we tend to forget that the first *protein* mutagenesis (Ser→Cys) was in fact demonstrated using a chemical method.[44–46] Given the potential strategic structural freedom of chemistry, then such methods have been unusually neglected and might come to prove powerful once again – to be applied now in the creation of a new generation of therapeutic molecules.

References

1. EU Commission, *New and Emerging Scientific Technologies*, 2005.
2. C. A. Lipinski, F. Lombardo, B. W. Dominy and P. J. Feeney, *Adv Drug Del Rev*, 1997, **23**, 3–25.

3. G. Walsh, ed., *Post-Translational Modification of Protein Biopharmaceuticals*, Wiley VCH, Weinheim, 2009.
4. A. Cuff, O. C. Redfern, L. Greene, I. Sillitoe, T. Lewis, M. Dibley, A. Reid, F. Pearl, T. Dallman, A. Todd, R. Garratt, J. Thornton and C. Orengo, *Structure*, 2009, **17**, 1051–1062.
5. C. A. Orengo and J. M. Thornton, *Annu Rev Biochem*, 2005, **74**, 867–900.
6. C. A. Orengo, T. P. Flores, W. R. Taylor and J. M. Thornton, *Protein Eng*, 1993, **6**, 485–500.
7. S. V. Taylor, K. U. Walter, P. Kast and D. Hilvert, *Proc Natl Acad Sci USA*, 2001, **98**, 10596–10601.
8. A. Rao, S. Chopra, G. Ram, A. Gupta and A. Ranganathan, *J Biol Chem*, 2005, **280**, 23605–23614.
9. G. L. Verdine and L. D. Walensky, *Clin Cancer Res*, 2007, **13**, 7264–7270.
10. A. Skerra, *Curr Opin Biotechnol*, 2007, **18**, 295–304.
11. S. D. Nuttall and R. B. Walsh, *Curr Opin Pharmacol*, 2008, **8**, 609–615.
12. L. K. Henchey, A. L. Jochim and P. S. Arora, *Curr Opin Chem Biol*, 2008, **12**, 692–697.
13. W. F. Goebel and O. T. Avery, *J Exp Med*, 1929, **50**, 521–531.
14. G. Ragupathi, R. Rao Koganty, D. Qiu, K. O. Lloyd and P. O. Livingston, *Glycoconj J*, 1998, **15**, 217–221.
15. T. A. Gilewski, G. Ragupathi, M. Dickler, S. Powell, S. Bhuta, K. Panageas, R. R. Koganty, J. Chin-Eng, C. Hudis, L. Norton, A. N. Houghton and P. O. Livingston, *Clin Cancer Res*, 2007, **13**, 2977–2985.
16. L. Holmberg and B. Sandmaier, *Expert Rev Vaccines*, 2004, **3**, 655–663.
17. V. Verez-Bencomo, V. Fernandez-Santana, E. Hardy, M. E. Toledo, M. C. Rodriguez, L. Heynngnezz, A. Rodriguez, A. Baly, L. Herrera, M. Izquierdo, A. Villar, Y. Valdes, K. Cosme, M. L. Deler, M. Montane, E. Garcia, A. Ramos, A. Aguilar, E. Medina, G. Torano, I. Sosa, I. Hernandez, R. Martinez, A. Muzachio, A. Carmenates, L. Costa, F. Cardoso, C. Campa, M. Diaz and R. Roy, *Science*, 2004, **305**, 522–525.
18. J. Harris and R. Chess, *Nat Rev Drug Discovery*, 2003, **2**, 214–221.
19. G. G. Kochendoerfer, *Curr Opin Chem Biol*, 2005, **9**, 555–560.
20. F. Veronese and G. Pasut, *Drug Discovery Today*, 2005, **10**, 1451–1458.
21. A. C. French, A. L. Thompson and B. G. Davis, *Angew Chem Intl Ed*, 2009, **48**, 1248–1252.
22. A. Abuchowski, J. R. Mccoy, N. C. Palczuk, T. Vanes and F. F. Davis, *J Biol Chem*, 1977, **252**, 3582–3586.
23. Y. Yamamoto, Y. Tsutsumi, Y. Yoshioka, T. Nishibata, K. Kobayashi, T. Okamoto, Y. Mukai, T. Shimizu, S. Nakagawa, S. Nagata and T. Mayumi, *Nature Biotechnol*, 2003, **21**, 546–552.
24. G. G. Kochendoerfer, S.-Y. Chen, F. Mao, S. Cressman, S. Traviglia, H. Shao, C. L. Hunter, D. W. Low, E. N. Cagle, M. Carnevali, V. Gueriguian, P. J. Keogh, H. Porter, S. M. Stratton, M. C. Wiedeke, J. Wilken, J. Tang, J. J. Levy, L. P. Miranda, M. M. Crnogorac, S. Kalbag, P. Botti, J. Schindler-Horvat, L. Savatski, J. W. Adamson, A. Kung, S. B. H. Kent and J. A. Bradburne, *Science*, 2003, **299**, 884–887.

25. P. E. Dawson and S. B. H. Kent, *Annu Rev Biochem*, 2000, **69**, 923–960.
26. S. Defrees, Z. G. Wang, R. Xing, A. E. Scott, J. Wang, D. Zopf, D. L. Gouty, E. R. Sjoberg, K. Panneerselvam, E. C. M. Brinkman-Van Der Linden, R. J. Bayer, M. A. Tarp and H. Clausen, *Glycobiology*, 2006, **16**, 833–843.
27. L. Wang, A. Brock, B. Herberich and P. G. Schultz, *Science*, 2001, **292**, 498–500.
28. A. P. Chapman, *Adv Drug Deliv Rev*, 2002, **54**, 531–545.
29. G. Mathe, T. B. Loc and J. C. R. Bernard, *Acad Sci*, 1958, **246**, 1626–1628.
30. J. D. Rodwell, V. L. Alvarez, L. Chyi, A. D. Lopes, J. W. F. Goers, H. D. King, H. J. Powsner and T. J. Mckearn, *Proc Natl Acad Sci USA*, 1986, **83**, 2632–2636.
31. A. Rosowsky, R. A. Forsch, J. H. Freisheim, J. Galivan and M. Wick, *J Med Chem*, 1984, **27**, 888–893.
32. R. A. Larson, E. L. Sievers, E. A. Stadtmauer, B. Lowenberg, E. H. Estey, H. Dombret, M. Theobald, D. Voliotis, J. M. Bennett, M. Richie, L. H. Leopold, M. S. Berger, M. L. Sherman, M. R. Loken, J. J. M. van Dongen, I. D. Bernstein, F. R. Appelbaum and Mylotarg Study Group, *Cancer*, 2005, **104**, 1442–1452.
33. E. Oflazoglu, K. M. Kissler, E. L. Sievers, I. S. Grewal and H. P. Gerber, *Br J Haematol*, 2008, **142**, 69–73.
34. C. M. Craig and G. J. Schiller, *Blood Rev*, 2008, **22**, 221–234.
35. S. O. Doronina, B. E. Toki, M. Y. Torgov, B. A. Mendelsohn, C. G. Cerveny, D. F. Chace, R. L. DeBlanc, R. P. Gearing, T. D. Bovee, C. B. Siegall, J. A. Francisco, A. F. Wahl, D. L. Meyer and P. D. Senter, *Nat Biotechnol*, 2003, **21**, 778–784.
36. J. M. Barnard and G. M. Downs, *Perspect Drug Discovery Des*, 1997, **7–8**, 13–30.
37. J. M. Stewart and R. J. Vavrek, US Patent Application No. 4801613, 1989.
38. D. S. Daniels, E. J. Petersson, J. X. Qiu and A. Schepartz, *J Am Chem Soc*, 2007, **129**, 1532–1533.
39. W. S. Horne, L. M. Johnson, T. J. Ketas, P. J. Klasse, M. Lu, J. P. Moore and S. H. Gellman, *Proc Natl Acad Sci USA*, 2009, **106**, 14751–14756.
40. H. Neumann, K. Wang, L. Davis, M. Garcia-Alai and J. W. Chin, *Nature*, 2010, **464**, 441–444.
41. B. G. Davis, *Science*, 2004, **303**, 480–482.
42. S. I. van Kasteren, H. B. Kramer, H. H. Jensen, S. J. Campbell, J. Kirkpatrick, N. J. Oldham, D. C. Anthony and B. G. Davis, *Nature*, 2007, **446**, 1105–1109.
43. J. M. Chalker and B. G. Davis, *Curr Opin Chem Biol*, 2010, **14**, 781–789.
44. K. E. Neet and D. E. Koshland, *Proc Natl Acad Sci USA*, 1966, **56**, 1606–1611.
45. L. Polgar and M. L. Bender, *J Am Chem Soc*, 1966, **88**, 3153–3154.
46. C. Zioudrou, M. Wilchek and A. Patchornik, *Biochemistry*, 1965, **4**, 1811–1822.

CHAPTER 6

Recent Advances in Antibody–Drug Conjugates

EDMUND I. GRAZIANI* AND L. NATHAN TUMEY

Oncology Medicinal Chemistry, Pfizer Global Research, 445 Eastern Point Road, Groton CT 06340, USA
*Email: edmund.graziani@pfizer.com

6.1 Introduction

While it has long been recognized that to extend the lives of cancer patients, effective treatments must eradicate tumors, especially those that cause metastatic disease, the means to achieve this end have proven limited. Traditional chemotherapeutic agents, while effective in eradicating tumors, generally have a narrow therapeutic index and can be poorly tolerated.[1,2] While efforts are underway to design new classes of small molecules that are more selective for protein targets implicated in aberrant cell types but down-regulated in healthy cells,[3] there remain significant challenges in achieving efficacy in heterogeneous populations of proliferating cells within tumors.[4] In contrast, cancer therapy with monoclonal antibodies (mAbs) has proven to be well tolerated and modestly efficacious in a number of tumor types, but this treatment is rarely curative.[5]

The concept of an antibody–drug conjugate (ADC) in which a highly potent (but poorly tolerated) chemotherapeutic agent is covalently attached to a mAb would appear to bridge the best aspects of each modality.[6] That is, while bound to the antibody, the chemotherapeutic "payload" no longer circulates systemically and is therefore well tolerated by healthy cells. However, if the mAb targets an antigen that is selectively expressed in tumor cell types, and if

RSC Drug Discovery Series No. 36
Biotherapeutics: Recent Developments using Chemical and Molecular Biology
Edited by Lyn H. Jones and Andrew J. McKnight

Published by the Royal Society of Chemistry, www.rsc.org

the mAb is rapidly internalized via endocytosis subsequent to antigen binding, it can deliver its cytotoxic cargo inside the tumor cell without affecting adjacent healthy tissue. Once inside the tumor cell the ADC can release its payload, much like a Trojan horse discharging its cohort of Greeks once inside the city walls, via "programmed" immolation of the linker or complete catabolism of the mAb. In this way, the highly potent general cytotoxin is converted to a more targeted therapy by linking to an antibody, and the mAb is made more potent via its attached payload. Figure 6.1 illustrates the ADC concept: a monoclonal antibody is covalently conjugated, in this example via sulfhydryl groups revealed by reduction of interchain disulfide bonds, to a proteolytically cleaved linker (see section 6.2.1) with a potent cytotoxic payload attached.

With the approval in the United States of the first ADC for human therapy, gemtuzumab ozogamicin (Mylotarg),[7] in 2000 and the subsequent approval of brentuximab vendotin (Adcetris) in 2012, the therapeutic potential of the ADC concept appears to be gaining traction. A number of excellent reviews have addressed the history of ADCs from early research to clinical development,[8,9] the specific small molecule payloads and linker combinations that have proven most effective,[10] and the current clinical pipeline of ADCs in development.[11,12] In light of the wealth of scholarship available, this review will focus on a number of recent advances in ADC modality: advances in linker technology, new methods for evaluating the potential benefits of site-specific ADCs, and advances in our understanding of the fate of the ADC upon *in vivo* exposure.

6.2 Recent Developments in ADC Linker Technology

ADC linkers serve two primary functions. First, the linker must anchor the cytotoxic payload to the antibody such that the ADC is benign in general circulation and does not release the payload prematurely. Premature release of the payload is likely to result in toxicity (due to increased exposure of normal tissue to the free payload) and/or reduced ADC efficacy (because the unconjugated antibody now competes with the ADC for the same cell surface antigen). Second, the linker must allow for the release of an active payload upon ADC internalization. This can be accomplished through some sort of linker-cleavage event (a "cleavable linker") or through complete ADC catabolism resulting in the release of the entire payload-linker moiety in an active form (a "noncleavable linker"). In addition to these two primary functions, a linker can play a variety of roles in modulating the biophysical properties of the ADC, the activity of the payload, and the stability of the ADC.

6.2.1 Cleavable Linkers

The term "cleavable linker" typically refers to a linker that is degraded as part of the payload release mechanism. This degradation usually occurs in the lysosome upon antigen-dependent internalization Some of the earliest cleavable linkers took advantage of the low lysosomal pH ($\sim$4.5) to facilitate the hydrolysis of hydrazone-linked payloads. This strategy was successfully

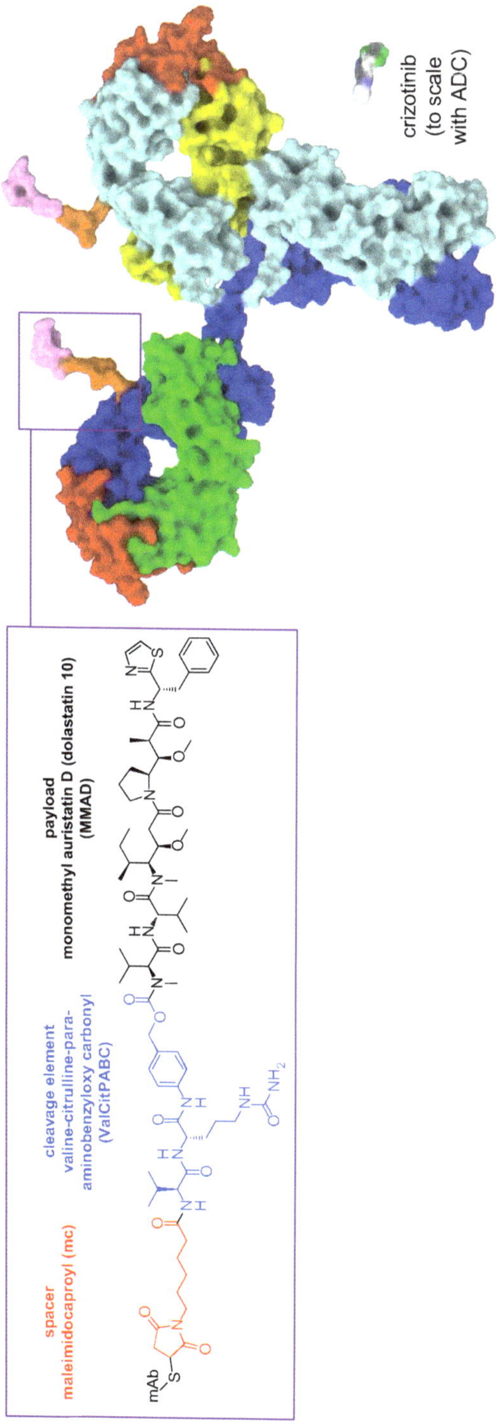

Figure 6.1 Representative model of an antibody drug conjugate (ADC), showing detail of linker elements and an auristatin payload.

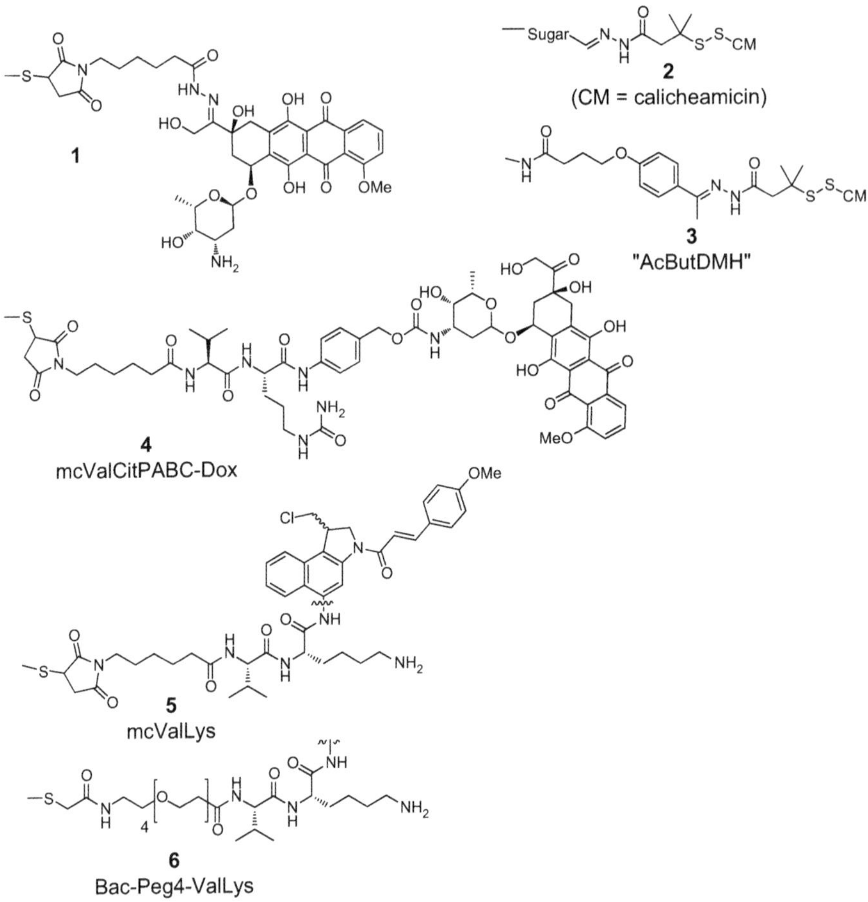

Figure 6.2 Doxorubicin, Calicheamicin, and CBI Linkers.

employed for doxorubicin conjugates, resulting in ADCs such as in Figure 6.2, **1**.[13] Unfortunately, these doxorubicin hydrazone conjugates have reportedly been abandoned due to limited efficacy, which is likely related to the observation that the hydrazone is slowly hydrolyzed in circulation.[14–16] A related linker (**2**) has been used to anchor calicheamicin to various antibodies such as anti-CD33, initially via a hydrazone to an oxidized carbohydrate on the mAb. Interestingly this payload requires the cleavage of a secondary cleavage element (a disulfide) in order to release the active species that causes DNA double-strand cleavage. A variation of this dual linkage was eventually incorporated into the commercialized calicheamicin conjugate gemtuzumab ozogamicin (Mylotarg, **3**). This linker, known as acetylbutyldimethylhydrazone (AcButDMH), was demonstrated to be stable at pH 7.4 but was hydrolyzed quickly at pH 4.5, the estimated lysosomal pH.[7] However, subsequent pharmacokinetic (PK) studies have demonstrated that this linker is slowly degraded in plasma thereby releasing the calicheamicin payload prematurely.[17]

Significant clinical and preclinical interest continues in the AcButDMH linker in spite of the voluntary withdrawal of Mylotarg in 2011.[18–20]

A major advance in cleavable linker technology was realized by Bristol-Myers Squibb in 2002 with the report, building upon earlier prodrug chemistry,[21] of the traceless linker valine-citrulline-para-aminobenzyloxy carbonyl (ValCitPABC, **4**) initially reported in the context of camptothecin[22] and doxorubicin[23] conjugates, but then quickly applied to auristatin payloads (Figure 6.3, 7) by Seattle Genetics in 2003.[16] This linker relies on the cleavage of a short peptide sequence (typically Val–Cit or Phe–Lys) by cathepsin B, resulting in a para-amino benzyl carbamate (PABC) that undergoes a self-immolation thereby releasing an amine containing payload, as shown in Scheme 6.1. The linkage was shown to have superior mouse and human plasma stability as compared to hydrazone linkers.[16] Since its disclosure in 2002, this linker has played a major role in the development of various ADCs, including the recently approved brentuximab vendotin (Adcetris). Several other mid- and late-stage clinical ADCs also utilize this linker[11] and minor permutations of it have been used to release aniline, phenol, and alcohol containing payloads.[22,24] Variations of this linker have been reported in which the peptide was moved to the C-termini of the auristatin payload (**8–11**), eliminating the need for the self-immolative PABC moiety. Interestingly, some peptide sequences such as

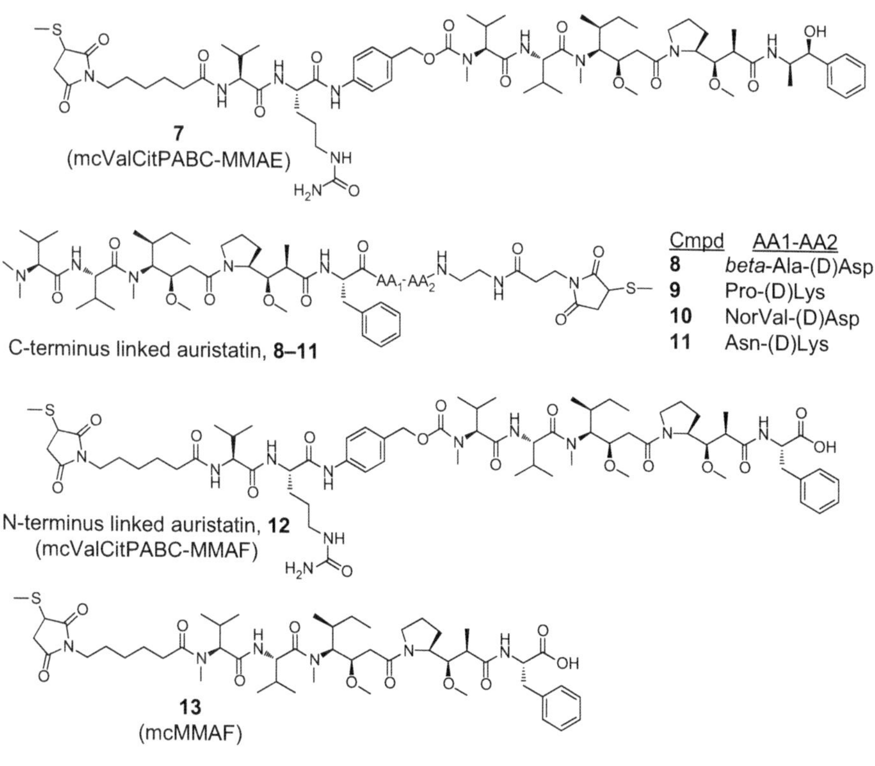

Figure 6.3 Auristatin-based linkers.

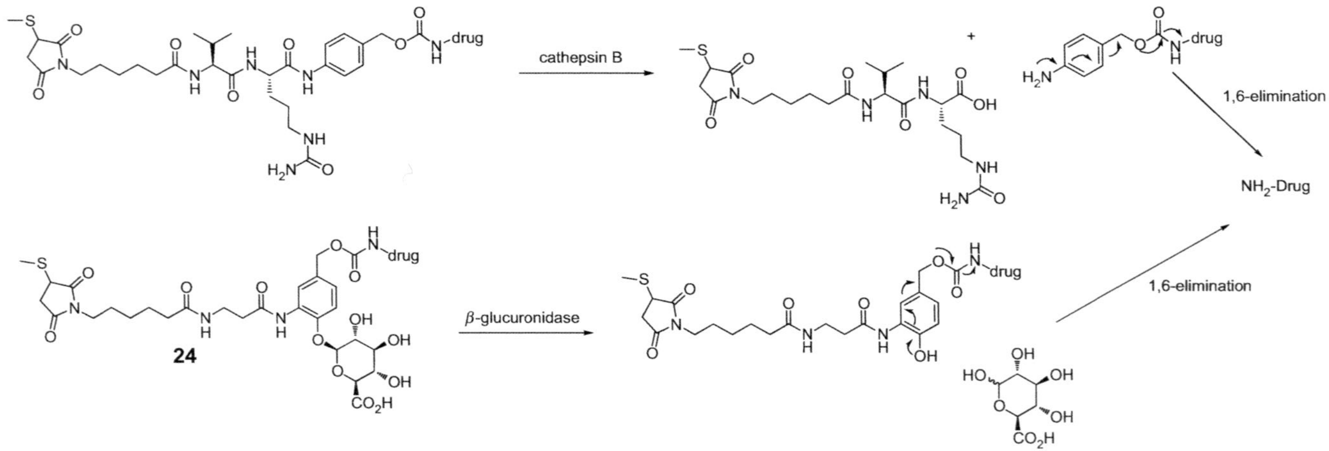

Scheme 6.1 Linkers containing the β-glucuronide cleavage element are cleaved by β-glucuronidase, resulting in a p-hydroxy benzyl carbamate that undergoes rapid 1,6-elimination and decarboxylation to release the amine-containing drug.

βAla-(D)Asp (**8**) and Pro-(D)Lys (**9**) resulted in ADCs that were essentially inactive against the antigen containing cell lines and had no *in vivo* efficacy in a tumor xenograft model. However, changing the peptide sequence to NorVal-(D)Asp (**10**) or Asn-(D)Lys (**11**) resulted in ADCs that were exceptionally potent both *in vitro* and *in vivo*. The latter sequence was incorporated into an ADC that showed superior efficacy and safety to mcValCitPABC-MMAF (**12**) in mouse models.[25] The development of these peptide-based linkers has been largely driven by cell-based cytotoxicity assays and therefore the exact mechanism of cleavage remains somewhat unclear. In fact, the above report unfortunately does not present any conclusive evidence demonstrating that the auristatin payload is released via proteolytic degradation. An additional complication for the design of peptide-based cleavable linkers is the putative role of enzymes other than cathepsin B in payload release, such as the work described by Jeffrey showing that particular peptide-linked conjugates showed excellent cellular potency but were not cathepsin B substrates.[24]

Concurrent with the development of the peptide-based cleavable linkers, a new class of disulfide-based cleavable linkers (Figure 6.4, **14** and **15**) utilizing a maytansinoid payload was emerging, primarily driven by the research efforts of ImmunoGen. The early maytansinoid disulfide linkers were structurally related to the first generation calicheamicin linkers, which incorporated a sterically hindered disulfide that is cleaved intracellularly thereby releasing the thiol-containing payload.[26,27] The release mechanism for these linkers relies on the high intracellular glutathione concentration, frequently estimated in the millimolar range, compared to the low micromolar levels of glutathione typically found in plasma. Interestingly, this release mechanism seems to be quite robust in spite of a report showing that the endocytotic pathway has a strong oxidizing potential.[28] A recent report demonstrated that SPP-DM1 (**14**) ADCs showed excellent cytotoxicity on a wide variety of antibody backbones, even in cases such as anti-CD21 where the antigen is known to be poorly internalized.[29] Unfortunately, this linker has also been shown to slowly release its payload prematurely due to the presence of glutathione in plasma.[30,31] This problem was partially ameliorated by the use of a modified linker, SPDB-DM4 (**15**), which incorporates additional steric hindrance near the disulfide. In mouse xenograft models this linker provides a significant boost in efficacy over SPP-DM1 that is attributed to improved *in vivo* stability of both the conjugate and the resulting free payload.[27,32,33] Further studies have shown that increasing the steric bulk around the maytansine-disulfide is a double-edged sword. Increases in bulk have consistently showed improved stability and exposure, but sometimes show decreased efficacy that is attributed to impaired reductive release of payload.[33] As such, optimization of this linker requires balancing high-stability and efficient release. A more thorough discussion of the cleavage mechanism will be presented in sections 6.2.1 and 6.2.2.

Recent study of this linker has focused on noncleavable (thioether) variants (*vide infra*) and on hydrophilic analogs. Zhao and colleagues have recently reported the incorporation of hydrophilic moieties into the SPDB chain, resulting in linkers such as **16** and **17**. Interestingly, they found that the

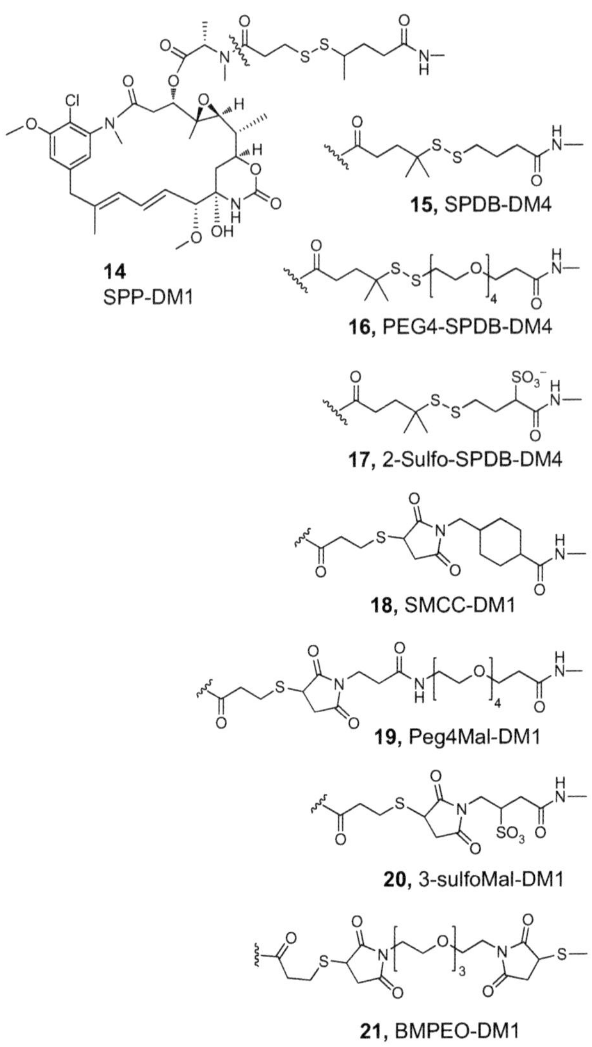

Figure 6.4 Maytansine linkers.

sulfo-linker **17** has nearly equivalent potency against MDR (multi-drug resistant) and non-MDR cell lines. In contrast, SPBD-DM4 (**15**) and PEG4-SPDB-DM4 (**16**) were observed to be 10–50 times less potent against the MDR cell lines. Moreover, **17** was shown to be far more effective in a mouse xenograft study against multidrug resistant COLO205-MDR tumors. The authors postulate that the metabolites of 2-sulfo-SPDB-DM4 ADCs may be poor substrates for MDR transporters such as P-glycoprotein (Pgp).[34] This hypothesis is based on the frequently noted preference of Pgp for hydrophobic substrates. Interestingly, it has been shown that metabolism of SPP-DM1 ADCs results initially in a lysine-linked maytansinoid with subsequent slow cleavage of the disulfide linkage (see section 6.4.1). In the case of 2-sulfo-SPDB-DM4, the corresponding intermediate (lysine-sulfo-SPDB-DM4) is presumably a poor Pgp substrate and therefore is more fully retained in MDR cell lines than the corresponding SPDB-DM4 intermediates.[30]

A new class of linkers that utilize a β-glucuronide cleavage element was reported in 2006 by Jeffrey and colleagues.[35] This class of linkers takes advantage of an enzyme, β-glucuronidase, that is found in high levels in lysosomes but is virtually absent in plasma. The cleavage of the glycosyl bond results in a *p*-hydroxy benzyl carbamate that undergoes rapid 1,6-elimination and decarboxylation to release the amine-containing drug. (Scheme 6.1) This linker has been shown to release the intended payload upon exposure to bacterial β-glucuronidase and to be stable in plasma for >7 days.[35–37] In fact, one report suggests that the β-glucuronide linker is more stable in plasma than the corresponding mcValCitPABC linker.[35] One intriguing property of this linker that differentiates it from its analogous peptide-based linkers is its hydrophilicity. As will be discussed below, hydrophilic linkers tend to allow for higher loading with lower levels of protein aggregation. As such, the β-glucuronide linker has proven to be particularly useful for the conjugation of hydrophobic payloads such as camptothecin,[36] cyclopropylbenz[e]indoles (CBIs),[37] and irciniastatin A.[38]

6.2.2 Noncleavable Linkers

In early 2006, nearly simultaneous reports from Seattle Genetics and ImmunoGen described a new class of linkers (such as **13** and **18**) that lacked any sort of designed lysosomal cleavage element. These ADCs were initially made as "control" ADCs that were expected to have minimal cytotoxicity and efficacy. Surprisingly, these "noncleavable" ADCs showed excellent *in vitro* cytotoxicity. The mechanism of payload release, for both the cysteine and lysine conjugates, was elucidated in an ingenious set of experiments using radiolabeled payloads to identify the major metabolites in cell culture (see section 6.4.1) In both cases, it was found that metabolism of the ADC resulted in the release of the amino acid-linked payload, i.e. lysine-SMCC-DM1 or cysteine-mcMMAF.[30,39,40]

In contrast to the cleavable linker SPP-DM1 (**14**), the activity of the noncleavable thioether SMCC-DM1 (**18**) has proven to be significantly target-dependent. For example, early reports showed that anti-CanAg ADCs of SPP-DM1 and SMCC-DM1 were equally cytotoxic but only the SPP-DM1 conjugate showed *in vivo* efficacy.[39] In contrast, scientists at Genentech report that trastuzumab conjugates of SMCC-DM1 have superior efficacy to the corresponding SPP-DM1 conjugates.[41] Likewise, a recent report elegantly illustrated the differences between these payload-linkers by conjugating both SMCC-DM1 and SPP-DM1 to antibodies against seven distinct antigens expressed on malignant B-cells. All seven of the SPP-DM1 conjugates showed *in vivo* efficacy while only two of the seven SMCC-DM1 conjugates showed efficacy.[29] In spite of the variable activity of this noncleavable linker, it has been repeatedly reported to have superior PK and lower toxicity than corresponding disulfide-linkers.[29,30,41] This is implicit in the fact that trastuzumab-SMCC-DM1 (known as "T-DM1" or "trastuzumab emtansine") has advanced to late stage clinical trials.[42] However, it has also been noted that disulfide-linked maytansinoid ADCs may have a therapeutic advantage in their ability to release cytotoxic payloads that can permeate to adjacent cells (the "bystander effect") thereby killing in an antigen independent manner.[43] This property may be particularly important in a clinical setting where antigen expression may be highly heterogeneous and is a probable explanation for why these disulfide linkers (SPP-DM1 and SPDP-DM4) persist in development pipelines in spite of their potential stability and toxicity problems.[11]

The success of the SMCC linker has prompted the development of 2nd generation noncleavable maytansinoid linkers that may evade common resistance mechanisms such as drug efflux. It has been noted that a variety of the cytotoxic agents used in ADC therapeutics such as calicheamicin, auristatins, and maytansinoids are subject to Pgp (MDR1) mediated efflux. While it might be speculated that noncleavable ADCs could overcome this resistance mechanism, a recent report nicely illustrates that this is not the case. Kovtun and colleagues show that cyclosporine A (a potent Pgp inhibitor) enhanced the potency of anti-EpCAM-SMCC-DM1 against Pgp-expressing cell lines such as HCT-15 but had no effect on the potency against Pgp-negative cell lines such as COLO205.[44] This, of course, suggests that the released entity (Lysine-SMCC-DM1) is a Pgp substrate and that the resistance could be overcome by evading the transporter-mediated efflux. In two reports, the authors went on to design linkers that replaced the greasy cyclohexyl group of SMCC with more polar linkers such as a PEG chain (**19**) and a sulfated alkyl group (**20**). These more polar linkers result in ADCs with vastly improved activity against the MDR cell lines, both *in vitro* and *in vivo*. The authors postulate that these new linkers result in far more polar released species that are either poor substrates for Pgp or are unable to diffuse into the membrane in order to access the transporter.[34,44] In addition, the PEG4Mal-DM1 (**19**) system was shown to have equal tolerability to SMCC-DM1 on the anti-CanAg platform and was shown to improve activity against Pgp-expressing cell lines on three different antibody platforms.[44] Aside from their influence on efflux, these more polar linkers also have a beneficial effect on the biophysical properties of the ADC, as will be discussed in section 6.2.3.

Yet another noncleavable maytansinoid linker was reported in 2010, this time enabling the conjugation of DM1 to engineered cysteine residues (see section 6.3.1). Known as bis-maleimido-trioxyethylene glycol (BMPEO,**21**), this linker was used to link DM1 to the engineered cysteine residue at position 114 of trastuzumab. The MBPEO-DM1 conjugate showed equal efficacy and superior tolerability to the conventionally conjugated SMCC-DM1. The reason for the improved tolerability is unclear given that plasma stability is unaffected by the change in linker. The authors suggest that the elimination of the high-DAR species present in the conventional conjugate may play a role in the improved safety of this conjugate.[45] This hypothesis is consistent with the reports of an improved therapeutic index observed for site-specific auristatin conjugates[46] (see section 6.3.1).

Noncleavable linkers have also become a major focus of efforts for auristatin-based ADCs. Doronina reports that anti-CD30 ADCs of mcMMAF (**13**) and ValCitPABC-MMAF (**12**) were equally potent *in vitro* and *in vivo*. Moreover, the mcMMAF ADC was tolerated at >3 times the maximum tolerated dose (MTD) of the ValCitPABC-MMAF conjugate. Similar to the SMCC-DM1 linker, study of radiolabeled mcMMAF ADCs showed that the released species was cysteine-mcMMAF, presumably resulting from complete ADC catabolism. It is noteworthy that, in contrast to mcMMAF, mcMMAE was essentially inactive *in vitro*.[40] This is reminiscent of the reports[29] from Immunogen showing that their noncleavable linkers (i.e. SMCC) are more target-dependent than the corresponding cleavable linkers. Likewise, this data from Genentech suggests that the potency of payload on a cleavable linker may not be indicative of its potency in a noncleavable context. This may reflect payload-dependent differences in ADC catabolism or lysosomal trafficking. In spite of this, the noncleavable technology has proven sufficiently robust to justify advancement of multiple mcMMAF conjugates into early clinical development.[11] A variation of the mc linker in which the maleimide was replaced by a bromoacetamide (bac) has also been reported. This linker did prove to be more stable than the mc linker (see section 6.2.1), but unexpectedly the bac-MMAF and mc-MMAF conjugates were found to have indistinguishable *in vitro* and *in vivo* efficacy.[47] Noncleavable oxime linkers have also been reported for auristatins conjugated to antibodies engineered to contain the unnatural amino acid pAcPhe.[48]

6.2.3 Impact of Linkers on Biophysical Properties

The two previous sections primarily focused on the impact that linkers play in ADC stability and payload release. In addition to these roles, it has long been recognized that the linker can play a role in the biophysical properties of the ADC. Recent clinical ADCs have generally coalesced around a small set of well-behaved linker-payloads and have typically aimed for loading in the range of 3–4 drugs per antibody.[11,42] However, there are cases in which more hydrophobic drugs and/or higher loading may be desirable. Various reports have shown that in some cases high-loaded ADCs provide superior activity over low-loaded ADCs, even when IC_{50} values are "normalized" for loading. For example, a loading increase from 2.5 to 6.6 has been reported to increase a maytansinoid ADC activity from 1500 nM to 27 nM.[34] Such "super-stoichiometric" increases in potency are not unique and have been observed for other ADC programs.[36] In these cases a highly loaded ADC is likely necessary in order to overcome limitations in ADC internalization, perhaps due to a low intrinsic rate of internalization or a limited number of antigen binding sites.

It is sometimes difficult to obtain higher loading while simultaneously avoiding protein aggregation, particularly for hydrophobic payloads. One of the earliest reports of this phenomenon came from efforts to synthesize high-loaded ValCitPABC doxorubicin conjugates. A dendritic linker was employed to generate cysteine-linked ADCs with a maximal loading of 16 drugs per mAb.

However, the resulting ADCs showed nearly complete aggregation through a noncovalent mechanism. It was found that the addition of short PEG chains to the payload resulted in ADCs that showed low levels of aggregation, even when the loading approached the desired 16 drugs per antibody.[49]

The above report illustrates that the mcValCit linker, while very versatile and widely utilized, can be prone towards aggregation, particularly when used in conjugation with hydrophobic payloads such as doxorubicins, CBIs, and camptothecins. Groups from Seattle Genetics have reported two strategies to overcome this problem. First, Jeffrey described a fascinating set of data in which a short polyethylene glycol (PEG) linker between the antibody and the cleavage element significantly decreased the level of aggregation observed for a series of CBI conjugates. By switching from a mcValLys linker (**5**) to a bac-PEG4-ValLys linker (**6**) the level of aggregation for an 8-loaded conjugate was reduced from ~100% to 5%.[24] Second, Burke reported that replacement of the hydrophobic cleavage element (ValCitPABC) with a more polar glucuronide cleavage element (Figure 6.5, **24**) resulted in greatly reduced aggregation for a set of camptothecin ADCs.[36] Maytansinoid ADCs have also been reported to aggregate and precipitate when the loading exceeds 4–5 drugs per antibody. This problem was overcome by the addition of polar groups to the linker element.[34] Another related approach was reported recently in which a phenolic moiety of a very hydrophobic CBI payload was phosphorylated, resulting in an ADC with good loading and low aggregation.[50] Such strategies will almost certainly continue to be employed as the field searches for novel payloads and cleavage strategies.

6.3 Recent Developments in Site Specific Conjugation of ADCs

The majority of ADCs that have been approved or are currently in late stage clinical trials[11] are conjugates in which the linker-payload is conjugated randomly to either: (1) the nucleophilic amine on the side chains of endogenous lysines[51] via an activated ester to form an amide linkage with the mAb or (2) the free sulfhydryls liberated upon reduction of the hinge and inter-strand disulfide bonds via a maleimide coupling to generate a thioether succinimidyl linkage to the mAb.[16] (Figure 6.5) Underlying each of these approaches (so-called "conventional" conjugates) is an essentially random conjugation process that yields a mixture of conjugates that can be best characterized as having an

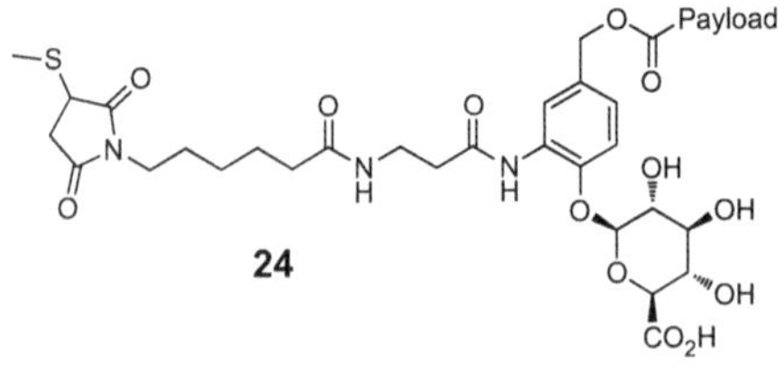

24

Figure 6.5 Miscellaneous linkers.

average drug–antibody ratio (DAR). There is some evidence[52] that for conventional ADCs conjugated via reduced disulfides an average DAR between 2 and 4 are better tolerated than related ADCs with higher loadings (average DAR = 6–8). This observation illustrates the importance of conjugation methodology that reproducibly achieves a defined DAR. Moreover, the simplified bioanalysis of advancing a single, homogenous drug substance with a very well-defined structure, loading, and impurity profile versus a mixture of ADCs would simplify PK-PD studies, metabolite identification, manufacturing, and perhaps regulatory approval. For these reasons, considerable interest in the field has shifted to the development of conjugation methods that are highly specific in both the residues involved in conjugation and the resulting ADC.

6.3.1 Engineered Cysteines

The first report[53] of engineering a cysteine residue at the surface of an antibody with the express purpose of introducing a sulfhydryl group for subsequent conjugation to a linker-payload came from a group at Celltech. Their approach involved a mutagenesis that introduced a cysteine onto the heavy chain of chimeric tumor binding anti-TAG72 mAb cB72.3. Molecular modeling was employed to identify putative sites that were (i) isosteric with cysteine, (ii) not involved in any hydrogen bonds (nor should the substitution should generate a new H-bond), (iii) not involved with antibody function, and (iv) accessible to small molecules at the mAb surface. Initial calculations identified five residues for substitution, T153, S156, S168, T173, and T207 by Kabat numbering. The resulting mAbs were titrated with 4,4′-dithiodipyridine to determine the number of free thiols present in each (with a theoretical maximum of 2 per mAb because each has two heavy chains). Results of this titration showed that while the S156C and T173C mutants gave approximately one free sulfhydryl/mAb, the other three site-mutants yielded free thiol levels that were identical to the control mAb. The authors concluded that the engineered cysteines were blocked (presumably as disulfides with cysteine). While mild reduction could yield free sulfhydryls, under these conditions the hinge disulfides were also reduced making site-specific conjugation problematic.

Recognizing the potential of this approach, workers at Glaxo Wellcome engineered[54] a cysteine at position S442 in a number of mAbs and were able to selectively reduce the engineered thiol (that was blocked with cysteine) with mercaptoethylamine to yield 2.0 free sulfhydryl/mol mAb (as determined by Ellman's reagent, 5,5′-dithiobis-(2-nitrobenzoic acid, DTNB). Reduction with a resin immobilized reductant also yielded 1.7 mol SH/mol mAb. Selectivity of the reduction at C438 (corresponding to S442) was determined via subsequent radiolabeling, digestion, and sequencing to identify labeled peptides.

Interestingly, researchers at Seattle Genetics took a complementary approach, preventing unwanted conjugation reactions from occurring at hinge sulfhydryls via mutagenesis of selected cysteines to serines.[55] They previously showed[52] that anti-CD30 ADCs having well-defined DARs (obtained via separation of individual isoforms from a conventional conjugate) enabled the optimization of efficacy and safety. They observed that while higher loading improved potency, this effect was offset by faster clearance and reduced tolerability. This suggests that control of specific DAR is highly desirable over and above improvements in homogeneity. Thus, by replacing solvent accessible cysteines with serine, variants with either two or four reactive cysteines were generated that yielded anti-CD30 ADCs with defined DAR and site of conjugation. The results obtained with these ADCs were similar to those observed previously with purified fractions having specifics DARs, though with improved conjugation yields and reduced heterogeneity. In a follow up experiment, Sun *et al.*[56] demonstrated that careful control of the reduction of hinge sulfhydryls gave populations of ADCs highly enriched in specific two-loaded species. This method employed either partial (selective) reduction of the native disulfides by dithiothreitol (DTT), tris(2-carboxyethyl)phosphine (TCEP), or aminoethanethiol (AET) or alternately complete reduction followed by partial reoxidation resulting in populations of ADCs highly enriched in specific two-loaded species. Nevertheless, HIC (hydrophobic interaction chromatography) was required to fully resolve individual DAR species.

In a key report[46] that brought together much of the previous work in this area Junutula *et al.* revived the concept of introducing an engineered cysteine to generate site-specific ADCs. Sites for engineering were chosen in the constant domains of the Fab region so as not to disrupt effector function or antigen binding. Reactive cysteines in those regions were identified via phage display.[57] Encouraging results from two anti-Her2 Fab mutants generated in this study, V110C and A114C, resulted in the generation of engineered cysteine-containing mAbs, or so-called thiomabs, on an anti-MUC16 antibody. As had been observed previously with engineered thiols, the resultant mAbs from fermentation were mixed disulfides capped with cysteine or glutathione and were resistant to initial attempts at conjugation. A two-step approach was developed whereby all the disulfides in the engineered thiomab were reduced with TCEP and the interchain disulfides were then allowed to re-oxidize using air oxidation or dehydroascorbate (Figure 6.6). Surprisingly, the V110C mutant gave only 25–50% of the expected loading following reaction of the reduced/reoxidized mAb with biotinyl-3-maleimido-propionamidyl-3,6-dioxaoctainediamine, prompting additional screening of other cysteinyl variants. While a number of sites were identified that gave >90% conjugation efficiencies, the A114C mutant emerged as a model system on which they further characterized the properties of the thiomab-drug-conjugates (TDCs).

Importantly, the anti-MUC16 TDC conjugated with the linker payload maleimide-caproyl-valine-citrulline-para-aminobenzyloxy carbonyl-[23] mono-methyl auristatin E (mcValCitPABC-MMAE, **7**) was shown to have antigen binding, specificity and efficacy comparable to the conventional conjugate, even with the reduced DAR of the TDC (DAR = 2 versus 4). Interestingly, the TDC showed improved tolerability upon administration to both rodents and cynomolgus monkeys, as measured by an absence of adverse events resulting from

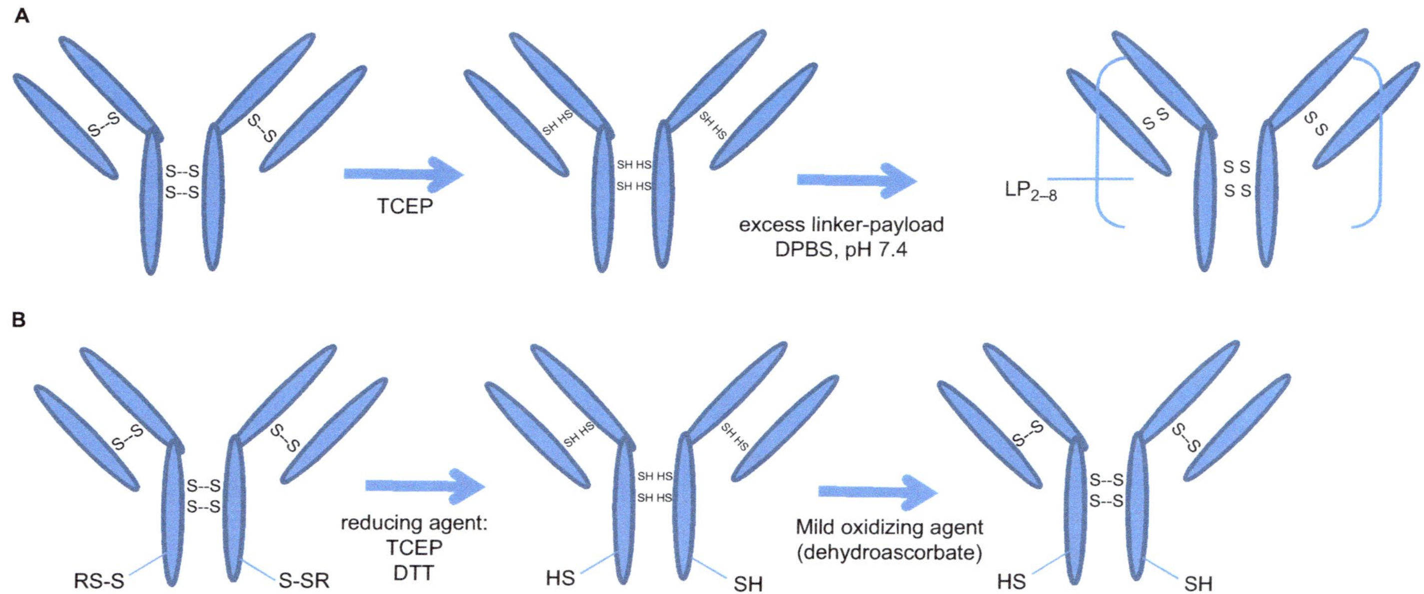

Figure 6.6 (A) Two-step preparation of reducing interchain disulfides to liberate free sulfhydryls for bioconjugation. Reaction with linker payloads bearing an electrophile yield mixtures of conjugates with 2–8 drugs per antibody. (B) Engineered cysteines on the antibody backbone require reduction followed by reoxidation and yield homogeneous conjugates with two drugs per antibody.

administration of the comparable ADC, even when normalized for payload exposure. Two hypotheses were presented to account for this difference, namely that (1) the absence of high DAR species in the TDC that are known to contribute to toxicity or (2) the kinetics of payload release are sufficiently slower on the engineered cysteines, due perhaps to payloads on these sites being less prone to proteolysis so that the TDC releases less free payload into the circulation than does the ADC, leading to decreased toxicity.

In a follow-on study that expanded the scope of sites used for incorporation of cysteine residues for site-specific conjugation, Shen *et al.*[58] explored the role of conjugation site on conjugate stability and biological activity. Cysteines were incorporated separately via engineering into the light chain (V205C), heavy chain Fab (A114C) and heavy chain Fc (S396C) regions of the anti-Her2 mAb trastuzumab and were conjugated with mcValCitPABC-MMAE. While antigen binding, internalization, and *in vitro* potencies were comparable across the three sites, the *in vivo* efficacy in a MMTV-HER2 Fo5 mammary tumor model varied significantly, with the V205C conjugate showing better efficacy than the A114C ADC, and S396C having little to no activity *in vivo*. This difference in activity was ultimately ascribed to significant differences in the plasma stability of the LP when conjugated to specific cysteines. Premature proteolysis of the ValCitPABC linker was ruled out as the major contributing mechanism to LP instability because free MMAE plasma levels were comparable across conjugates. Moreover, in a series of elegant experiments employing *in vivo* or *in vitro* incubation followed by affinity capture of ADC and subsequent LC-MS analysis,[59] the increased linker stability observed at specific cysteines was attributed to succinimide ring hydrolysis that prevents deconjugation in a retro-Michael fashion[47] (Scheme 6.2), with ring hydrolysis presumably catalyzed by the local positively charged residues surrounding sites like V205C and that are absent at sites such as 396C. This phenomenon will be discussed in more detail in section 6.4.2.

As mentioned, the thiomab technology has also been applied to maytansinoid ADCs[45] by conjugating a bis-maleimido-trioxyethylene glycol (BMPEO) spacer to an A114C variant of trastuzumab and then subsequent reaction with the maytansinoid payload DM1 (Figure 6.4, **21**). The resultant homogeneous conjugate (DAR = 1.8) showed improved tolerability in cynomolgus monkeys compared to a conventional lysine DM1 conjugate (DAR = 3.3).

6.3.2 Unnatural Amino Acids as Reactive Handles for Antibody Conjugation

One of the key challenges in preparing antibody–drug conjugates, and perhaps with bioconjugation reactions in general, is the absolute requirement that the covalent bond forming reactions be performed under mild aqueous conditions at or around physiological pH, and most especially under conditions that are bioorthogonal to the functional groups present on its component amino acids and carbohydrates.[60] The discovery that amino acids beyond the canonical set

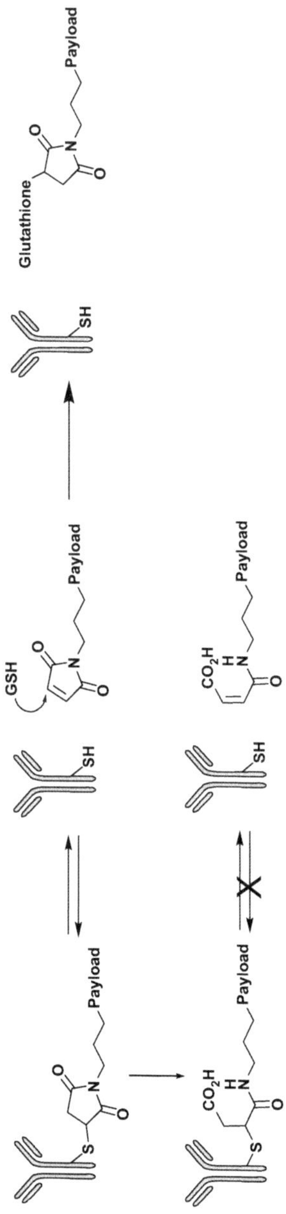

Scheme 6.2 Maleimide linkers conjugated to cysteines are known to undergo maleimide-exchange via retro-Michael reaction with exogenous thiols such as glutathione, cysteine, and perhaps serum albumin. Conjugation of maleimides to specific cysteines can undergo hydrolysis to give a "ring opened" succinimide that is resistant to retro-Michael exchange.

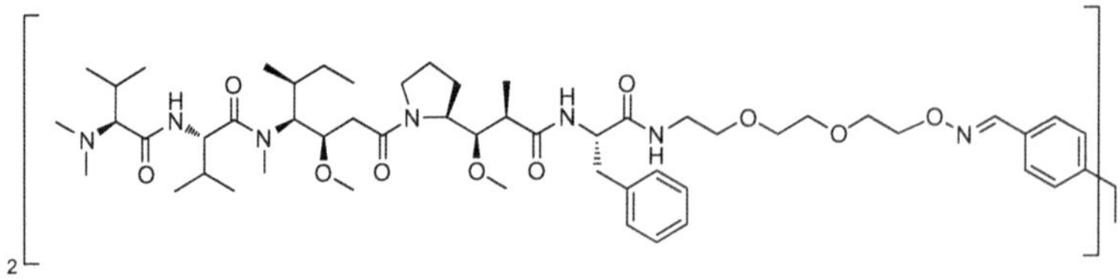

Figure 6.7 A noncleavable hydroxylamine bearing linker attached to monomethylauristatin F (MMAF) was conjugated via an oxime-forming reaction at pH 4.5 to yield site-specifically conjugated ADCs with DAR = 2.

of 20 (so-called "unnatural" amino acids) can be incorporated into the backbone of polypeptides via genetic engineering[61,62] opened up the possibility of installing functional groups into a protein that are present nowhere else in the biomolecule, notably aldehydes, ketones,[63,64] and selenothiols.[65,66]

The first published report[48] of an ADC made via conjugation to an unnatural amino acid employed an amber suppressor tRNA/aminoacyl-tRNA synthetase pair to incorporate p-acetylphenylalanine (pAcPhe) into both Fab domains and full length IgG1 variants of an anti-Her2 trastuzumab backbone. Two sites on the light chain(K169X and S202X) as well as one site on the heavy chain(A121X) were expressed, with reported yields up to 300 mg/L. As was the case in related approaches to determining sites for conjugation, proximity away from antigen binding or Fc effector function as well as solvent accessibility were used to guide selection. A noncleavable hydroxylamine bearing linker attached to monomethylauristatin F (MMAF) was conjugated via an oxime-forming reaction at pH 4.5 to yield site-specifically conjugated ADCs with DAR = 2 for the full IgG case (Figure 6.7). The three IgG conjugates showed similar cytotoxicity in a $HER2^+$ SK-BR-3 cell line (reported IC_{50}s ranged from 1.8 to 8.3 nM) and potency remained unchanged following a three day incubation in mouse serum at 37 °C, suggesting that the oxime linkage is stable in serum. While the pharmacokinetics of the A121X IgG conjugate were evaluated for antibody exposure and shown to be comparable to the unconjugated antibody, the ADC exposure (i.e. measurement of DAR change over time or evaluation of conjugated payload using an anti-payload ELISA) was not evaluated. The A121X ADC showed complete tumor regression in an MDA-MB-435 mouse xenograft model following a single injection at 5 mpk, but no assessment of safety was reported other than weight loss and qualitative histopathology of liver and kidney in mice, which are known to be relatively insensitive to auristatin toxicity.[67]

6.3.3 Chemical and Enzymatic Methods for Site-Specific Protein Conjugation

While cysteine engineering and the incorporation of unnatural amino acids have been applied successfully to the preparation of ADCs, it is worth mentioning a number of chemical and enzymatic methods that have been employed to modify antibodies site-specifically, though none of these methods have yet been extensively reported on for the production of ADCs for oncology.

One of the earliest reports[68] of site-specific modifications of mAbs was the use of periodate oxidation of the oligosaccharide moieties on the mAb to provide an aldehyde handle for covalent attachment of chelating groups for radionuclide binding. In this study, the average loading was reported to be 5 mol/mol of ^{125}I per mAb, and antigen binding was shown to be superior to a comparable mAb with chelating groups attached randomly via lysine conjugation.

More recently, other oxidative methods have been applied specifically to the bioconjugation of antibodies. Early work on the site-specific modification of N-terminal residues on proteins[69,70] led the Francis group to develop a biomimetic transamination reaction that was successfully applied to the modification of an anti-FLAG IgG.[71,72] Their approach employed pyridoxal 5′-phosphate (PLP) to selectively oxidize the amino group of any N-terminal amino acid to an imine that, upon hydrolysis, reveals an aldehyde or ketone functionality available for bioorthogonal conjugation with hydroxyl amines and hydrazines. Scheme 6.3 illustrates the underlying mechanism for the selectivity of this reaction for the N-terminal amino group, namely, that upon formation of imine **25** a proton having a relatively low pKa is uniquely available (in contrast to amino groups on amino acid side chains) for tautomerization to the glyoxyl imine (**26**), which readily undergoes hydrolysis to yield an aldehyde or ketone (**27**). In related work, Witus *et al.*[73] demonstrated that the specific sequence of amino acids can have an effect on the efficiency of conjugation using PLP oxidation. Systematic screening of short peptide libraries using a colorimetric assay wherein the most reactive sequences are labeled with an oxime forming dye identified AKT as a particularly efficient tripeptide tag for oxidative coupling at the N-terminus of a protein.

Ban *et al.*[74] have also reported a method for site-specific modification of tyrosine residues and demonstrated that it can be used to attach a targeting

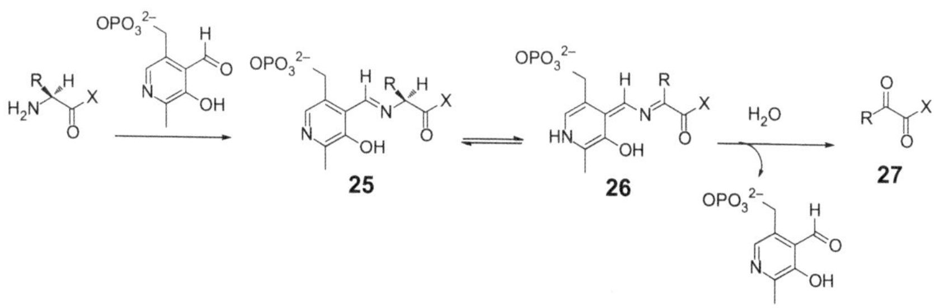

Scheme 6.3 A biomimetic transamination reaction that employed pyridoxal 5′-phosphate (PLP) to selectively oxidize the amino group of any *N*-terminal amino acid to an imine that, upon hydrolysis, reveals an aldehyde or ketone functionality available for bioorthogonal conjugation with hydroxyl amines and hydrazines.

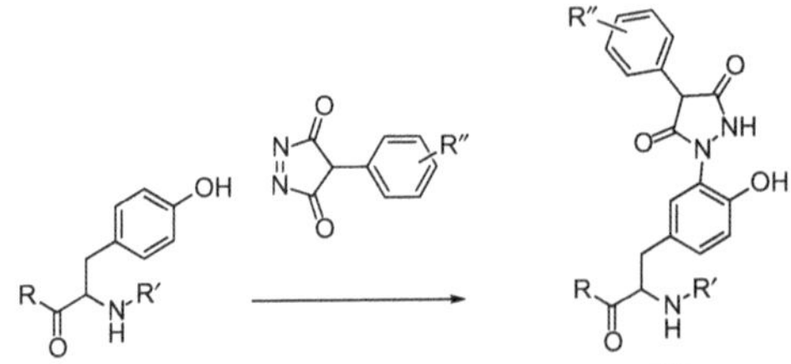

Scheme 6.4 Site-specific conjugation to tyrosine via an ene reaction with a phenyl-triazole-dione.

peptide to trastuzumab. A phenyl-triazole-dione reacts selectively in an ene-type reaction at the position ortho to the phenol group of tyrosine to yield the adduct **28** (Scheme 6.4). The stability of the new linkage was tested in model systems and shown to be stable to both extremes of pH and temperature, though the stability of the trastuzumab conjugate was not evaluated. In related work,[75] tyrosine was modified via reaction with 4-formylbenzene diazonium hexafluorophosphate to install an aldehyde for functionalization. This method was employed to generate a biotinylated trastuzumab via formation of an oxime of the modified tyrosine(s).

Enzymes have likewise been employed to introduce site-specific modifications to a antibodies. For example, transglutaminases that catalyze the formation of stable isopeptidic linkages between the side chains of lysine and glutamine have been used to modify IgGs for many years, either randomly using non-engineered mAbs or site-specifically by employing specific tags that are engineered into the N- or C- terminus of the protein.[76–78] More recently, Jeger *et al.*[79] reported the site-specific labeling of an N297Q variant of the anti-L1-CAM mAb chCE7 using bacterial transglutaminase (BTG). Interestingly, this site had been engineered to prevent N-glycosylation of the mAb, and therefore the deglycosylated mAb reveals another glutamine, Q295 that would otherwise be inaccessible. Surprisingly, although there are additional solvent accessible glutamines on the mAb, only Q295 and Q297 were modified by BTG, yielding a conjugate with DAR = 4.

6.4 Recent Developments in ADC Metabolism and PK

The same improvements in bioanalytical techniques that have driven the recent explosion in site-specific conjugation technologies have also revolutionized our understanding of payload release, ADC stability, and mechanism of action. Antibodies (IgG1 in particular) typically have a rather long half-life in serum, often in the range of 1–3 weeks in humans. It is generally considered imperative that the drug–antibody linkage be stable over this time period in order to minimize serum levels of free payload and unconjugated antibody, which may lead, respectively, to toxicity and lack of efficacy. While a high degree of serum stability is required, it is also essential that the payload be released efficiently during the process of antigen-mediated internalization. Recent advances in bioanalysis have enabled a more thorough understanding of these processes and the discussion below will highlight a few studies that illustrate this progress.

6.4.1 Payload Release In Targeted Tissues

A number of recent studies have elucidated the mechanism of antigen-mediated cytotoxicity for cleavable and noncleavable maytansinoid ADCs.[30,80,81] It has been widely reported that the first step of payload release for both thioether and disulfide-linked maytansines is lysosomal catabolism to the lysine-linked payload, as shown in Scheme 6.5. This lysine-linked species is toxic when released intracellularly, but is relatively benign in circulation due to its poor permeability.[39] In some cases, this lysine-linked maytansine is thought be subject to Pgp efflux, thereby limiting the potency of the conjugate. As discussed previously, polar linkers have been developed to evade efflux and the resulting lysine catabolites have been shown to accumulate in higher levels in tumor cells resulting in superior efficacy.[44] There has been some suggestion that thioether-linked maytansinoids may undergo an oxidative release of free drug, but it is unclear at present whether this is a physiologically relevant mechanism of action.[82]

Following lysosomal catabolism, disulfide-linked maytansine ADCs undergo a subsequent slow reduction of the disulfide to release the potently cytotoxic maytansinoid. This species permeates nearby cells resulting in some non-antigen

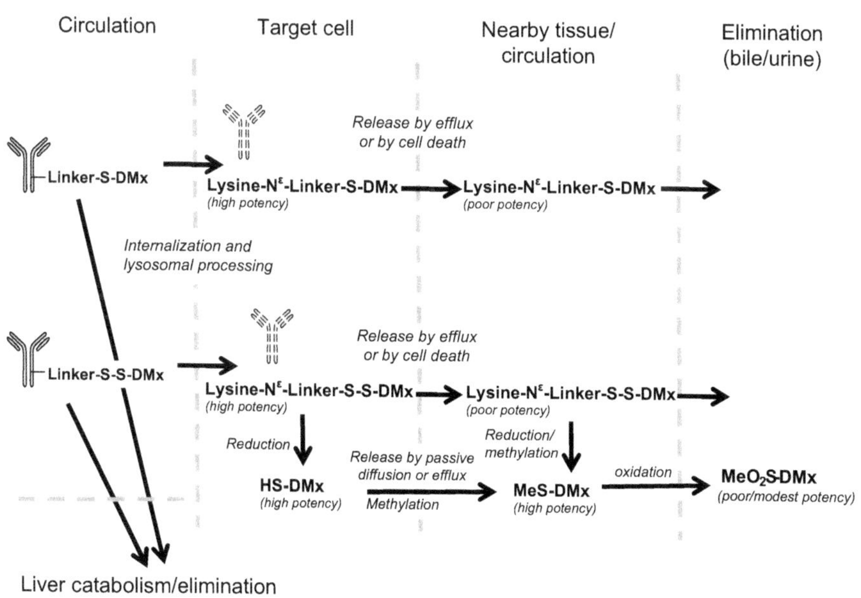

Scheme 6.5 Pathways for ADC metabolism and clearance.

dependent cytotoxicity (bystander effect).[43] The reduction of the disulfide likely takes place in the reducing environment found in the cytoplasm or nucleus rather than the lysosome, which has recently been shown to have a generally oxidizing milieu.[28] The resulting thiol-containing payload is metabolized in the tumor to S-methyl-DMx by the ubiquitously expressed S-methyl transferase. The rate of S-methylation is much greater for DM4 than for DM1, but this is likely inconsequential because S-methylation has little effect on the cytotoxicity.[80,81] The S-methyl DMx is rapidly oxidized in the liver to a far less toxic sulfoxide or sulfone.

Like the noncleavable lysine-linked maytansines, the noncleavable cysteine-linked auristatins have been shown to release an amino acid adduct of the payload.[40] This suggests that complete protein catabolism is a fairly ubiquitous process and can be a mechanism of payload release for a variety of linkers. Interestingly, there have been no reports of peptide cleavable conjugates resulting in an amino acid adduct of the payload. Presumably the rate of linker cleavage is comparable to (or faster than) the rate of protein catabolism. For example, the only payload-derived species that has been observed upon ADC internalization of radiolabeled SGN-35 (anti-CD33-mcValCitPABC-MMAE) is the free payload, MMAE. Like its maytansinoid counterparts, it was shown to diffuse into adjacent cells resulting in non-antigen-dependent cell killing.[83]

6.4.2 Premature Payload Release In Circulation And Non-Targeted Tissues

Premature cleavage of the payload from the antibody has been a cause of concern from the earliest reports of doxorubicin and calicheamicin conjugates.[15,19] It was generally assumed that moving away from hydrazone-based linkages would decrease the amount of drug released in serum. While this has been demonstrated by several studies,[16] it is also clear that the current cohort of cysteine and lysine-linked ADCs continue to suffer from premature payload release.

Pharmacokinetic analysis of huC242-SPP-DM1 (**14**) shows that the clearance of bound-DM1 from mouse plasma samples is approximately three times the rate of clearance for the antibody alone, strongly suggesting that the payload is being prematurely cleaved in circulation.[31] The more sterically hindered linker, SPDB-DM4 (**15**), provides a significant boost in efficacy that is attributed to improved *in vivo* stability of the conjugate.[27,32] The improvement in stability by moving from the disulfide SPP-DM1 to the thioether SMCC-DM1 (**18**) is similarly striking. The total antibody clearance for both ADCs is approximately equal. However, the clearance rate of antibody-bound DM1 from the thioether conjugate is approximately half of that of the disulfide conjugate. Interestingly, however, even the thioether conjugate T-DM1 (i.e. SMCC-DM1, **18**) is observed to have a significantly higher clearance than the total antibody clearance.[30] This implies that there must be some other mechanism by which maytansine is being cleaved prematurely from the ADC. While an oxidative mechanism has been proposed,[82] it seems possible that the maleimide moiety of the SMCC linker may be prone to a "maleimide exchange" similar to what has recently been observed for maleimide–cysteine ADCs (*vide infra*).

A fascinating report in 2008 clearly illustrated that noncleavable maleimide-based ADCs can slowly loose payload *in vivo*. Two mcMMAF ADCs were shown to slowly release their auristatin payload ($T\frac{1}{2}$ ~7 days) over the course of a 2-week PK study in mice. Plasma stability studies showed that the payload was being transferred to the cysteine-34 residue of serum albumin. Replacement of the maleimide with a thioether derived from a bromoacetamide (bac) linker resulted in a dramatic improvement in stability.[47] Likewise, a group from Genentech reported that mcMMAF bound to a particular engineered cysteine site (LC-V205C) gave improved *in vivo* stability over a conventional mcMMAF conjugate. A careful mass spectrometric analysis of ADC recovered from plasma showed that the mcMMAF bound at these sites underwent a rapid hydrolysis to give a "ring opened" succinimide, as shown in Scheme 6.2. Conversely, mcMMAF bound at sites such as HC-S395C were resistant to hydrolysis and the payload was rapidly lost from the ADC. The authors postulate that favorable sites, such as LC-V205C, undergo a rapid ring-opening, which renders them resistant to maleimide exchange with exogenous thiols such as glutathione, cysteine, and perhaps serum albumin.[58] Such ring-opened succinimides have recently been shown to be resistant to forced maleimide-exchange and retro-Michael reactions.[84] A similar loss of payload-linker is thought to occur for all maleimide-linked payloads, including mcValCitPABC. Another recent study demonstrated that loss of mcValCitPABC-MMAE occurs at a similar rate in human, rat, and monkey plasma. Perhaps unexpectedly, the loss does not occur via cleavage of the peptidic linker. Rather, it appears to be complete loss of mcValCitPABC-MMAE followed by the addition of an exogenous moiety such as cysteine.[59] This is consistent with the above retro-Michael mediated mechanism of payload loss. One clear message that emerges from the above reports is that minimizing the premature payload loss has a beneficial effect on efficacy and/or toxicity.[47,58,85]

6.4.3 Pharmacokinetics/Pharmacodynamics

While there is widespread experimental agreement on the mechanism of ADC catabolism and the mechanism of payload release, it has been far more difficult to use this data in a precise understanding of the pharmacokinetics/pharmacodynamics (PK/PD) of various ADCs. For instance, a noncleavable (SMCC-DM1) anti-CanAg ADC shows significantly inferior efficacy to a corresponding cleavable (SPP-DM1) ADC in spite of the fact that the noncleavable ADC has better plasma exposure and higher levels of released drug in the tumor.[80] One possible reason for this discrepancy is that catabolism of the disulfide-linked ADC results in permeable, highly cytotoxic species, which can go on to affect adjacent (bystander) tumor cells. However, trastuzumab conjugates show the opposite trend. The noncleavable linker (SMCC-DM1) gives superior efficacy as compared to the cleavable

SPP-DM1.[41] In this case, the levels of released drug are comparable suggesting that the bystander effect is not playing a significant role in efficacy.[30]

Two possible explanations have been put forward to explain the lack of good PK/PD correlations. First, PK modeling studies suggest the sensitivity of tumors to various agents may be impacted by tumor cell growth rate.[86] This variable is often not discussed when comparing efficacy in various tumor models. Second, the kinetics of ADC internalization may not be fully understood. For example, the improved stability of T-SMCC-DM1 (T-DM1) resulted in greater plasma exposure and tumor exposure as compared to T-SPP-DM1. However, the increased tumor exposure unexpectedly did not result in increased tumor levels of maytansinoid catabolites. The reason for this discrepancy is unclear but the authors suggest that it may be due to the kinetics of exposure. The plasma levels of both ADCs are highest and closest together early in the experiment (≤24 hours) and tumor localization reaches a maxima at ~24 hours. It is possible that most tumor internalization and c-tabolism takes place in this time window, minimizing the effect of later differences in plasma exposure.[30] Needless to say, a great deal remains to be learned about ADC PK/PD relationships.

6.4.4 ADC Clearance and Elimination

A seminal study by Hamblett in 2004 clearly illustrated that drug loading can have a significant impact on ADC exposure and clearance. It was found that a low loaded mcValCitPABC-MMAE ADC (DAR ~2) had similar exposure to unmodified antibody. However, the corresponding 4-loaded and 8-loaded ADCs had ~35% and ~80% lower exposure, respectively. Likewise, the 4-loaded and 8-loaded ADC were cleared, respectively, 1.6 and 5 times faster than the unmodified antibodies while the 2-loaded ADC was cleared at approximately the same rate as unmodified antibody.[52] Boswell and colleagues reported a similar finding in 2012. They found that the total antibody of a 2-loaded site-specific conjugate was cleared at approximately the same rate as unmodified antibody while a 3-loaded conventional conjugate was cleared ~35% faster than its unconjugated counterpart. Since the analysis was performed based on total antibody level, the difference is unlikely to be due to differential rates of deconjugation. Rather, the authors infer that the observed difference in clearance is due to elevated hepatic uptake of higher-loaded ADCs, possibly due to the increased hydrophobicity of the conjugates.[85] Similar observations of increased clearance of ADC as compared to unmodified antibody have been reported in clinical studies of ADCs.[87] The mechanistic basis for this increased hepatic uptake is unclear, but one plausible explanation is impaired FcRn-dependent recycling.[88,89]

A recent report from Genentech confirms that hepatic uptake is the major route of ADC elimination. Shen and colleagues followed the fate of radiolabeled T-DM1 (SMCC-DM1) in rats and show that ~80% of the isotopically labeled material was recovered in feces. A separate study with bile duct cannulated rats showed that ~50% of the radiolabeled material was recovered in bile while <10% was recovered in urine. Moreover, the vast majority of the DM1-derived material was found in the acetonitrile-soluble fraction, indicating that the ADC is hepatically metabolized and subsequently eliminated via the biliary system. Cumulatively this data strongly indicates that the vast majority of the payload is lost hepatically rather than systemically.[90] This raises an interesting dilemma: if the majority of payload is released (and then eliminated) in the liver, why is more liver toxicity not observed? Recent work by Sun and colleagues offer two suggestions. First, they conjecture that tubulin targeting payloads (such as auristatins and maytansines) may not be particularly toxic to the fully differentiated, nondividing cells of the liver. Second, their work shows that maytansines are quickly methylated and oxidized to give derivatives that are generally far less toxic than the parent compounds.[81]

Finally, it should be pointed out that there is emerging data that strongly suggests that the net charge (pI) of the ADC can play a major role in the PK.[91] The effect of antibody charge on PK has been previously noted, but was systematically demonstrated by Igawa in 2010.[92] This study notes that increasing the pI (catonization) of an antibody via Fc mutations results in a significant increase in clearance and decrease in half life. Conversely, decreasing the pI (anionization) results in lower clearance and extended half life. While these results have not been specifically observed on ADCs, this is a phenomenon likely to receive increasing attention in the future and may help explain some of the discrepancies between ADC clearance and mAb clearance. As can be easily seen from this discussion, many of the payloads and linkers being explored for ADCs are charged and are therefore likely to have an impact on the ADC pI. Moreover, many conjugation methodologies employed today utilize acylation of lysine residues, which potentially could increase the ADC pI and therefore have an adverse effect on clearance.

6.5 Summary and Future Prospects

The recent approval of a new antibody conjugate for oncology (Adcetris, 2012) combined with encouraging clinical data for a number of other ADCs in late stage development represent the fruition of an idea that has been investigated for decades. That the time is now ripe for fully realizing the goal of targeting a highly potent cytotoxin to a tumor while greatly reducing systemic exposure via conjugation to an antibody is perhaps a function of recent advances in antibody engineering, medicinal chemistry, and bioanalytical methodologies. That said, it is also vital to acknowledge that these recent advances relied heavily on much of the pioneering early work in this field that pointed the way to current success. Looking forward, the plethora of clinical ADCs now in early development will no doubt provide a rich set of data that will inform the next generation on ADCs. These conjugates of the future will likely rely on advances in small molecule chemistry and its related disciplines as well as protein science. The field can look forward to the discovery of new classes of payloads that work as ADCs, advances in the ability to design and synthesize linkers that give stable conjugates and that allow for efficient trafficking and delivery of ADC payloads to their target cells, new ways of enabling site-specific conjugates, as well as improvements in our ability to monitor metabolism of ADCs and better understand their pharmacokinetics. It will be by continuing to build on the advances in the areas outlined in this review that the targeting power of monoclonal antibodies and the cell-killing power of small molecules will form a more perfect union.

References

1. B. A. Chabner and T. G. Roberts Jr, *Nat. Rev. Cancer*, 2005, **5**, 65–72.
2. V. T. DeVita and E. Chu, *Cancer Res.*, 2008, **68**, 8643–8653.
3. J. J. Cui, M. Tran-Dubé, H. Shen, M. Nambu, P.-P. Kung, M. Pairish, L. Jia, J. Meng, L. Funk, I. Botrous, M. McTigue, N. Grodsky, K. Ryan, E. Padrique, G. Alton, S. Timofeevski, S. Yamazaki, Q. Li, H. Zou, J. Christensen, B. Mroczkowski, S. Bender, R. S. Kania and M. P. Edwards, *J. Med. Chem.*, 2011, **54**, 6342–6363.
4. M. Gerlinger, A. J. Rowan, S. Horswell, J. Larkin, D. Endesfelder, E. Gronroos, P. Martinez, N. Matthews, A. Stewart, P. Tarpey, I. Varela, B. Phillimore, S. Begum, N. Q. McDonald, A. Butler, D. Jones, K. Raine, C. Latimer, C. R. Santos, M. Nohadani, A. C. Eklund, B. Spencer-Dene, G. Clark, L. Pickering, G. Stamp, M. Gore, Z. Szallasi, J. Downward, P. A. Futreal and C. Swanton, *NEJM*, 2012, **366**, 883–892.
5. L. M. Weiner, R. Surana and S. Wang, *Nat. Rev. Immunol.*, 2010, **10**, 317–327.
6. S. DeCarvalho, H. J. Rand and A. Lewis, *Nature*, 1964, **202**, 255–258.
7. P. R. Hamann, L. M. Hinman, I. Hollander, C. F. Beyer, D. Lindh, R. Holcomb, W. Hallett, H.-R. Tsou, J. Upeslacis, D. Shochat, A. Mountain, D. A. Flowers and I. Bernstein, *Bioconj. Chem.*, 2002, **13**, 47–58.
8. P. J. Carter and P. D. Senter, *Cancer J.*, 2008, **14**, 154–169.
9. P. D. Senter, *Curr. Opin. Chem. Biol.*, 2009, **13**, 235–244.
10. J. A. Flygare, T. H. Pillow and P. Aristoff, *Chem. Biol. Drug. Des.*, 2013, **81**, 113–121.
11. P. Sapra, A. T. Hooper, C. J. O'Donnell and H.-P. Gerber, *Expert Opin. Invest. Drugs*, 2011, **20**, 1131–1149.
12. P. R. Hamann and R. G. Dushin, *New Frontiers in Chemical Biology: Enabling Drug Discovery*, ed. M. E. Bunnage, RSC Publishing, 2011, pp. 224–257.
13. P. A. Trail, D. Willner, S. J. Lasch, A. J. Henderson, S. Hofstead, A. M. Casazza, R. A. Firestone, I. Hellstrom and K. E. Hellstrom, *Science*, 1993, **261**, 212–215.
14. G. Casi and D. Neri, *J. Controlled Release*, 2012, **161**, 422–428.
15. P. A. Trail, D. Willner, J. Knipe, A. J. Henderson, S. J. Lasch, M. E. Zoeckler, M. D. TrailSmith, T. W. Doyle, H. D. King, A. M. Casazza, G. R. Braslawsky, J. Brown, S. J. Hofstead, R. S. Greenfield,

R. A. Firestone, K. Mosure, K. F. Kadow, M. B. Yang, K. E. Hellstroem and I. Hellstroem, *Cancer Res.*, 1997, **57**, 100–105.
16. S. O. Doronina, B. E. Toki, M. Y. Torgov, B. A. Mendelsohn, C. G. Cerveny, D. F. Chace, R. L. DeBlanc, R. P. Gearing, T. D. Bovee, C. B. Siegall, J. A. Francisco, A. F. Wahl, D. L. Meyer and P. D. Senter, *Nat. Biotechnol.*, 2003, **21**, 941.
17. E. R. Boghaert, K. M. Khandke, L. Sridharan, M. Dougher, J. F. DiJoseph, A. Kunz, P. R. Hamann, J. Moran, I. Chaudhary and N. K. Damle, *Cancer Chemother. Pharmacol.*, 2008, **61**, 1027–1035.
18. A. D. Ricart, *Clin. Cancer Res.*, 2011, **17**, 6417–6427.
19. P. R. Hamann, L. M. Hinman, C. F. Beyer, L. M. Greenberger, C. Lin, D. Lindh, A. T. Menendez, R. Wallace, F. E. Durr and J. Upeslacis, *Bioconj. Chem.*, 2005, **16**, 346–353.
20. P. Sapra, J. DiJoseph and H.-P. Gerber, *Drug Delivery Oncol.*, 2012, **2**, 395–410.
21. P. L. Carl, P. K. Chakravarty and J. A. Katzenellenbogen, *Journal of Medicinal Chemistry*, 1981, **24**, 479–480.
22. M. A. Walker, G. M. Dubowchik, S. J. Hofstead, P. A. Trail and R. A. Firestone, *Bioorg. Med. Chem. Lett.*, 2002, **12**, 217–219.
23. G. M. Dubowchik, R. A. Firestone, L. Padilla, D. Willner, S. J. Hofstead, K. Mosure, J. O. Knipe, S. J. Lasch and P. A. Trail, *Bioconj. Chem.*, 2002, **13**, 855–869.
24. S. C. Jeffrey, M. Y. Torgov, J. B. Andreyka, L. Boddington, C. G. Cerveny, W. A. Denny, K. A. Gordon, D. Gustin, J. Haugen, T. Kline, M. T. Nguyen and P. D. Senter, *J. Med. Chem.*, 2005, **48**, 1344–1358.
25. S. O. Doronina, T. D. Bovee, D. W. Meyer, J. B. Miyamoto, M. E. Anderson, C. A. Morris-Tilden and P. D. Senter, *Bioconj. Chem.*, 2008, **19**, 1960–1963.
26. L. M. Hinman, P. R. Hamann, R. Wallace, A. T. Menendez, F. E. Durr and J. Upeslacis, *Cancer Res.*, 1993, **53**, 3336–3342.
27. W. C. Widdison, S. D. Wilhelm, E. E. Cavanagh, K. R. Whiteman, B. A. Leece, Y. Kovtun, V. S. Goldmacher, H. Xie, R. M. Steeves, R. J. Lutz, R. Zhao, L. Wang, W. A. Blaettler and R. V. J. Chari, *J. Med. Chem.*, 2006, **49**, 4392–4408.
28. C. D. Austin,X. Wen,L. Gazzard,C. Nelson,R. H. Scheller and S. J. Scales, *Proc. Natl. Acad. Sci. U. S. A.*, 2005, **102**, 17987–17992.
29. A. G. Polson, J. Calemine-Fenaux, P. Chan, W. Chang, E. Christensen, S. Clark, F. J. de Sauvage, D. Eaton, K. Elkins, J. M. Elliott, G. Frantz, R. N. Fuji, A. Gray, K. Harden, G. S. Ingle, N. M. Kljavin, H. Koeppen, C. Nelson, S. Prabhu, H. Raab, S. Ross, J.-P. Stephan, S. J. Scales, S. D. Spencer, R. Vandlen, B. Wranik, S.-F. Yu, B. Zheng and A. Ebens, *Cancer Res.*, 2009, **69**, 2358–2364.
30. H. K. Erickson, G. D. Lewis Phillips, D. D. Leipold, C. A. Provenzano, E. Mai, H. A. Johnson, B. Gunter, C. A. Audette, M. Gupta, J. Pinkas and J. Tibbitts, *Mol. Cancer Ther.*, 2012, **11**, 1133–1142.

31. H. Xie, C. Audette, M. Hoffee, J. M. Lambert and W. A. Blaettler, *J. Pharmacol. Exp. Ther.*, 2004, **308**, 1073–1082.
32. L. Ducry and B. Stump, *Bioconj. Chem.*, 2010, **21**, 5–13.
33. B. A. Kellogg, L. Garrett, Y. Kovtun, K. C. Lai, B. Leece, M. Miller, G. Payne, R. Steeves, K. R. Whiteman, W. Widdison, H. Xie, R. Singh, R. V. J. Chari, J. M. Lambert and R. J. Lutz, *Bioconj. Chem.*, 2011, **22**, 717–727.
34. R. Y. Zhao, S. D. Wilhelm, C. Audette, G. Jones, B. A. Leece, A. C. Lazar, V. S. Goldmacher, R. Singh, Y. Kovtun, W. C. Widdison, J. M. Lambert and R. V. J. Chari, *J. Med. Chem.*, 2011, **54**, 3606–3623.
35. S. C. Jeffrey, J. B. Andreyka, S. X. Bernhardt, K. M. Kissler, T. Kline, J. S. Lenox, R. F. Moser, M. T. Nguyen, N. M. Okeley, I. J. Stone, X. Zhang and P. D. Senter, *Bioconj. Chem.*, 2006, **17**, 831–840.
36. P. J. Burke, P. D. Senter, D. W. Meyer, J. B. Miyamoto, M. Anderson, B. E. Toki, G. Manikumar, M. C. Wani, D. J. Kroll and S. C. Jeffrey, *Bioconj. Chem.*, 2009, **20**, 1242–1250.
37. S. C. Jeffrey, M. T. Nguyen, R. F. Moser, D. L. Meyer, J. B. Miyamoto and P. D. Senter, *Bioorg. Med. Chem. Lett.*, 2007, **17**, 2278–2280.
38. S. C. Jeffrey, J. De Brabander, J. Miyamoto and P. D. Senter, *ACS Med. Chem. Lett.*, 2010, **1**, 277–280.
39. H. K. Erickson, P. U. Park, W. C. Widdison, Y. V. Kovtun, L. M. Garrett, K. Hoffman, R. J. Lutz, V. S. Goldmacher and W. A. Blaettler, *Cancer Res.*, 2006, **66**, 4426–4433.
40. S. O. Doronina, B. A. Mendelsohn, T. D. Bovee, C. G. Cerveny, S. C. Alley, D. L. Meyer, E. Oflazoglu, B. E. Toki, R. J. Sanderson, R. F. Zabinski, A. F. Wahl and P. D. Senter, *Bioconj. Chem.*, 2006, **17**, 114–124.
41. G. D. Lewis Phillips, G. Li, D. L. Dugger, L. M. Crocker, K. L. Parsons, E. Mai, W. A. Blaettler, J. M. Lambert, R. V. J. Chari, R. J. Lutz, W. L. T. Wong, F. S. Jacobson, H. Koeppen, R. H. Schwall, S. R. Kenkare-Mitra, S. D. Spencer and M. X. Sliwkowski, *Cancer Res.*, 2008, **68**, 9280–9290.
42. B. A. Teicher and R. V. J. Chari, *Clin. Cancer Res.*, 2011, **17**, 6389–6397.
43. Y. V. Kovtun, C. A. Audette, Y. Ye, H. Xie, M. F. Ruberti, S. J. Phinney, B. A. Leece, T. Chittenden, W. A. Blaettler and V. S. Goldmacher, *Cancer Res.*, 2006, **66**, 3214–3221.
44. Y. V. Kovtun, C. A. Audette, M. F. Mayo, G. E. Jones, H. Doherty, E. K. Maloney, H. K. Erickson, X. Sun, S. Wilhelm, O. Ab, K. C. Lai, W. C. Widdison, B. Kellogg, H. Johnson, J. Pinkas, R. J. Lutz, R. Singh, V. S. Goldmacher and R. V. J. Chari, *Cancer Res.*, 2010, **70**, 2528–2537.
45. J. R. Junutula, K. M. Flagella, R. A. Graham, K. L. Parsons, E. Ha, H. Raab, S. Bhakta, T. Nguyen, D. L. Dugger, G. Li, E. Mai, G. D. Lewis Phillips, H. Hiraragi, R. N. Fuji, J. Tibbitts, R. Vandlen, S. D. Spencer, R. H. Scheller, P. Polakis and M. X. Sliwkowski, *Clin. Cancer Res.*, 2010, **16**, 4769–4778.

46. J. R. Junutula, H. Raab, S. Clark, S. Bhakta, D. D. Leipold, S. Weir, Y. Chen, M. Simpson, S. P. Tsai, M. S. Dennis, Y. Lu, Y. G. Meng, C. Ng, J. Yang, C. C. Lee, E. Duenas, J. Gorrell, V. Katta, A. Kim, K. McDorman, K. Flagella, R. Venook, S. Ross, S. D. Spencer, W. L. Wong, H. B. Lowman, R. Vandlen, M. X. Sliwkowski, R. H. Scheller, P. Polakis and W. Mallet, *Nat. Biotechnol.*, 2008, **26**, 925–932.
47. S. C. Alley, D. R. Benjamin, S. C. Jeffrey, N. M. Okeley, D. L. Meyer, R. J. Sanderson and P. D. Senter, *Bioconj. Chem.*, 2008, **19**, 759–765.
48. Y. Axup Jun, M. Bajjuri Krishna, M. Ritland, M. Hutchins Benjamin, H. Kim Chan, A. Kazane Stephanie, R. Halder, S. Forsyth Jane, F. Santidrian Antonio, K. Stafin, Y. Lu, H. Tran, J. Seller Aaron, L. Biroc Sandra, A. Szydlik, K. Pinkstaff Jason, F. Tian, C. Sinha Subhash, B. Felding-Habermann, V. Smider Vaughn and G. Schultz Peter, *Proc. Natl. Acad. Sci. U. S. A.*, 2012, **109**, 16101–16106.
49. H. D. King, G. M. Dubowchik, H. Mastalerz, D. Willner, S. J. Hofstead, R. A. Firestone, S. J. Lasch and P. A. Trail, *J. Med. Chem.*, 2002, **45**, 4336–4343.
50. R. Y. Zhao, H. K. Erickson, B. A. Leece, E. E. Reid, V. S. Goldmacher, J. M. Lambert and R. V. J. Chari, *J. Med. Chem.*, 2012, **55**, 766–782.
51. J. M. Lambert, *Drugs Future*, 2010, **35**, 471–480.
52. K. J. Hamblett, P. D. Senter, D. F. Chace, M. M. C. Sun, J. Lenox, C. G. Cerveny, K. M. Kissler, S. X. Bernhardt, A. K. Kopcha, R. F. Zabinski, D. L. Meyer and J. A. Francisco, *Clin. Cancer Res.*, 2004, **10**, 7063–7070.
53. A. Lyons, D. J. King, R. J. Owens, G. T. Yarranton, A. Millican, N. R. Whittle and J. R. Adair, *Protein Eng.*, 1990, **3**, 703–708.
54. J. B. Stimmel, B. M. Merrill, L. F. Kuyper, C. P. Moxham, J. T. Hutchins, M. E. Fling and F. C. Kull, *J. Biol. Chem.*, 2000, **275**, 30445–30450.
55. C. F. McDonagh, E. Turcott, L. Westendorf, J. B. Webster, S. C. Alley, K. Kim, J. Andreyka, I. Stone, K. J. Hamblett, J. a. Francisco and P. Carter, *Protein Eng., Des. Sel.*, 2006, **19**, 299–307.
56. M. M. C. Sun, K. S. Beam, C. G. Cerveny, K. J. Hamblett, R. S. Blackmore, M. Y. Torgov, F. G. M. Handley, N. C. Ihle, P. D. Senter and S. C. Alley, *Bioconj. Chem.*, 2005, **16**, 1282–1290.
57. J. R. Junutula, S. Bhakta, H. Raab, K. E. Ervin, C. Eigenbrot, R. Vandlen, R. H. Scheller and H. B. Lowman, *J. Immunol. Methods*, 2008, **332**, 41–52.
58. B.-Q. Shen, K. Xu, L. Liu, H. Raab, S. Bhakta, M. Kenrick, K. L. Parsons-Reponte, J. Tien, S.-F. Yu, E. Mai, D. Li, J. Tibbitts, J. Baudys, O. M. Saad, S. J. Scales, P. J. McDonald, P. E. Hass, C. Eigenbrot, T. Nguyen, W. A. Solis, R. N. Fuji, K. M. Flagella, D. Patel, S. D. Spencer, L. A. Khawli, A. Ebens, W. L. Wong, R. Vandlen, S. Kaur, M. X. Sliwkowski, R. H. Scheller, P. Polakis and J. R. Junutula, *Nat. Biotechnol.*, 2012, **30**, 184–189.
59. K. Xu, L. Liu, O. M. Saad, J. Baudys, L. Williams, D. Leipold, B. Shen, H. Raab, J. R. Junutula, A. Kim and S. Kaur, *Anal. Biochem.*, 2011, **412**, 56–66.

60. J. Kalia and R. T. Raines, *Curr. Org. Chem.*, 2010, **14**, 138–147.
61. C. J. Noren, S. J. Anthony-cahill, M. C. Griffith, P. G. Schultz and I. April, *Science*, 1989, **244**, 182–188.
62. L. Wang, a. Brock, B. Herberich and P. G. Schultz, *Science*, 2001, **292**, 498–500.
63. L. Wang and P. G. Schultz, *Ang. Chem. Int. Ed. Engl.*, 2004, **44** , 34–66.
64. P. Wu, W. Shui, B. L. Carlson, N. Hu, D. Rabuka, J. Lee and C. R. Bertozzi, *Proc. Natl. Acad. Sci. U. S. A.*, 2009, **106**, 3000–3005.
65. T. Hofer, J. D. Thomas, T. R. Burke and C. Rader, *Proc. Natl. Acad. Sci. U. S. A.*, 2008, **105**, 12451–12456.
66. J. D. Thomas, H. Cui, P. J. North, T. Hofer, C. Rader and T. R. Burke, *Bioconj. Chem.*, 2012, **23**, 2007–2013.
67. J. C. Mirsalis, J. Schindler-Horvat, J. R. Hill, J. E. Tomaszewski, S. J. Donohue and C. a. Tyson, *Cancer Chemother. Pharmacol.*, 1999, **44**, 395–402.
68. J. D. Rodwell, V. L. Alvarez, C. Lee, a. D. Lopes, J. W. Goers, H. D. King, H. J. Powsner and T. J. McKearn, *Proc. Natl. Acad. Sci. U. S. A.*, 1986, **83**, 2632–2636.
69. K. Geoghegan and J. G. Stroh, *Bioconj. Chem.*, 1992, **3**, 138–146.
70. R. Wetzel, S. R. Halualani, J. T. Stults and C. Quan, *Bioconj. Chem.*, 1990, **1**, 114–122.
71. J. M. Gilmore, R. a. Scheck, A. P. Esser-Kahn, N. S. Joshi and M. B. Francis, *Angew. Chem.*, 2006, **118**, 5433–5437.
72. R. a. Scheck and M. B. Francis, *ACS Chem. Biol.*, 2007, **2**, 247–251.
73. L. S. Witus, T. Moore, B. W. Thuronyi, A. P. Esser-Kahn, R. a. Scheck, A. T. Iavarone and M. B. Francis, *J. Am. Chem. Soc.*, 2010, **132**, 16812–16817.
74. H. Ban, J. Gavrilyuk and C. F. Barbas, *J. Am. Chem. Soc.*, 2010, **132**, 1523–1525.
75. J. Gavrilyuk, H. Ban, M. Nagano, W. Hakamata and C. F. Barbas, *Bioconj. Chem.*, 2012, **23**, 2321–2328.
76. a. Josten, L. Haalck, F. Spener and M. Meusel, *J. Immunol. Methods*, 2000, **240**, 47–54.
77. N. Kamiya, T. Takazawa, T. Tanaka, H. Ueda and T. Nagamune, *Enz. Microb. Technol.*, 2003, **33**, 492–496.
78. T. L. Mindt, V. Jungi, S. Wyss, A. Friedli, G. Pla, I. Novak-Hofer, J. Grünberg and R. Schibli, *Bioconj. Chem.*, 2008, **19**, 271–278.
79. S. Jeger, K. Zimmermann, A. Blanc, J. Grünberg, M. Honer, P. Hunziker, H. Struthers and R. Schibli, *Angew. Chem. Int. Ed.*, 2010, **49**, 9995–9997.
80. H. K. Erickson, W. C. Widdison, M. F. Mayo, K. Whiteman, C. Audette, S. D. Wilhelm and R. Singh, *Bioconj. Chem.*, 2010, **21**, 84–92.
81. X. Sun, W. Widdison, M. Mayo, S. Wilhelm, B. Leece, R. Chari, R. Singh and H. Erickson, *Bioconj. Chem.*, 2011, **22**, 728–735.
82. N. Fishkin, E. K. Maloney, R. V. J. Chari and R. Singh, *Chem. Commun.*, 2011, **47**, 10752–10754.

83. N. M. Okeley, J. B. Miyamoto, X. Zhang, R. J. Sanderson, D. R. Benjamin, E. L. Sievers, P. D. Senter and S. C. Alley, *Clin. Cancer Res.*, 2010, **16**, 888–897.
84. A. D. Baldwin and K. L. Kiick, *Bioconj. Chem.*, 2011, **22**, 1946–1953.
85. C. A. Boswell, E. E. Mundo, C. Zhang, D. Bumbaca, N. R. Valle, K. R. Kozak, A. Fourie, J. Chuh, N. Koppada, O. Saad, H. Gill, B.-Q. Shen, B. Rubinfeld, J. Tibbitts, S. Kaur, F.-P. Theil, P. J. Fielder, L. A. Khawli and K. Lin, *Bioconj. Chem.*, 2011, **22**, 1994–2004.
86. N. L. Jumbe, Y. Xin, D. D. Leipold, L. Crocker, D. Dugger, E. Mai, M. X. Sliwkowski, P. J. Fielder and J. Tibbitts, *J. Pharmacokinet. Pharmacodyn.*, 2010, **37**, 221–242.
87. R. A. Herbertson, N. C. Tebbutt, F.-T. Lee, D. J. MacFarlane, B. Chappell, N. Micallef, S.-T. Lee, T. Saunder, W. Hopkins, F. E. Smyth, D. K. Wyld, J. Bellen, D. S. Sonnichsen, M. W. Brechbiel, C. Murone and A. M. Scott, *Clin. Cancer Res.*, 2009, **15**, 6709–6715.
88. D. Bumbaca, C. A. Boswell, P. J. Fielder and L. A. Khawli, *AAPS J.*, 2012, **14**, 554–558.
89. K. Lin and J. Tibbitts, *Pharm. Res.*, 2012, **29**, 2354–2366.
90. B.-Q. Shen, D. Bumbaca, O. Saad, Q. Yue, C. V. Pastuskovas, S. C. Khojasteh, J. Tibbitts, S. Kaur, B. Wang, Y.-W. Chu, P. M. LoRusso and S. Girish, *Curr. Drug Metab.*, 2012, **13**, 901–910.
91. C. A. Boswell, D. B. Tesar, K. Mukhyala, F.-P. Theil, P. J. Fielder and L. A. Khawli, *Bioconj. Chem.*, 2010, **21**, 2153–2163.
92. T. Igawa, H. Tsunoda, T. Tachibana, A. Maeda, F. Mimoto, C. Moriyama, M. Nanami, Y. Sekimori, Y. Nabuchi, Y. Aso and K. Hattori, *Protein Eng., Des. Sel.*, 2010, **23**, 385–392.

CHAPTER 7

MicroRNA Biotherapeutics: Key Challenges from a Drug Development Perspective

STERGHIOS ATHANASIOS MOSCHOS

Department of Molecular and Applied Biosciences, School of Life Sciences, University of Westminster, 115 New Cavendish Street, London, W1W 6UW, UK
Email: s.moschos@westminster.ac.uk

7.1 Introduction

The description of the RNA interference (RNAi) phenomenon is widely considered to be one of the most disruptive scientific discoveries in biology and medicine. Initially proposed in *Caenorhabditis elegans* in 1998[1] and confirmed in *Homo sapiens* within 3 years,[2,3] RNAi is generally accepted to involve post-transcriptional, pre-translational inhibition of gene expression by double stranded RNA (dsRNA) molecules ~21 nucleotides (nt) long. Most life scientists are familiar with RNAi through the widely adopted technology of small interfering RNA (siRNA). The mode of action of siRNAs is commonly accepted to involve sequence-specific targeting and endonucleotic cleavage of mRNA strands.[4,5] Thus, with the knowledge of the mRNA sequence of a particular protein and through careful design, substantial and specific reduction of transcript and protein levels (also referred to as knock-down [KD]) can be achieved. Less commonly known, on the other hand, are the endogenous mediators of RNAi, termed microRNAs (miRNA).

RSC Drug Discovery Series No. 36
Biotherapeutics: Recent Developments using Chemical and Molecular Biology
Edited by Lyn H. Jones and Andrew J. McKnight

Published by the Royal Society of Chemistry, www.rsc.org

Like siRNAs, the mature form of the vast majority of these molecules also involves ~21 nt-long dsRNA, albeit with one or more sites of imperfect complementarity in the duplex (Figure 7.1). Moreover, both siRNAs and miRNAs load onto the same effector complexes.[6] However, in stark contrast to siRNAs, miRNA are generally accepted to drive translational repression at a network level through imperfect sequence recognition of multiple mRNA targets over their untranslated regions. In addition, these targets are not actively cleaved, but are rather sequestered from the translational machinery in the cytosol, to be eventually degraded via a relatively poorly understood mechanism.[7,8] Thus, whilst structurally and functionally very similar,

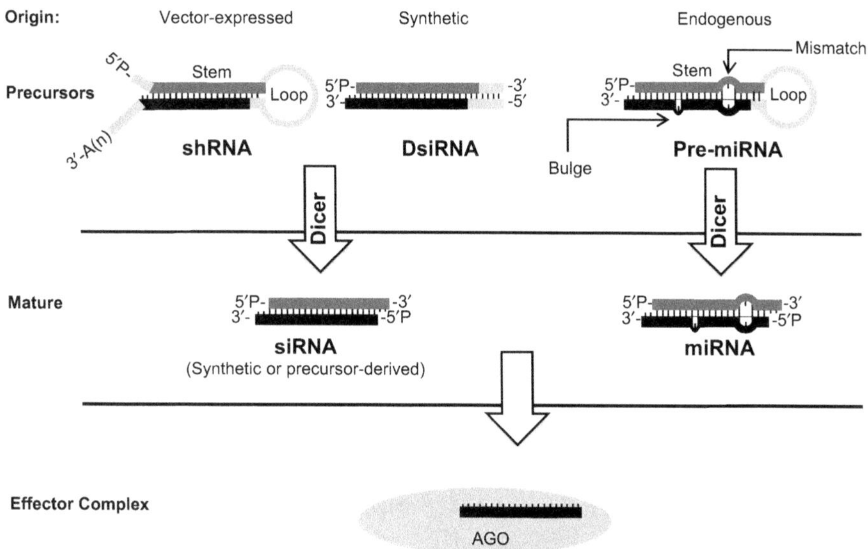

Figure 7.1 A structural comparison of mature siRNA and miRNA and shRNA. The endogenous (miRNA) and exogenous (siRNA) mediators of RNAi possess considerable similarities in structure and function. The main difference in synthetic siRNA structure to mature miRNA pertains to the almost universal presence of one or more mismatched bases or bulges in the miRNA stem. Like miRNA, shRNA are also processed from expressed hairpin precursors into mature siRNA by the cytosolic enzyme Dicer, which excises accessory sequences (light grey). Both siRNA and miRNA are loaded into an Argonaute protein (AGO) as double stranded RNA molecules, leading to the selection of the active or guide strand (black), which is antisense to the RNA target, and the removal of the passenger strand (dark grey). Some synthetic siRNA precursors like Dicer substrate siRNA (DsiRNA) or even long, synthetic RNA hairpins also rely on Dicer-mediated maturation to become active. shRNA can also be designed to adopt the structures of miRNA precursors (pre-miRNA). These miRNA scaffold systems use the same loop sequences as the parent miRNA, and replicate mismatches at the same positions.

siRNAs and miRNAs can be both used to address aberrant gene expression causal to disease.

Historically discovered first, siRNAs quickly became a versatile research tool as well as one of the most highly promising candidate therapeutic modalities. With the human genome sequencing project complete, the design of effective tool siRNAs quickly became a virtually bioinformatic exercise for any target of interest, including so-called undruggable genes (i.e. molecular targets with no obvious small molecule approach to intervention).[9] Moreover, simple, automated synthesis platforms allowed for rapid production and financially accessible, widespread use of siRNAs by the research community. The net result was a newfound capacity to complete functional *in vitro* assays within a few months of project initiation at a fraction of the cost previously associated to such work. Equally accelerated was also the progress of siRNAs from the research bench to the clinical bedside. In an unprecedented fashion for any novel therapeutic modality, the COBALT phase III siRNA trial against age-related macular degeneration was initiated within 9 years of the discovery of RNAi. However, this apparently robust success was to be cut short by a chain of publications implicating misinterpretation of off-target effects,[10] many of which were immune related,[11–14] and a lack of effective delivery to the appropriate cytosolic compartments.[12,14] The sum impact of these reports was a refocusing of discovery efforts to mitigating immunostimulation risks and the development of drug delivery solutions. Whilst chemical modification of siRNAs has been shown to circumvent some of their immunostimulatory potential,[15,16] delivery has yet to be conclusively resolved. Given the structural and physicochemical similarity of various synthetic miRNA mimics to siRNA, many of these limitations apply to miRNA replacement therapy too.

Interestingly, the same research group that published the conservation of the RNAi mechanism for synthetic RNAi,[2] also documented the conservation of the endogenous mediators of RNAi.[3] Nevertheless, interest in miRNAs bloomed only after their implication in cancer[17,18] and other diseases.[19] Moreover, research *in vivo* mainly assessed their utility as biomarkers[20] and potential diagnostics. Tellingly, successful *in vivo* manipulation of miRNA levels was achieved at the same time as the first in class siRNA clinical studies were being initiated.[21] Initial work examined the ablation of miRNA levels through the use of single-stranded antisense oligonucleotide (ASO) molecules. Since then, efforts have been led by investigators using a simple nucleoside modification called locked nucleic acid (LNA) aiming to inhibit miRNAs.[22,23] Presently, the most advanced applications of this technology appear to successfully accumulate in targeted tissues and achieve on-target activity in the absence of delivery systems.[24–26] These ASOs have been progressed to phase II trials with some setbacks around safety (SPC5001, http://www.clinicaltrials.gov ID: NCT01350960).[27] Notwithstanding the promise of miRNA depletion therapies, loss of miRNA expression has been proposed to be causative of many diseases,[19] especially cancer,[18,28] creating a clear case for attempting miRNA replacement therapy. Nevertheless, whether in the form of synthetic analogues or as a biotherapeutic modality, such a solution has yet to be attempted in man. In contrast, several diagnostic assays based on miRNA signatures have been commercialized (e.g. miRview® assays by Rossetta Genomics).[29]

A third class of RNAi mediators that straddles the siRNA and miRNA fields involves the molecular engineering of vector-expressed, short hairpin RNA (shRNA) duplexes, named so on account of their hairpin-like, stem-loop structure, which is highly reminiscent of miRNA precursors (Figure 7.1). Originally developed in the wake of the description of RNAi in man,[30,31] shRNAs are typically expressed either off plasmid DNA (*in vitro*) or recombinant viral vectors (primary cells and *in vivo*). Essentially a form of gene therapy, the RNA transcripts of shRNA systems utilize the microRNA maturation pathway to process shRNA into mature siRNAs.[32] Indeed, mimicry of microRNA precursor (pre-miRNA) structures, so called 'miRNA scaffolds', were proposed quickly after the development of the first systems to achieve siRNA expression *in vitro* and *in vivo*.[33] Thus, many lessons from the shRNA therapy field translate to miRNA biotherapeutics development. This was best exemplified when the most commonly used promoter system for both shRNA and ectopic miRNA expression was shown to overwhelm the miRNA maturation pathway: processing of the highly overexpressed shRNA transcript resulted in depletion of cellular miRNAs, and death.[34]

The 2012 European Medicines Authority approval of Glybera®, the first virus-based gene therapy marketed in the West, coupled to the penetrance of viral miRNA expression systems, indicates a potentially bright future for miRNA biotherapeutics. The three pillars of drug discovery require that a compelling drug candidate (i) must be exposed at the site of action at sufficient amounts to (ii) demonstrably engage it's target and (iii) result in a functional change in man.[35] In the age of intracellular molecular therapeutics R&D, such as miRNA biotherapeutics, 'site of action' no longer means tissue, but cell subtype and even subcellular compartment.[36] This chapter will therefore take a careful, critical look into key aspects of miRNA biosynthesis and function from a pharmacological perspective, and how these impact on biologic approaches to miRNA replacement therapy and inhibition. Although many aspects of RNAi are conserved across plants, lower animals and man, the information in this chapter will focus entirely on processes convincingly documented in *H. sapiens* and commonly used laboratory species. Emphasis will also be placed on interpreting findings and proposed approaches from a drug discovery and development perspective, presenting challenges, risks and opportunities. To achieve this, examples of the major accomplishments and mishaps across the many facets of the RNAi field will be used to highlight both best standards of practice as well as key knowledge gaps remaining to be addressed. Whilst the delivery approach, viral or otherwise, is expected to play a key role in the success of the anticipated solutions, this aspect will be covered only in the form of naturally occurring viral miRNA (vmiRNA), as potential recombinant hosts for exogenous miRNA expression.

7.2 MicroRNA Biogenesis

The location from which a mature miRNA originates tends to be overlooked; after all, most functional studies interrogate the properties of the mature miRNA, which is the potentially active component perceived as pharmacologically valuable. Nonetheless, over the years it has become clear that the genomic and transcriptomic organization of miRNAs is a much more subtle and important matter than previously anticipated. This is especially true when considering options around recombinant miRNA expression. Historically, mammalian miRNAs were described through small RNA cloning experiments, with little attention paid on the processing of the transcript. However, many miRNAs are organized in the genome in clusters of 2 or more miRNA (Figure 7.2A).[37–39] In addition, a significant fraction of these arises from 'junk' DNA regions, i.e. intergenic, non-protein-coding loci, which often feature vague transcriptional start sites.[40]

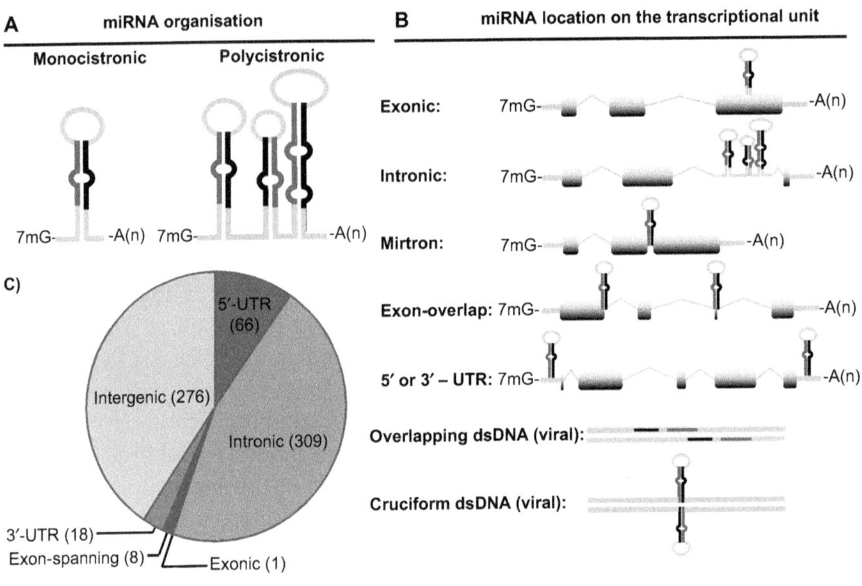

Figure 7.2 Genomic organisation of miRNA. (A) miRNA hairpins are organised in the genome individually or in clusters, giving rise to mono- or polycistronic transcripts (light grey lines). (B) miRNA can be located either in exons (grey boxes) or introns (light grey angled lines). Mirtrons make up the entire intron sequence and have so far been described only in protein-coding transcriptional units. Exon-overlap miRNA extend into, or span entire short exons. Some miRNA are also found in 5′ or 3′ UTR regions, though they tend to be classified as intergenic miRNA. Double-stranded DNA (dsDNA) viral genomes like those of the herpesvirus family can feature overlapping miRNA found on either DNA strand, or imperfect palindromes that organise into cruciform-like structures of 2 miRNAs. (C) Distribution of miRNA across UCSC known protein-coding genes in the human genome (assembly 18) according to miRGen.

Interrogation of the genomic context of a particular miRNA can be easily carried out using any of the popular genome browsers (e.g. NCBI Pubmed, Ensembl, or University of California at Santa Cruz (UCSC) Genome Browser) or dedicated miRNA genomic analysis tools such as miRGen (http://www.diana.pcbi.upenn.edu/miRGen.html)[38,39] and miRbase (http://www.mirbase.org).[41] The standard genomic annotation of miRNAs is in the format of a species designator, followed by the acronym miR and a number indicating it's order of discovery and annotation (e.g. hsa-miR-16 is the 16th miRNA annotated in *H. sapiens* – *'hsa'*). Although some databases might feature tools of varied interest and complexity, the reader is strongly advised to consider individual database characteristics, such as miRNA cluster binning parameters, genome assembly versions, use of predicted and/or experimentally validated miRNA sequences, automated or manual curation algorithms, etc. Moreover, transcriptional features, particularly splicing patterns, 5′ and 3′ untranslated region (UTR) sequences, are highly variable between databases and depend on tissue type, cell type and experimental conditions accepted by database curators.

Given transcript origin influences substantially the miRNA maturation process, this should not be ignored when engineering ectopic miRNA expression. Presently, endogenous mammalian RNA transcripts that are processed into miRNAs can be generally grouped in two ways: by transcriptional unit type and miRNA location on the transcriptional unit.[42,43] The major categorization depends on whether the transcript is protein-coding or not. The minor categorization pertains to whether the miRNA locus is located inside or outside an exon.

7.2.1 Genomic Organization of Endogenous miRNAs

Intergenic transcripts, also referred to as non-coding RNA (ncRNA) genes, are the hosts of a large fraction (>30%) of miRNAs and were the first class of miRNA to be ever described.[44] The precursors of these miRNAs, also known as primary miRNA (pri-miRNA), feature 5′ and 3′ end structures common to mRNA[45] and can contain either individual miRNAs or multiple miRNAs in clusters (Figure 7.2A).[46] Moreover, these non-coding transcripts can also be subject to splicing and the associated regulatory mechanisms, with the miRNA mono- or polycistron located either in the spliced-out (intronic miRNA) or the spliced-in (exonic miRNA) section of the transcript (Figure 7.2B).[42] Presently, it is assumed that the sequences immediately flanking miRNA precursors are generally non-functional and dispensable. Perplexingly, many of these are several kilobases long.

MicroRNA harboured within protein-coding genes (intragenic miRNA) are found principally within introns. As with intronic miRNAs processed from ncRNAs, this subset can be organized as a subcomponent of an intron hundreds to thousands of bases long, either mono- or polycistronically.[42,47] Unlike ncRNA, however, protein-coding genes can also harbour miRNA precursors that span an entire, miRNA-sized intron. These so-called mirtrons harbour only single miRNAs, and have so far been described exclusively within protein coding genes (Figure 7.2B).[48–50] Protein-coding genes can also harbour exonic miRNA. Interestingly, a small subset of these miRNAs span both introns and exons of a gene, or entire small exon sequences (exon-overlap miRNA).[38] As with exonic miRNA from non-coding transcriptional units, production of exonic miRNA from protein-coding genes can be dependent on host gene splicing.[42] Another unusual subset of miRNAs is processed out of UTRs of mRNAs either at the 5′ or the 3′ end of the transcript (Figure 7.2B,C).[51] However, many of these tend to be classed as intergenic miRNAs. A mIRGen summary of experimentally validated human miRNA and their genomic distribution with respect to UCSC genome browser known genes is provided in Figure 7.2C (accurate as of January 2013).

7.2.2 Virus-Encoded miRNAs

7.2.2.1 Herpesviruses

Regulatory RNAs, including miRNAs, have also been described in numerous viral genomes, particularly those of DNA viruses: release 19 of miRbase lists 436 putative mature vmiRNA sequences. These are of significant interest as they represent evolutionarily optimized solutions for engineering ectopic miRNA expression. The most notable examples can be found among the herpesvirus family, where most members are known to contain miRNAs.[52] The genomic organization of herpesvirus vmiRNAs reflects in many respects that of the human genome. Thus, Epstein-Barr Virus (EBV; 44 miRNA),[53–55] the best-studied herpesvirus representative, features three miRNA clusters in two protein-coding genes. These have been identified both computationally and experimentally from clinical samples and cell line studies.[55–57] Curiously, the loss of EBV loci apparently redundant under culture conditions prevented the detection of a large proportion of the EBV vmiRNAs,[52] highlighting genomic regions of EBV inappropriate for vector engineering. Dispensability of viral miRNAs has also been observed in murine viruses.[58] In contrast, differential detection of vmiRNAs between clinical[59] and experimental[60] setups has been reported in human cytomegalovirus (hCMV; 17 miRNA), despite minimal genetic drift in these loci.[60] hCMV is also interesting with respect to the genomic organization of vmiRNAs, given it features 12 precursors in both intragenic and intergenic regions, as well as introns.[61]

Other notable, miRNA-containing members of the herpesvirus family includes Kaposi Sarcoma-associated Herpes Virus (KHSV; also known as human herpesvirus 8; 25 miRNA),[55,62–66] the neurotropic Herpes Simplex Virus (HSV) 1 and 2 (26 and 24 miRNA respectively),[67–69] and Herpes B Virus (BV; 3 mature miRNA).[70,71] Like EBV, KSHV is characterized by encoding vmiRNAs intronically, but also features an exonic vmiRNA, miR-K10, which is part of the kaposin protein open reading frame (Figure 7.2C).[65] Another feature presently exclusive to KSHV is the presence of overlapping miRNA transcripts among the sense and antisense strands of the dsDNA viral genome (Figure 7.2B).[66,72] Likewise, many of the miRNAs unique to BV are found in regions not conserved in other herpesviruses, and is characteristic of long sequence repeats[71], as with some human intergenic miRNAs. Notably, of these vmiRNAs, many were convincingly shown to package into infectious virions in their mature format, with interesting implications for engineering mature miRNA delivery.[71] The drivers behind mature miRNA inclusion in virion packaging are presently unknown. In HSV, on the other hand, there appear to be cruciform-like vmiRNAs encoded in large palindrome-like sequences (Figure 7.2B),[68,73] some of which are located highly proximal to the viral origin of replication.[68] Thus, despite their evolutionary relationship, the genomic organization of vmiRNAs in herpesvirus appears to be quite divergent and relates to differential tissue specificity,[68,71] presenting interest opportunities for tailored ectopic miRNA expression beyond virion packaging of mature miRNAs.

7.2.2.2 Adenovirus

Original reports on the impact of Adenovirus on RNAi, implicated the highly expressed, virus-associated RNA transcript I (VAI) in blockade of the canonical miRNA maturation pathway.[74] However, it was later shown that this highly structured RNA can be processed into small RNAs, associate with functional components of the RNAi pathway, and enact RNAi in both a reporter system[75] and against specific endogenous targets.[76] Thus, it is debatable if Adenovirus blocks or hijacks miRNA biogenesis (or potentially both) to promote vmiRNA production and the viral life cycle. Curiously, the transcriptional control of VAI impacts substantially on vmiRNA maturation. Similarly, KHSV features two promoters that regulate the expression of one 12-vmiRNA cluster. The primary promoter yields a 12-hairpin pri-miRNA. During the productive phase of the viral life cycle the secondary promoter, which is located upstream of the last two vmiRNAs in the cluster, is also activated generating a secondary 2-hairpin pri-miRNA, enhancing the production of these two vmiRNAs exclusively.[63,77]

7.2.2.3 RNA Viruses: an Unlikely, yet Flexible Group of Vectors

Unlike DNA viruses, RNA viruses appear at first instance[78] unlikely candidates for hosting miRNA genes. After all, why would an RNA virus encode a molecule that, if loaded onto RISC, could potentially block viral expression or destroy the viral genome? West Nile Virus is such a +RNA virus that has been proposed to encode miRNAs in 3′-UTRs,[79] though studies in mammalian cells and with native viral genomes is presently lacking. Moreover, emerging evidence indicates potentially alternative roles for this viral ncRNA.[80] Retroviruses feature dsRNA genomes and are primarily represented in research by Human Immunodeficiency Virus (HIV)-1, which allegedly contains 4 vmiRNA.[81,82] However, the veracity of these vmiRNAs remains a controversial subject.[83–86] Detection of some putative HIV-1 vmiRNAs has been independently reproduced only from *in vivo* samples,[81,87] proposed to arise from the processing of the cis-acting, transactivation response (TAR) element, a ncRNA structure responsible for HIV-1 expression regulation. However, these putative vmiRNAs are comparatively shorter than most endogenous miRNAs[87] and are found in low yields.[88] Thus, a statistical analysis of their average per-cell prevalence compared to the endogenous miRNA pool, coupled to the assumption that more miRNA means stronger potential for function,[86] has challenged their biological significance.[85,86] Given that, like with other viruses,[89] the vast majority of T cells *in vivo* do not harbour HIV-1 viral genomes,[90] the vmiRNA signal would be expected to be dilute relative to the endogenous miRNA pool. Moreover, most viruses, including HIV, can have vastly different expression kinetics for small RNAs among cell types, let alone *in vivo*.[91,92]

In addition, like other reports on vmiRNA abundance,[93] our own studies on virus-encoded shRNA maturation have shown that more does not necessarily mean better in terms of RNAi mediator bioactivity (see section 7.4). Further investigation of the RNAi potential of ncRNA transcripts of HIV-1 is therefore relevant to developing alternative approaches to ectopic miRNA expression engineering. In support of this tenet is a report that retroviral vmiRNAs are possible, as such molecules have been described in bovine leukemia virus.[94] Curiously, this bovine retrovirus mimics a well-described oncogenic endogenous miRNA, underpinning the premise that endogenous miRNAs can be expressed off viral vectors naturally hosting vmiRNAs. Similarly, small viral RNAs encoded in hepatitis C virus (HCV), highly reminiscent of miRNAs, have been shown to associate with RNAi effector proteins.[93]

Elsewhere, laboratory efforts to introduce miRNA expression cassettes in RNA viruses have been reported as successful in a number of examples both *in vitro*[95–98] and *in vivo*.[99] Thus, an EBV vmiRNA introduced into the 3′ non-coding region of the tick-born encephalitis virus (TBEV) +RNA genome was found to be well tolerated by the vector *in vitro* and not to be subjected to rapid excision or mutation.[97] Similarly, Sindbis Virus (SV) was reported to tolerate well the subcloning of an entire murine pri-miRNA in its own 3′ non-coding region as well as allow it's function, albeit somewhat reducing viral genome levels.[98] Influneza A and vesicular stomatitis virus (VSV), on the other hand, were shown to tolerate and allow the function of the same pri-miRNA as SV, this time in an intronic[86] and intragenic[100] context, respectively, and without impact on virus turnover. The biological mechanisms driving differential tolerance of ectopic miRNA expression off RNA vectors is presently unknown.

7.2.3 Canonical miRNA Maturation

Given the vast diversity of miRNA location in the genome, it should come as no surprise that endogenous miRNAs are transcribed by both RNA polymerase II[101] and III. The latter, however, has only been associated to the biogenesis of miRNA loci in the context of Alu repeats,[102] and in the context of some vmiRNAs.[74–76] Overall, miRNA transcription is mediated in the same fashion as other RNA transcripts and is driven by many commonly studied promoter and transcription factor systems.

7.2.3.1 Excision of miRNA Precursors from Primary Transcripts

Post-transcriptionally, the most well-studied pathway in miRNA biogenesis is a minimally 7-step, Ran guanosine triphosphate (RanGTP)-dependent process (Figure 7.3). After transcription, *in silico* simulations suggest that miRNA-containing transcripts assume secondary structures featuring the stem-loop

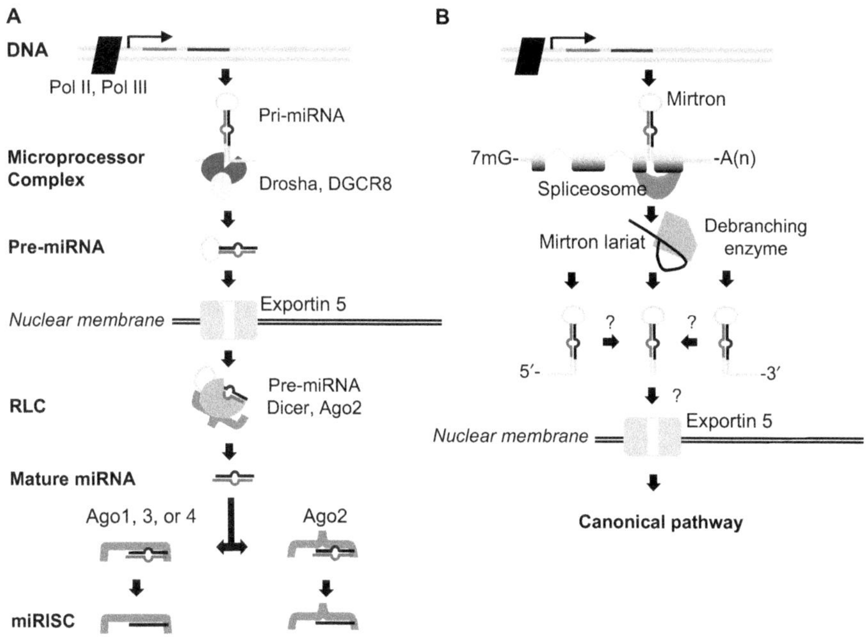

Figure 7.3 Main miRNA Biogenesis Pathways. (A) The canonical miRNA biogenesis pathway is a minimally 7-step process involving pre-miRNA excision from pri-miRNA transcripts by the microprocessor complex. These are exported into the cytosol by the karyopherin Exportin 5, leading to the loading of the pre-miRNA onto the RISC-loading complex (RLC). The RLC component Dicer eliminates the loop section, whereupon the duplex is loaded into one of four Argonaute proteins. Only Ago2 features a slicer function. After elimination of the passenger strand the RNAi-bioactive miRNA-induced silencing complex (miRISC) is formed. (B) Mirtrons are matured via the action of the spliceosome, which yields mature mRNA and an excised intron lariat containing the mirtron. The debranching enzyme subsequently releases the intron, which is presumed to form a pre-miRNA-like structure that might feature a 5′ or 3′ tail originating from the splice junction sequence. After the tail is trimmed, the mirtron is matured into a pre-miRNA structure, which is assumed to cross into the cytosol and enter the canonical miRNA maturation pathway by engaging with Exportin 5.

regions destined to become mature miRNAs. These primary miRNA-containing transcripts (pri-miRNA) are recognized by the so-called micro-processor complex,[103–105] which removes the miRNA-containing stem-loop hairpins to generate miRNA precursors (pre-miRNA).

The minimal components of the microprocessor complex are the class 2 RNA endonuclease III enzyme Drosha, which carries out pre-miRNA excision, and the dsRNA binding protein DiGeorge syndrome critical region gene 8 (DGCR8). Drosha yields stem-loop hairpins of ~70 nt length featuring 5′ phosphate groups and 3′ dinucleotide overhangs characteristic of RNAse III action.[46] Key components in the pre-miRNA excision process are the 'accessory' ssRNA sequences surrounding the pre-miRNA, at what is termed the single-stranded RNA (ssRNA)-double stranded RNA (ssRNA-dsRNA) junction.[106,107] Thus, mechanistic *in vitro* studies in a commonly observed stem-loop assembly indicated that the loop structure and sequence are not crucial, but structural alteration of the 'accessory' sequences either side of the pre-miRNA could abolish pre-miRNA excision. In addition, the length of the single-stranded sections in these accessory sequences was shown to be crucial in directing the Drosha cleavage site some 11 bp from the ssRNA-dsRNA junction into the hairpin stem.[106] More recently, however, it was observed that the loop sequence impacts on Drosha processing efficiency in at least one, highly conserved miRNA.[108] Whether these rules apply to other pre-miRNA structures has not been elucidated.

7.2.3.2 *Nuclear Export of Pre-miRNAs*

The pre-miRNAs are then transferred to Exportin 5 (EXP5), which is sensitive to the size (>16 nt) and structure[109] of the Drosha-generated ends and, in a RanGTP-dependent manner, translocates the pre-miRNAs from the nucleus into the cytosol.[110,111] The same karyopherin is also responsible for nuclear export of recombinantly expressed shRNAs.[111] The exact mechanics of EXP5 function, as well as the hand-over processes between EXP5 and it's upstream (microprocessor) and downstream partners have yet to be defined in man. Moreover, whether pre-miRNA transport across the nuclear membrane is a one-way or a two-way process is presently unclear. This question arises from studies on ectopic vmiRNA expression in RNA vectors that should function exclusively cytosolically. Thus, in some cases, their maturation was found to be Drosha-dependent, indicating that pre-miRNAs might (re-)enter the nucleus.[97]

7.2.3.3 *Removal of the Hairpin Loop*

Upon pre-miRNA translocation into the cytosol, the class 3 RNA endonuclease III enzyme Dicer recognizes the 3′ dinucleotide overhang of the pre-miRNA hairpin and directs cleavage of some 20–22 nucleotides from this end, thereby eliminating the loop sequence and generating an RNA duplex with 5′ phosphate groups and 3′ dinucleotide overhangs at both ends.[112] This process has been shown to be ATP-independent *in vitro*,[113,114] and to involve an

~0.5 MDa macromolecular protein complex[115] comprised of Argonaute2 (Ago2),[113,114] the HIV-1 TAR RNA-binding protein (TRBP)[114,116] and PACT.[117] Although PACT[117] and TRBP[118] are not essential to Dicer processing of pre-miRNAs, participation of PACT greatly enhances Dicer function.[117] Recent data suggests that the loop sequence might affect Dicer processing of pre-miRNAs, in a similar fashion to the way it affects Drosha.[108] Beyond these roles, it is presently unknown if the miRNA loop sequence engages in other biological functions.

7.2.3.4 Selection and Loading to Effector Protein Complexes

At this stage, mature miRNAs have been generated. Differing to siRNA's only on account of the mismatched bases (Figure 7.1), these duplexes also feature two potentially active strands. The two strands are loaded into one of four Ago proteins found in the human genome. Whether and how exactly miRNAs are specifically sorted into each of the four human Argonaute proteins is still a subject of debate.[6] This is complicated by the almost complete lack of differential expression data between tissues and cell types for the four Argonautes, let alone their involvement in disease.[119] Early work indicated that unlike lower organisms, mammalian species did not sort miRNA into specific Argonautes.[120–123] More recently, however, a few examples involving mature miRNA strands processed out of the same pre-miRNA, but with minor differences on their 5′ or 3′ ends due to RNA editing or other processing mechanisms (also known as isomiR's), have been put forward as examples of differential and preferential sorting.[122,124,125] In mice, the absence of mismatches has been shown to direct loading to Ago1 and 2 as opposed to Ago3 and 4, which exhibit preference towards miRNA-like structures.[126] In human cell line studies, this has not been observed.[127] Whether this differential preference is ubiquitous between cell types of primary or immortalized nature and similarly conserved in man has yet to be conclusively reported; however, it remains a crucial knowledge gap given that a host of miRNA functional validation studies rely on fully base-paired, siRNA-like miRNA mimics assessed *in vitro* and in rodents.[18,128]

Cell-free studies have suggested Dicer does not participate in loading siRNA duplexes onto Argonautes,[113] though later work indicated that Dicer, in conjunction with TRBP, interacts with exogenously delivered siRNAs to, at the very least, assess their 5′ end thermodynamic stability.[129] Whether this interaction is retained with pre-miRNAs is still unclear. In accordance, more recent cell culture experiments have indicated that Dicer is a dispensable,[130] yet useful component in selecting efficacious molecules for Ago2 loading, sensitive to the sequence, thermodynamic and 3′ end features of the duplexes.[131] Argonaute 2, on the other hand, exhibits substantially lower (×30) K_D values for adenosine and uracil compared to cytosine and guanine, a feature naturally selected for in endogenous miRNA sequences.[132] Moreover, some evidence indicates that ATP might play a contributing factor to the process.[133] Alternatively, a synthetic siRNA duplex can be engineered to function exclusively in a Dicer-dependent manner, through use of polarized 3′ overhang design (left or right end of the duplex) and judicious placement of deoxynucleotides in the antipodal blunt end (Figure 7.1).[134,135] These designs are closer to the experimental evidence of Dicer involvement in duplex thermodynamic recognition.[129] Examination of the utility of D-siRNA scaffolds for miRNA mimicry has yet to be described. Similarly, data on D-siRNA Argonaute preference is also lacking. In the case of shRNAs, association with all four Argonaute members has been reported, albeit only in murine studies.[136] In contrast to synthetic siRNAs,[126] preferential loading of shRNA-derived siRNAs to Ago members has not been determined.

7.2.3.5 Guide Strand Selection

After duplex loading onto an Argonaute protein, one of the two strands is selected to become biologically active, driving the ensuing RNAi mechanism. This strand is commonly referred to as the guide strand (also as antisense strand for siRNA's, with reference to the targeted RNA being 'sense'). This strand, in conjunction with the Argonaute protein that is loaded on, forms the final, effector RNAi complex: the RNA-Induced Silencing Complex (RISC; siRISC if loaded with siRNA or miRISC if loaded with miRNA). Guide strand selection has been long considered to be a process dependent on the relative stability of the two 5′ ends of the duplex, on account of bioinformatics analysis of miRNA thermodynamic profiles, *in vitro* biochemical assays and siRNA efficacy studies in cell culture.[127,137,138] Alternatively, non-permissive chemical end-modification can also be used to drive strand selection.[16] Of note, the process has been shown to differ considerably between species, even for the same, conserved miRNAs.[139] Moreover, increasing numbers of observational reports on human disease are further challenging the accepted models of strand selection.[140] However, a 5′ phosphate has been elegantly demonstrated to be a requirement for efficient miRNA guide strand loading onto at least Ago1.[141] The other strand, referred to as the passenger strand or the star strand (designated by a star symbol (*)), is released and/or degraded. However, it is now clear that many strands presumed to be of passenger nature, can indeed form bioactive guide strands.

The precise fate of the strand not loaded onto RISC depends on the Argonaute protein in the particular RISC complex. Thus, Ago2 is the only member of the mammalian Argonaute family to feature an endonuclease activity, also referred to as 'slicer', directed against RNAi targets or, in this case, the passenger strand. As a result, Ago2-loaded miRNA or siRNA duplexes direct the cleavage of the passenger strand at the phosphodiester bond between the nucleotides opposite positions 10–11 from the 5′ end of the guide strand, yielding a 9 nt 5′ fragment and a 12 nt 3′ fragment. This event, determined *in vitro*[4] and in cells,[5] requires complementarity with the passenger strand over the slicer active site. Once the passenger strand is cleaved, the component 3 promoter of RISC (C3PO) complex engages with RISC to proactively remove the cleaved passenger strand in an Mg^{2+}-dependent, endonucleotic manner.[142,143] Importantly, both the slicer and C3PO functions have been shown to require scissile phosphodiester bonds on the passenger strand to maintain appreciable efficacy. These findings pose interesting challenges to synthetic siRNA and miRNA mimic strategies involving heavily stabilizing chemical modifications designed to prevent phosphodiester backbone breakdown. In contrast, passenger strand cleavage does not occur with Ago1-, 3- and 4-loaded duplexes because the slicer activity is absent. Thus, as with heavily modified siRNAs loaded onto Ago2, the removal of passenger strands from Ago1, 3 and 4 appears to involve a slicer-independent, low turnover, strand unwinding mechanism.[4] In this case, the presence of mismatches has been shown to improve the unwinding rate of the Ago-loaded duplex and, in contrast to the 'common' miRNA geometry featuring a central mismatch, unwinding efficiency is improved if the mismatch is proximal to the ends of the duplex.[133,144] In fact, the structural elements shown in cell-free studies to favour Dicer loading and Argonaute loading appear to be mutually opposing. Moreover, limited evidence indicates that ATP does not participate in the unwinding process.[133]

7.2.4 EXP5 and Safety of Ectopic RNAi Mediator Expression

Studies initially performed on shRNA therapeutics *in vivo* and subsequently on adenovirus *in vitro* have shown that one of the most crucial steps in the canonical miRNA maturation pathway is nuclear export of expressed RNAi precursors. Thus, lethality was observed in both mice[34] and non-human primates[145] on account of powerful promoters driving shRNA expression in the liver. Toxicity was also reported in similar studies attempting shRNA expression in the rat brain[146] and dog heart.[147] In all cases, perturbation of tissue-specific miRNA homeostasis and miRNA-controlled gene dysregulation was documented under toxic shRNA expression levels, which may be alleviated through the use of either less powerful[148] or mutation-tuned promoters.[145] The culprit for this unexpected risk was found to be the core protein involved in both miRNA[111] and t-RNA[149] nuclear export, EXP5. In the initial report[34] overexpression of EXP5 alleviated the shRNA-induced lethality *in vivo*, whereas elsewhere[150] EXP5 saturation was shown to also negatively impact on Dicer protein levels through disrupted interaction between EXP5 and Dicer mRNA.

Given Dicer also processes other structured endogenous transcripts,[151] overexpressed hairpins can have a diverse and compounding negative effect on RNA homeostasis. Notably, the highly overexpressed ($>10^8$copies) VAI non-coding transcript of Adenovirus has been demonstrated to engage EXP5[152] and Dicer,[74] disrupting ectopic miRNA maturation and Dicer function, even when introduced at lower levels than those encountered in infection.[74] Yet down-tuning levels of shRNA expression comparable to those of VAI in the mouse and non-human primate liver, though not directly shown to rescue competition with endogenous miRNA for EXP5, also alleviated lethality.[145] Notwithstanding the differences in the models used in these studies, there appears to be a yet undescribed threshold of competitive binding for EXP5 below which ectopic hairpin expression might indeed be safe. Given the impact of promoter strength in this process, tissue and perhaps *in vivo* cell type-specific assessment[14] of EXP5 levels and competition tolerance are strongly advised; after all EXP5 levels are not ubiquitous across tissues. Thus, comparatively low levels of EXP5 have been documented in the mouse brain,[153] challenging the promise of miRNA and shRNA biotherapeutics in neurological disease. Of note, at least one report involving ectopic miRNA expression off an RNA vector claims to achieve ectopic miRNA maturation and function at levels comparable to those of endogenous miRNAs without impact on their turnover or efficacy.[100] Nevertheless, this platform also yields unexpectedly high levels of apparently bioactive passenger strand, which normally is suppressed.

7.2.5 Alternative Maturation Pathways

For many years miRNA biogenesis was believed to occur strictly in accordance with the canonical biogenesis pathway. Notably, many parameters of the process were determined using highly artificial approaches, such as cell-free systems, ectopic miRNA expression, polyploid cell lines (e.g. HeLa), modified proteins or through minimal validation studies in human cell lines after mechanism scrutiny in other organisms (e.g. EXP5 and Dicer).[46,111–113,130,133,138,143,154] Moreover, the so-called dogma of RNAi was established using only a handful of miRNAs.[46,111,112] Despite the distance of these systems from molecular physiology, the definition of various such 'rules' of RNAi continues to be developed in a similar fashion.[10] However, with the advent of more advanced techniques such as next generation sequencing, and the realization that many details differ between species,[139] the biochemistry of the RNAi world is becoming increasingly clearer. Through observation of biochemical events under physiologically relevant conditions, or indeed primary/tissue sample analysis rather than highly artificial experimental setups, an ever-increasing list of alternative maturation pathways is being continuously described. The following examples serve as a reminder that the RNAi field has repeatedly suffered extrapolations between systems with minimal cross-validation, and substantial impact on therapeutics development.

7.2.5.1 Mirtrons and Simtrons

Initially described in invertebrate studies[155,156] involving small RNA next generation sequencing, mirtrons are a subfamily of miRNAs that arise from comparatively short intronic sequences (Figure 7.2B). Unlike pri-miRNAs, these do not feature long stems cleaved by the microprocessor complex some 11 nt from the dsRNA-ssRNA junction. Rather, the spliceosome is responsible for their production along with the lariat debranching enzyme, which confers to mirtrons 5′ monophosphates and 3′ hydroxyls.[156] On the basis of these invertebrate studies, mirtrons are also commonly believed to have unique maturation fates in that the 3′ arm of the hairpin is generally favoured to harbour the guide strand, on account of splicing reaction constrains on the RNA sequence. Thus, sequence permitting, they should present an opportune system in directing guide strand selection. The remaining biogenesis pathway is still EXP5/Dicer dependent (Figure 7.3B).

In mammals, the existence of mirtrons has been validated in man, chimpanzee, macaque and laboratory rodent species in a study that also documented evolutionarily distinct mirtron subsets to invertebrate organisms.[48] Although the fraction of mirtrons was initially thought to be <1% in the human miRNA repertoire,[48] experimentally validated mirtrons are now considered to consist at least 12% of the human miRnome.[157] The most important finding of these surveys was also a stark contrast to the strand selection mechanism found in invertebrates; highly expressed mammalian mirtrons produce an excess of guide strands from the 5′ arm of the hairpin, though 3′-arm guide strands are also generated. Crucially, bioinformatic prediction of strand preference was also contrasted by observational data,[48] highlighting that such rules are, at best, limited in their scope. Present understanding of mirtron organization in mammals is sub-categorized into whether these feature 5′ or 3′ tails connecting the hairpin to the splice donor or acceptor signal,[157] or if their excision from primary transcripts is both microprocessor and spliceosome-independent (simtrons).[158] Although the enzymatic processing of 5′ and 3′ tails is elucidated in invertebrates, this remains to be determined in man. On the other hand, indirect evidence in human cell studies indicates that EXP5 might not be responsible for the translocation of all mirtrons into the cytosol, as modulation of free EXP5 levels does not universally affect mature mirtron guide strand levels.[159] Importantly, not all genome or miRNA databases are up to date with these subcategorisations.

Simtrons (Figure 7.4) are notable exceptions in that their bioprocessing appears to be Drosha-dependent but DGCR8, EXP5 and Ago2-independent.[158] These conclusions were derived through ectopic simtron expression studies in human cell lines or mouse embryonic stem cells featuring RNAi-mediated protein KD or gene knock-outs (KO). Moreover, for at least two of the simtrons tested, the flanking exons appeared to have no influence on simtron maturation, indicating the intronic sequences directed the bioprocessing fate of the hairpin. In terms of function, mature simtrons were co-immuniprecipated with all four human Argonautes in overexpression studies. Therefore, if these data are corroborated in studies without ectopic simtron expression, this class of miRNA might represent an entirely EXP5-independent means of miRNA expression.

At least one viral genome has been also proposed to express Drosha-independent miRNA precursors (Figure 7.4).[160] Ectopic expression studies indicate that these arise off a Pol III promoter transcript encoding for a tRNA; the processing of the tRNA precursor by tRNaze Z also yields a pre-miRNA. Moreover, the transcript is believed to be tolerant to various tRNA structures, miRNA or shRNA sequences, making it an interesting framework for ectopic miRNA expression. These findings are engaging as they are corroborative to

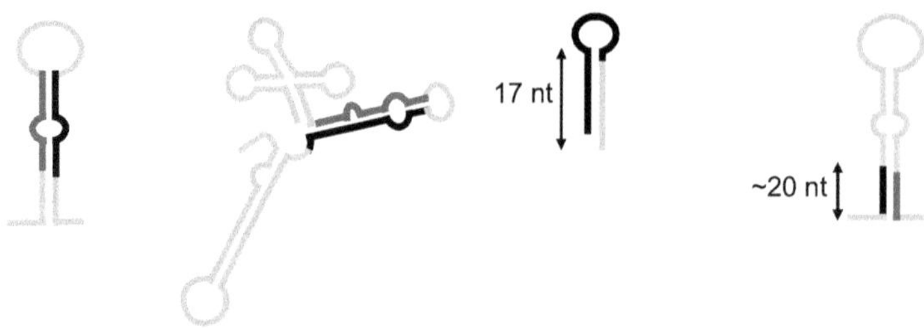

	Simtrons	Murine γ Herpesvirus 68 miR-m1-7	Mammalian miR-451	Mammalian moRNA
Spliceosome	–	–	–	–
Drosha	+	–	+	+
tRNAse Z	–	+	–	–
Exportin 5	–	?	–	?
Dicer	–	?	–	?
Ago2	–	?	+	?
RNAi	✔	✔	✔	?

Figure 7.4 Unusual miRNA and their biogenesis determinants. Key features of unique miRNAs or miRNA subsets are indicated, as well as the proteins involved in their maturation, where evidence towards this has been accrued.

previously described RNAi-acting tRNA::shRNA fusion constructs.[161] Whether these sidestep the EXP5 system or present associated cytotoxicity risks has not been elaborated.

7.2.5.2 miRNAs Encoded in Small Nucleolar RNAs

On the basis of deep sequencing data, it has been proposed that endogenous tRNAs and other small nucleolar RNAs (snoRNA) in man might also give rise to miRNAs.[161] Many of these propositions feature computational RNA 2ry structure predictions that are highly reminiscent of pri-miRNA transcript organization.[162,163] Moreover, some examples also indicate that components of the RNAi pathway such as Dicer might be needed for their maturation, alluding to engagement with the RNAi pathway,[164] a potentially evolutionarily ancient feature.[165] Although some of these studies benefit from performing deep sequencing after immunoprecipitation of Argonaute proteins, they do not examine the subcellular fate of the mature putative guide strands. Thus, though a pre-miRNA-like maturation process might engage these transcripts, and elaborate ectopic expression system evidence might allude to an RNAi function, the presently accumulated body of evidence remains unclear as to their actual roles in cell biology. Given the RISC complex has been implicated in epigenetic regulatory pathways in the nucleus, elucidating the functions of these RNAs is even more so challenging.

7.2.5.3 miR-451

The critical role of Ago2 in embryogenesis and development has been well described in Ago2-null rodent studies. By developing heterozygous mice featuring a slicer inactive Ago2 allele,[166] homozygous Ago2 slicer negative mice were shown to survive to post-gestational stages of development, but to die soon thereafter with overt signs of anaemia. This was found to be on account of arrested erythroid progenitor maturation in the foetal liver on account of maturation impairment of a single miRNA: miR-451. Unlike other miRNAs, pre-miR-451 has a 17 nt stem which is non-permissive to Dicer processing and features a 5′ arm-derived guide strand that extends into the tetranucleotide hairpin loop and possibly into the first two nucleotides of the 3′ arm of the hairpin. Moreover, pre-miR-451 is unique in exhibiting a perfect stem structure, i.e. it is highly reminiscent of an shRNA. Ectopic overexpression of miR-451 in various conditional mutant cell lines[166–168] resolved it's maturation to be a Drosha-dependent, Dicer-independent process. In the homozygous Ago2 slicer negative animal livers, the miR-451 precursor was found to co-precipitate with Ago2. In addition, *in vitro* biochemical assays indicated pre-miR-451 maturation was dependent on Ago2 slicer activity. Moreover, sequence modification to express other miRNAs or shRNAs without interruption of the structural elements of mir-451 indicated that direction of the precursor to this Ago2-dependent maturation pathway was not a sequence-dependent phenomenon, and of use in directing potentially strand-selective ectopic miRNA expression.[166]

However, evidence towards this has been criticized as circumstantial due to the lack of explicit proof for Dicer-independent maturation for these constructs.[169] More recently, detailed investigation of the structural features that drive miR-451 biogenesis have indicated that (i) a single mismatch close to the centre of the stem results in increased processing of the mature guide strand, (ii) Argonaute stoichiometry can affect pre-miR-451 maturation, (iii) G–C content in the pri-miRNA influences guide strand maturation in a cell type-specific manner and (iv) a pre-miR stem length of 17 nt decides Dicer-independent maturation.[170] Thus, whilst clear opportunities might exist in strand-selective shRNA hairpin design, extrapolations to other shRNA cassettes are ill advised. It is presently assumed that miR-451 matures in an EXP5-dependent manner, and it is the only mammalian miRNA that uses this biogenesis pathway. Limited studies in a single cell line indicate that this might not be the case after all.[151]

7.2.5.4 miRNA Offset RNAs

A recent attempt to characterize the miRNA gamut of the simple chordate *Ciona intestinalis* revealed that ~20 nt long ssRNA were generated in a controlled fashion from the stem sequences flanking a pre-miRNA: these were termed as miRNA offset RNA's (moRNA).[171] Re-examination of small RNA library reads generated from human prefrontal cortex samples indicated that moRNAs might indeed be found in man too: the moRNAs originating from the flanking sequences of at least one miRNA (miR-219) appear to be conserved between *C. intestinalis* and man, although it appears to be expressed in relatively small quantities.[172] MoRNAs have been also documented *in vitro*,[173,174] in murine brain samples,[175] in human tumours,[176] in herpesvirus genomes,[66,68,177] and even in *in vitro* transfected PCR amplicons encoding miRNA hairpins purported to direct self transcription without the need for a promoter.[178] Evidence from mice indicates their maturation is generally reflective of miRNA guide strand selection,[175] though it is unclear whether the moRNA or the miRNA drives the process. The mechanism of moRNA biogenesis, cellular distribution, precise function and putative engagement with RNAi components is presently unclear, but of capital scientific interest. It is expected that as a minimum, a substantial role will be described for moRNAs in miRNA biogenesis, if not gene expression regulation.

7.2.5.5 Viral Lessons on RNAi Precursor Maturation

Another as yet uncharacterized route of miRNA maturation has been proposed from experiments using cytosolic RNA viral vectors. Studies on TBEV indicated that ectopic vmiRNA expression might involve a nuclear translocation step during maturation, requiring Drosha activity.[97] On the other hand, the processing of ectopic pri-miRNA encoded in Sindbis virus was shown to be a Dicer-dependent, DGCR8- and EXP5-independent, i.e. essentially cytosolic, process.[98] In follow-up studies with recombinant vesicular stomatitis virus, the unexpected generation of canonically eliminated passenger strand was also put forward as evidence of alternative maturation processes. One interpretation of these findings pertains to nuclear membrane breakdown during cell division under culture conditions.[86] Alternatively, these data might be evidence of recruitment of the microprocessor components to the cytoplasm.[100] A simpler interpretation would involve residual microprocessor activity in the cytosol before nuclear translocation of the proteins after synthesis. This explanation would require either a high, or a preferential Drosha turnover rate for these pri-miRNAs in the cytoplasm. Unfortunately, the exact enzyme kinetics of Drosha and it's sequence/structural motif preferences have yet to be determined. As for the higher prevalence of the ectopic passenger strand in transduced cells, this might relate to the entirely out-of-context expression of the ectopic miRNA under investigation: identical strand selection between different cell types is a commonly held assumption. In the specific example,[100] a neuronal miRNA was expressed in diverse non-neuronal cell types and tissues both *in vitro* and *in vivo*.

On the other hand, one curious yet highly instructive intricacy in miRNA maturation has been reported in the context of Adenoviral vmiRNAs. Thus, the processing of VAI-encoded hairpins is directed by the nature of the primary transcript. VAI has been long known to feature two transcription initiation sites, only 3 nucleotides apart.[179] Given Dicer processes substrates using the hairpin 3′ end as a guide,[112] one would expect such a difference to be inconsequential. Nonetheless, the two VAI transcripts, differing in length by 3 nt's at their 5′ end alone, give rise to two separate 5′ vmiRNA strands incorporated onto RISC, also differing in length at their 5′ end by 3 nucleotides. This feature was shown to affect both RISC incorporation efficiency and RISC function against an artificial target, probably due to the lack of a 5′ monophosphate in the shorter transcript. The 3′ strand of the vmiRNA, although consistent between both transcriptional variants, and more efficiently assembled onto RISC (>200-fold versus the 5′ variants), was shown to be ×20 less active than the 5′ strand.[93] Studies on Adenovirus vmiRNA maturation have also shown that the relative abundance of miRNA precursors does not direct mature miRISC assembly. Thus, the VAII transcript, which also appears to encode vmiRNAs, accumulates at levels ×20 lower than VAI, but results in twice as many vmiRNA-loaded RISC complexes.[180] Collectively, these data underscore the importance of accurate mapping of transcriptional start sites in ectopic shRNA or miRNA expression cassettes, the value of assessing 5′ end chemistry in maturation variants, and dispel the commonly held assumption[86] that more precursors or mature strands mean better or more RNAi.

7.2.6 Post-Transcriptional Processing of miRNAs

One of the most frequently sidelined yet of cornerstone therapeutic importance aspects of miRNA biosynthesis is the process of miRNA modification. As the name alludes, increasing evidence has amassed from both observational and biochemical studies to suggest that individual miRNAs arising from a single transcriptional unit might undergo considerable processing during their maturation. The net result is the production of a cohort of mature guide and passenger strands of differing 5′ starts and 3′ ends from a single pre-miR hairpin, also known as isomiRs. Importantly, this process is not to be confused with miRNA paralogues, i.e. mature guide strands arising from distinct genomic loci but with identical, or highly similar sequences (e.g. the miR-146 family, consisting of miR-146a, which originates from chromosome 5, and miR-146b, which originates from chromosome 10, differ only by 2 bases in their 3′ termini).[181]

Evidence towards the existence of isomiRs has been as old as the description of miRNA guide strands by northern blotting: anywhere between 1–5 bands can be typically observed using this method for a given mature strand. Interestingly, this feature is not exclusive to endogenous miRNA, but shRNA vectors too. Crucially, until very recently it was commonly assumed that the number of northern blot bands equalled the number of variants produced from a pre-miRNA, i.e. no more than a handful.[182] Elsewhere, isomiRs would be described, but discarded in favour of the most prevalent isoform, or the sequence already annotated in online repositories.[56,183] These variations are largely on account of imprecision of Drosha and Dicer processing (Figure 7.5).[182,184] In our own work on therapeutic shRNA maturation, we have also observed such variability. Using deep sequencing on three separate shRNA sequences, ranges of 4–10 nucleotides in strand variant length were

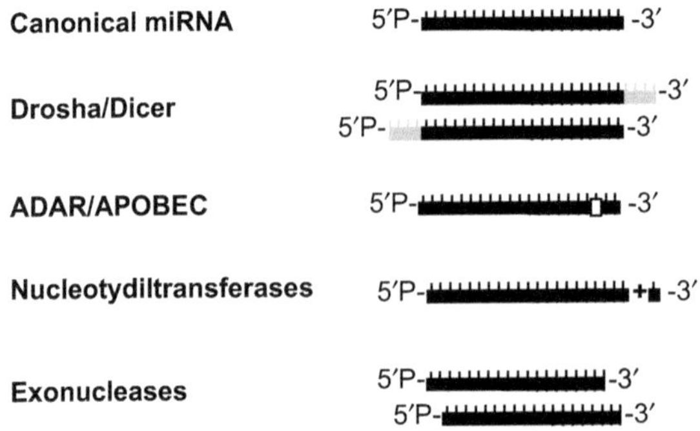

Figure 7.5 IsomiR structure and their production mechanism. Sequences arising from the pri-miRNA stem on account of imprecise Drosha/Dicer cleavage feature stem/loop nucleotides (grey bars) in the mature guide strand (black bar). Double-stranded RNA specific adenosine deaminases (ADAR) and cytosine deaminases, such as APOBEC family proteins, introduce base substitutions (boxed nucleotides), whereas nucleotydiltransferases introduce single nucleotides to the free 3′ ends of miRNAs. Finally, exonuclease-mediated sequence trimming is also believed to take place, removing one or more nucleotides from 5′ ends, 3′ ends or both ends of active strands.

encountered, with the most prevalent mature strand being 8 bases longer than the intended mature strand length (Figure 7.6). Interestingly, even though from an engineering perspective none of the three hairpins were expected to exhibit strand maturation preference, the third shRNA transcript in a tandem triple promoter-shRNA expression cassette exhibited exclusivity for the maturation of the 5′ hairpin arm (Figure 7.6C). It is still unclear if this is on account of 3′ arm RNase susceptibility, unconventional maturation processes or other mechanisms that eliminate this strand from the sequenced RNA pool. Importantly, 'passenger' strand elimination kinetics is a generally underexplored area.

Notwithstanding maturation imprecision, miRNA are believed to be subject to the action of a host of RNA modifying enzymes.[185] Thus, observational evidence based on next generation sequencing has reportedly documented adenosine deaminase-mediated adenosine-to-guanine or -inosine transitions, cytidine deaminase-mediated transitions to uracil[186,187] and mononucleotide 3′ extensions, possibly involved in miRNA targeting regulation.[188] Even Ago2 has been tentatively proposed to participate in miRNA 3′ end trimming.[189] Although the veracity of some of these findings is controversial on account of potential error misinterpretation in sequencing data or simply sample degradation,[190–193] evidence is steadily accumulating in support of non-random documentation of isomiR production.[194–201] Recently, through ultra-deep sequencing sample analysis, even moRNA editing has also been indicated.[201] Although in their vast majority these reports indicate stronger action on guide strand 3′ ends, at least one group have encountered miRNA editing to be focused preferentially to the 5′ end.[202] Curiously, many of these modified guide

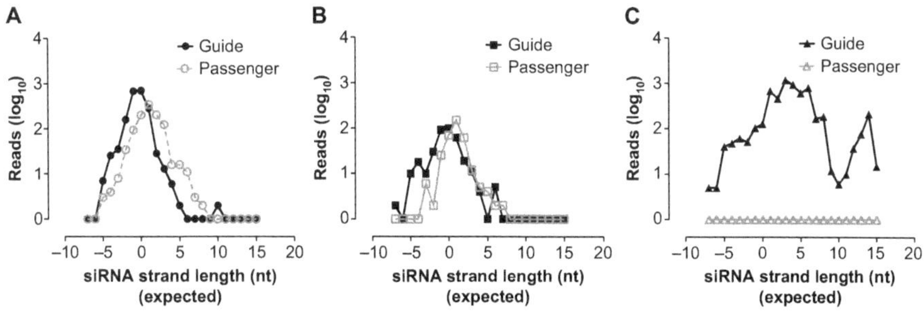

Figure 7.6 Strand maturation variability for an shRNA biotherapeutic. TT-034 is an adeno-associated virus-encoded shRNA therapeutic against HCV encoding 3 shRNA hairpins. Small RNA sequencing of the siRNA maturation products indicated that strands up to -5 nucleotides smaller or 3 nucleotides larger than the expected size (x-axes) were produced at substantial amounts (>10 reads; y-axes). Two of the siRNA precursors (A, B) produced comparable levels of guide and passenger strands, with the guide strands dominated by equivalent levels of the engineered (0) and slightly smaller (–1 nt) sequences. In contrast, one precursor (C) was found to be matured into only guide strands of a wider length range, typically larger (<5 nt) than the intended.

strands were reported to be dominant over their parent miRNAs, at least in terms of abundance[56,203] or Ago loading,[199] lending weight to hypotheses around functional relevance. In our studies on shRNA maturation variants, the 3rd most prevalent guide strand consisting only 1% of processed variants was also the most active.

Elsewhere, minor isomiR sequence differences have been proposed as a means of focusing targeting to a more concise network of mRNAs.[194] However, this tenet has been put forward with limited consideration as to how it would present an evolutionary advantage or evidence of systems-level integration through directed or stochastic editing processes. In at least one study aiming to rationalize isomiR function by examining reporter construct responsiveness to synthetic isomiR transfection, evidence of expanded rather than focused mRNA targeting was put forward.[202] However, in the case of at least one miRNA with a well defined 5′ isomiR, both of which are expressed in the brain, experimental evidence alludes to independent, specific functions.[204] Further insight into disease processes, cell type-specific events and differences between species will be necessary before the role of isomiRs begins to be defined.

From a pharmacological and drug development perspective, however, the existence of isomiRs is substantially impactful in a number of ways:

(1) Which of the mature miRNA (or indeed shRNA) strands is biologically active in the process under scrutiny?
(2) When expressing a miRNA recombinantly, is the biologically active strand produced at pharmacologically sufficient levels and able to overcome competition with co-matured strands for RISC complex formation?
(3) What are the safety implications of unintentionally produced strands?

With the shRNA or miRNA expression vector (or indeed synthetic miRNA mimic) being effectively a pro-drug, these questions carry considerable implications in validating the safety and efficacy of miRNA therapeutics. It was with these concerns in mind, after all, that the first miRNA backbone vectors for shRNA (sh-miRNAs) were proposed.[33] To the knowledge of this author, there is limited, if any, public domain data on therapeutic miRNA candidate editing either from *in vitro* studies or the therapeutically targeted tissue *in vivo*. This limitation extends to the vast majority of studies on miRNA functional validation, despite volumes of data generated with isomiR-agnostic methodologies (e.g. RT-PCR, microarrays, etc.). Notwithstanding these crucial knowledge gaps, evidence indicates that the same miRNA may not be processed into the same mature strands across different cell types.[140,182,203] These findings expand the scope for miRNA biotherapeutic isomiR characterization in most, if not all successfully transfected tissues, even at low levels of expression.

Can isomiR production be tailored? The structural drivers behind miRNA length variety have been biochemically studied for a number of human miRNAs.[182] Thus, through northern blot, end-labelling and mutational analyses of endogenous pre-miRNAs, structural symmetry and pre-miR length were shown to be capital drivers in putative guide strand diversity and principal guide strand length. Duplexes with minimal mismatches would yield 21-mer guide strands, whereas one or more bulges would direct the production of 22- or 23-mer strands. Although utility in engineering concise guide strand production was proposed by the authors, it was not explored in this manuscript. Moreover, these data are of preliminary nature given conclusion reliance on northern blotting assays.

7.3 miRNA Function

Upon formation of the mature RISC complex, the minimal requirements for the final, active RNAi machinery have been met. The RISC complex can then proceed to engage in gene expression regulation in a variety of ways (Figure 7.7). The best understood mechanisms involve action on mRNA translation. However, increasing amounts of evidence indicate that RISC is bioactive also in the nucleus, being intimately involved in transcriptional regulation at the epigenetic level. Moreover, endogenous miRNAs have been also shown to engage in viral life cycle regulation through entirely unconventional processes, with HCV being the most intensely scrutinized example.

7.3.1 Molecular Actions of miRNAs: Therapeutic Challenges and Opportunities

7.3.1.1 Ago2-Mediated Translational Repression

The most well described mechanism of action for RISC complexes involves the recognition of mRNA targets through Watson–Crick base pairing by the RISC-loaded guide strand. Complementarity across the slicing active site in an

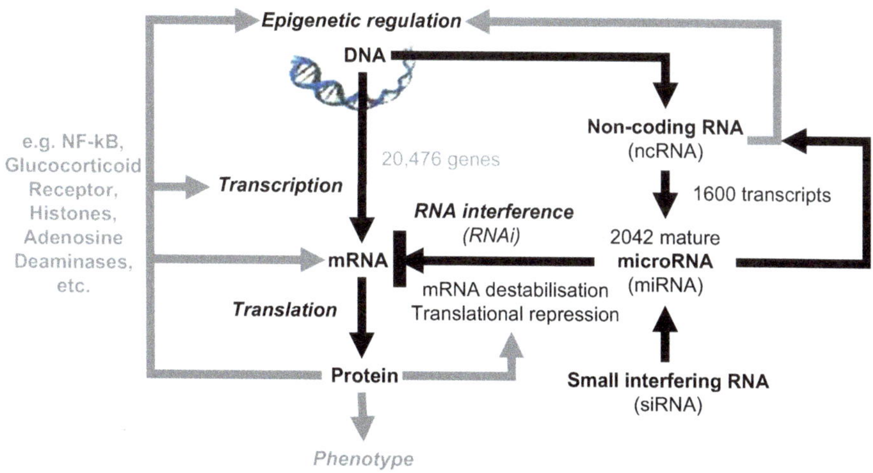

Figure 7.7 An updated version of the Central Dogma of Molecular Biology highlighting the role of miRNAs in human gene expression regulation. The Central Dogma of Molecular Biology indicates that DNA is transcribed into mRNA, which is translated into protein, which generally exerts phenotype. Proteins are now known to regulate this process at all levels, controlling epigenetic markers that permit or prevent transcription from taking place, undertaking and directing transcriptional processing and editing protein-coding mRNA as well as non-coding RNAs. Well-described functions of non-coding RNAs, besides ribosomal and transfer RNA, involve epigenetic regulation of gene expression. microRNAs are a sub-class of ncRNA processed out of at least 1600 ncRNA transcripts into over 2000 mature, active strands of ~21 nt in length. These molecules exert endogenous RNAi through Argonaute proteins, repressing mRNA translation through mRNA destabilisation. Alternatively, miRNA::mRNA interactions have been put forward where miRNA promote mRNA translation. Moreover, miRNA have been implicated in epigenetic regulation of DNA expression, as well as the regulation of ncRNA function. The widely used siRNAs are essentially synthetic analogues of miRNAs, which hijack the RNAi machinery to silence gene expression. Protein-coding gene and microRNA-encoding transcript statistics accurate according to the *Homo sapiens* Genome Reference Consortium assembly GRCh37 (Ensembl release 69, October 2012, www.ensembl.org) and miRbase v.19 (August 2012, www.mirbase.org).

Ago2 RISC context as a minimum, along with extensive base pairing along the rest of the guide strand sequence, is required to permit target RNA cleavage by Ago2.[120,121,205] This phosphate-dependent, ATP-independent function[114] is the core mode of action of siRNAs, and can take place across the entire length of spliced mRNAs, spanning 5′ to 3′ UTRs and including coding regions. Crucially, slicer activity is retained in Ago2 miRISCs:[113] miRNAs can therefore function as siRNAs, provided they are loaded onto Ago2 and a putative target is complementary across the slicer active site. This function is not trivial, as Ago2 appears indispensible in development. In mouse, Ago2 mutation is embryonic lethal[121] and not tolerated even in embryonic stem cell culture.[126]

Conversely, dual knockouts for Ago1, 3 or 4 have been reported as apparently viable and phenotypically normal, though results have been published only for Ago1/Ago3 dual knockouts.[206]

The slicer-permissive nature of miRNAs has promoted hypotheses that these biomolecules participate in antiviral biodefence in an RNAi context.[207] However, rapid evolution of resistance has been robustly reported in culture studies.[208] Moreover, RNA viruses seem to tolerate well vectoring their own miRNAs (see section 7.2.2.3) and Ago1/Ago3 knockouts rather than Ago2 mutation appears to increase viral lung disease susceptibility.[206] Together, these data point against an antiviral mechanism of miRNA action. With regards to other putative targets, such as endogenous RNAs (coding or otherwise), it is unknown what the roles of mismatch tolerance, target affinity (G/C richness) and miRNA expression levels might be in slicer-dependent processes.

The main reason for this is that the extent of base pairing required beyond the slicer active site for efficiently recognizing targets and permitting Ago2 slicer cleavage is an obscure point driven by a decade of siRNA algorithm design rather than empirical determination of slicer-active miRNAs. Depending on experimental context, various permissive and non-permissive features on both guide strands and target sites have been described in the pursuit of optimal siRNA engineering,[209] including, importantly, target RNA secondary structure.[210] It is unclear which, if any, of these rules have been confirmed in the context of their individual applications beyond algorithm construction. Thus, in our own work, an up to 3 mismatch tolerance rule was originally put in place for efficient siRNA design. However, experimental data on a number of targets showed these comfortably tolerated up to 5 consecutive mismatches at the 5′ end of a guide strand, without impact on slicer activity as confirmed by 5′ Rapid Amplification of cDNA Ends (5′ RACE). Analogously, use of siRNA design rules to ascertain putative miRISC slicer sites should be superseded by empirical approaches. Thus, a small subset of miRNA have been shown to be slicer-active against endogenous targets in experimental contexts.[211] Alas, present implementations of 5′ RACE are targeted and Sanger-based, therefore not suited to global transcriptomics for agnostic determination of off-targets of either siRNAs or slicer-active miRNAs.

7.3.1.2 Slicer-Independent Translational Repression

Lack of slicer activity either on account of slicer site mismatches in Ago2 or in an Ago1, 3 and 4 miRISC context can result in the prevention of mRNA translation. In the field of siRNA therapeutics these interactions are considered the principle source of non-immune off-target effects.[212] A hexa- or heptanucleic acid motif at the 5′ end of the guide strand starting at position 2 referred to as the 'seed region', is the feature considered to drive target identification;[213] however, examples where partial[214] or seed-independent target recognition occurs[211,215] have been put forward. How these are biologically promoted is presently poorly understood, though, like with siRNA target selection, target secondary structure plays an important role.[215] Unlike slicer endonucleotic target cleavage, slicer-independent translational repression is generally accepted to involve miRNA interactions with mRNA 3′ UTRs rather than exonic or intronic sequences. These interactions result in the formation of p-bodies in the cytosol,[216,217] usually proximal to endosomes,[218,219] wherein translational inhibition without mRNA cleavage takes place or mRNA degradation is promoted.[220] Crucially, this degradation mechanism has yet to be elucidated. The net result is a network-level suppression of protein production, with one or more miRNAs being the nodal element(s) that controls the translation of the targeted RNAs, orchestrating effector protein modulation individually or in concert, either linearly, cumulatively or rheostatically.[215]

On the basis of Watson–Crick base pairing with 3′-UTR's, putative miRNA target sites, also referred to as miRNA recognition elements (MREs), can be bioinformatically postulated across the entire genome. Various tools are publicly available to propose such targets, each with their own flair on what constitutes a more confident prediction.[221] How these predictions relate to cell type-specific transcriptomics is, however, obscure, and poorly supported by experimental data. Instead, the typical approaches taken for validating putative MREs involves one or more of a variety of methods (Figure 7.8). The aim of these methods is to show pharmacologically or molecularly either that modulation of miRNA expression affects putative target levels or that a predicted MRE can confer miRNA susceptibility to a transgene. Thus, whilst studying the function of miR-146a,[222] a miRNA induced in response to inflammatory stimuli, it was shown through reporter studies that in monocytic cells it would engage with the transcripts of two components of the NF-kappaB pathway, TRAF6 and IRAK1, to result in the suppression of their protein levels and thereby result in inhibition of inflammatory responses. However, transduction of miR-146a mimics or inhibitors in epithelial cells resulted in the expected effect on inflammation modulation but independently of TRAF6, IRAK1 or a host of other bioinformatically predicted mRNA targets.[128] Indeed, all the surveyed targets exhibited robust expression at the protein level. Similar problems in determining the miRNAs that interact with specific mRNAs in different cell type/culture systems abound.[18]

The main reason for this situation is the absence of methods that can conclusively prove specific miRNA::mRNA interactions within a given context. Rather, the techniques described previously provide suggestive evidence, albeit in highly artificial systems, substantially deviating from physiology. Thus, miRNA mimics tend to be fully hybridized, siRNA-like molecules rather than mismatch/bulge-containing structures,[223] a feature known to affect Ago loading.[126] Moreover, they usually lack any nucleotide modification[224] known to abolish TLR activation (e.g. LNA or 2'-O-methyl)[16] to avoid disruption of affinity/selectivity for mRNA targets or 'miRNA-like off target effects' as is want for siRNAs.[15,16,212] This introduces risk of misinterpretation of immune effects as RNAi, the bane of the first clinical siRNA candidate to reach phase III trials without success. Moreover, although siRNA editing has been shown to take place in the cytosol,[225] it is unclear whether miRNA mimics are

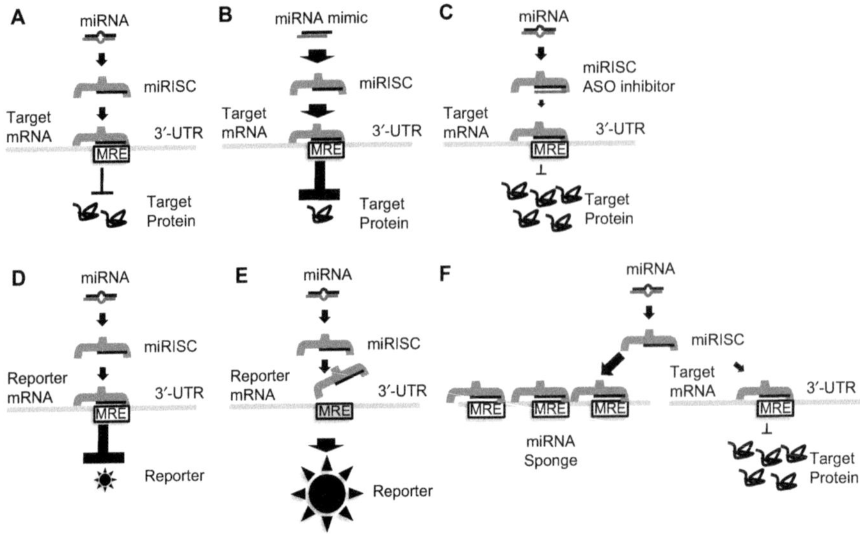

Figure 7.8 Current *in vitro* methods for the validation of miRNA targets. MicroRNA target validation is presently pursued by a variety of pharmacological or recombinant methodologies that aim to amplify or suppress (A) canonical miRNA::mRNA target interaction in cell culture experiments, resulting in observable changes in miRNA-targeted protein expression. Many of these methods are also explored for their therapeutic significance. In these examples, approaches to validating 3′-UTR-mediated translational repression are examined. Transfection of (B) a miRNA mimic expands (enlarged arrows and inhibition 'T' signs) translation suppression whereas (C) a miRNA antisense inhibitor oligonucleotide (ASO) restores target translation (smaller arrows and inhibition signs). Introduction of a 3′-UTR fragment containing a (D) wild type putative MRE downstream of a reporter gene (typically luciferase) should similarly result in inhibition of reporter gene expression (small 'sun' symbol) whereas (E) a mutated MRE (grey MRE box) should allow for stronger reporter gene expression (larger 'sun' symbol). Alternatively, the introduction of a (F) transgene with multiple MREs in it's 3′-UTR (miRNA sponge) can sequester the miRNA of interest.

processed similarly to their cognate counterparts or according to an exogenous siRNA editing mechanism.

Antisense inhibitors of miRNAs on the other hand are designed to be paralogue-sensitive, if possible through use of high T_M modifications such as LNAs,[226,227] but are not isomiR selective. Moreover, LNA-containing ASO's have been shown to infiltrate DNA duplexes,[228] to disrupt even pre-miR structures and, in cases, to demonstrate a 2- or at best 3-log selectivity for their targets.[14] Such features are rarely[229] controlled for in the literature. Alternatively, use of quantitative RT-PCR for a miRNA of interest[224] or northern blotting[24] is used to demonstrate that an inhibitor has eliminated the pool of free miRNA from mRNA interaction. Similar to other reports,[224] in our hands simple spiking of an ASO inhibitor in tissue prior to RNA extraction is sufficient to yield identical results to those put forward from *in vivo* experiments interpreting miRNA sequestration as proof of cytosolic delivery and a miRNA-specific effect. Given that both miRNA inhibitors and mimics require comparatively herculean concentrations (10^6–10^8 copies per cell) to exert their function, as opposed to the 10^4–10^5copies of individual miRNAs in the cytosol, it is likely that such circumstantial evidence of function necessitates a robust elimination of potential off-target modes of action. To the point, it is now clear that such overexpression approaches perturb cellular miRNA minicircuitry.[230]

Similarly, transgene expression, be it a miRNA sponge or a reporter-3′-UTR construct, is similarly overwhelmingly produced in the cell. This deviates miRNA::mRNA homeostasis beyond physiology to force interaction; the titrating (EC_{50}) effect of such inhibitory approaches is typically never investigated. Notwithstanding transgene expression intensity, although secondary structure has been demonstrated to affect MRE accessibility,[231–233] this is also sidelined in reporter –3′-UTR fusion approaches. Moreover, the 3′-UTR context of the MRE under scrutiny can vary considerably between investigators (a few tens to a few hundreds of bases) and is derived from database annotation rather than experimental determination in the system under investigation, ignoring 3′-UTR plasticity[234] within and between cell types.

In an attempt to avoid cell manipulation, transcriptome and/or proteome analysis can be carried out in parallel to miRNA expression analysis.[235–237] However, the false discovery rate of these approaches typically exceeds 50%,[214] on account of poor discrimination of primary (causal) and secondary (consequential) effects. Presently, the only method available to directly interrogate miRNA::mRNA interactions is high throughput sequencing after cross-inking and immunoprecipitation (HITS-CLIP; also referred to as CLIP-seq).[238] In this approach, sample irradiation chemically conjugates nucleic acids to their protein counterparts in ribonucleoprotein complexes such as a target-engaging miRISC. After immunoprecipitating the protein of interest, the RNA conjugated to it can be released and deep sequenced, eliminating potential false–positive interactions that can manifest by simple co-immunoprecipitation.[239] As a result, false discovery rates are reduced to <30%.[214] In the original study proposing the use of HITS-CLIP to decode miRNA::mRNA interaction maps[240] the 2A8 monoclonal antibody was used,[241] which does not discriminate between the four human Argonaute proteins. Although the study was also focused on the determination of the interaction of miRNAs with non-coding regions of transcripts typically interrogated for miRNA function, about a quarter of the interactions were found to be within coding regions of mRNAs, and about an eighth within introns. These findings indicate that a considerable element in miRNA function beyond UTR's has been overlooked, as is alluded by various experimental setups increasingly implicating functional accessibility of coding regions to miRISC.[242–249] Importantly, later studies utilizing HITS-CLIP substantiated the tenet that miRNA mimic transfection biases MRE detection,[221] whereas others showed that miRNA mimic transfection could also generate Argonaute interaction artefacts.[250] Moreover, it is now clear that HITS-CLIP is a static rather than dynamic method for MRE detection.[221]

In other words, it can only contextualize and validate MRE predictions within specific experimental contexts and lack of MRE confirmation does not mean that the MRE might not be available to miRISC under other circumstances.

Although a substantial improvement in the process of querying miRNA biology, the information provided by Argonaute HITS-CLIP is incomplete as it does not validate specific miRNA::mRNA interaction pairs: it provides a global protein::RNA interaction picture. As mismatch tolerance can be extensive and isomiR or paralogue sequence differences can be minimal, it can be challenging to confirm molecular interactions beyond statistical inferences around sequence prevalence. Alternatively, parallel analysis of RNA ends (PARE) can also be used to globally query unexpected 5′ end generation as a means of evidencing miRNA-mediated RNAi.[251] However, the technique suffers low sensitivity on account of sequencing read wastage to unrelated RNA sections. The sum knowledge around HITS-CLIP and PARE findings is presently archived in starBase,[252] with efforts around miRNA target validation collated in TARBASE.[253] In the last few years numerous bioinformatics and statistical data enrichment models have been proposed,[214,254–258] each relying on different approaches on experimental data integration with modelling parameters. However, a comprehensive, organized effort to promote orthogonal interrogation of these molecular interactions with a view of collectively improving mathematical prediction approaches is currently lacking.

7.3.1.3 Other Mechanisms of miRNA Action

An additional complicating factor in understanding miRNA activities is the continuous discovery of potential exceptions in the 'rules' around miRNA function. Thus, whilst most attention has focused on translational repression, miRNAs are now understood to also promote translation under some circumstances.[259] So far, such observations appear to relate to cell cycle phase,[260–263] with crucial implications in miRNA-mediated cancer therapy.[264] The diversity of miRNA under cell cycle control is unclear, given that only a handful of *in vitro* studies have involved synchronized cell culture. Moreover, such functions also appear to be regulated by subcellular localization of miRNAs, as at least one upregulatory mechanism is nuclear-specific.[262] Thus, it is presently understood that after cytosolic maturation of miRISC, the complex can enter the nucleus[265] in an Importin 8[266] and CRM1-dependent[267] fashion, a pathway that can be accessed by siRNAs[267,268] or shRNAs,[269] possibly in a sequence-dependent manner.[268] The extent of miRNA compartmentalization appears to be variable,[270] with one report indicating that isomiR length might correlate to sub-localisation.[271]

Once in the nucleus, the function of miRNAs (and siRNAs or, for intellectual property purposes, antigene RNAs (agRNAs)) appears to remain Argonaute dependent, but expands beyond the classical notions of RNAi to involve mechanisms of transcriptional gene silencing (TGS), activation (TGA) and epigenetic complex-mediated regulation, driving control of heterochromatin formation. Thus, the defining element in most cases appears to be interaction with promoter-associated transcripts.[272–278] This process then alters histone methylation status,[278–280] which controls the expression potential of the targeted gene. Importantly, these results are not restricted to *in vitro* studies.[281] Interestingly, some of the TGS/TGA-active miRNAs are indeed encoded in the promoters they regulate,[274] indicating additional feedback/forward loops in transcriptional regulation through miRNA::ncRNA interactions. Elsewhere, intronic targeting of a siRNA was shown to mediate splicing control in an Ago1- and antisense transcript-dependent fashion involving histone modifications,[282] whereas partial exon targeting siRNA or miRNA-like duplexes could either eliminate or partially subvert, respectively, targeted exon inclusion; in contrast to the other report on splicing-regulating siRNAs, these duplexes were shown to act in an Ago2-dependent mechanism.[283] Together, these functions expand further the potential off-targeting gamut of siRNAs (therapeutic or otherwise), yet present novel approaches to manipulating RNAi for the treatment of disease:[284]

The prospect of heritable or, at least prolonged effects through chromatin modification is of particular relevance. Interestingly, unlike their applications in cytosolic miRNA function and therapeutic targeting, LNA antisense use in querying miRNA function in the nucleus has indicated that such molecules might indeed preferentially bind DNA to ncRNA transcripts.[285] Conversely, modification tolerance studies in agRNAs guide strands indicated limited utility of nucleoside analogues.[286] MicroRNA tolerance to these modifications in a TGS/TGA context has yet to be assessed, though the agRNA studies are in agreement with the general findings on functional restriction of RNAi-active miRNAs after modification. Presently, there is only a single report on Argonaute protein selection for nuclear targeting involving promoter-targeting siRNAs against genome-integrated HIV, indicating the potential for controlled subcellular targeting of RISC complexes.[287] The most advanced approach to querying nuclear compartmentalization of RISC (or nRISC) involves fluorescence correlation spectroscopy and fluorescence cross-correlation spectroscopy[265] as well as single molecule microscopy.[288] However, these techniques should be used with caution as they rely on heavily modified constructs to enlighten intracellular RISC trafficking. Thus, nRISC accumulation was found to be a consequence of slicer-restricted interactions in the cytosol;[265] a minimum requirement to validation would involve similar characterization of the ncRNA partner to the nRISC interaction involved in TGS/TGA.

The nucleus is not the only organelle that might be involved in miRNA function because Ago2 or miRNA have been purified from mitochondrial fractions both *in vitro*[289–291] and *in vivo*.[292] Whether mitochondrial targeting is sequence-dependent, of functional relevance or a purification artefact is unclear: these findings are of exceptional interest, however, because manipulation of mitochondrial targeting might present novel opportunities for treating mitochondrial diseases, for which limited options exist. Tellingly, *in vitro* and *in silico* studies indicate that some miRNA genes might be encoded in the microbial genome.[293] Whether these are an evolutionary adaptation of the prokaryotic RNAi-like system[294,295] or a result of genetic exchange with the nucleus remains to be determined. Thus, it appears that appropriate spatiotemporal assessment of miRNA function extends beyond cell type-specific, kinetic analysis of expression and mRNA interaction, to subcellular localization and replicative state.

MicroRNA interactions are also believed not to be restricted to the 3′ UTR or ORF of an mRNA target. Thus miR-10a was the first miRNA shown to potentially mediate a 5′ UTR interaction involving most of the miRNA sequence, to drive translation enhancement even under starvation conditions such as those encountered in cancer.[296] In this respect, miR-10a is not unique: the liver-specific miR-122 is required for the Ago2-dependent[297,298] survival and replication of HCV in culture,[299–302] and appears to be intimately involved in permitting, rather than suppressing viral genome translation,[301] as well as protecting it from exonuclease degradation.[303–305] Indeed HCV might hijack more RNAi components than just Ago2 in its life cycle, such as other Argonautes[306] and p-body proteins.[306–308] Mechanistically, a long-range interaction between the HCV 5′ internal ribosomal entry site (IRES) where the miR-122 MRE's reside[309] and it's 3′ genome end have been proposed in mediating translational control.[300] However, it appears that cell lysate experiments, commonly used in other miRNA functional and particularly biochemical studies, may disproportionately emphasize the role of this phenomenon,[310] a tenet corroborated in studies involving novel oligonucleotide inhibitors designed to interrupt this interaction[229] and reporter studies.[306] Instead, despite the overlapping location of the two MRE's in the HCV IRES, recent studies indicate that both might be simultaneously occupied,[311] bearing interesting implications on the functional value of juxtaposed MREs in other RNAs. The clinical value of miR-122 targeting in HCV therapy is presently ambiguous according to clinical investigations on miR-122 levels in infected liver biopsy versus viral load from three independent groups.[312–314] Notwithstanding potential value in a personalized medicine context similar to the paradigm of cancer therapy,[315] multiple angles of evidence suggest loss of miR-122 expression in the liver might underpin fibrosis and hepatocellular carcinoma, conditions which are both related to late stage HCV chronic infection.[316–318]

5′ UTR interaction does not necessarily drive translational upregulation as experimental construct studies have revealed that 5′ UTR localization of a MRE can drive translational repression too.[319] This work also details investigations on the susceptibility to miRNA regulation for HCV IRES fused to artificial let-7 MRE's. These studies concluded that even such 5′-encoded MRE's can mediate translational repression. Notably, this finding is in stark contrast to the mode of action of miR-122 on HCV function, and emphasizes further the importance of transcript structure on ensuing miRNA activity. Moreover, this manuscript is of additional interest because it is among the few in the literature reporting variance in reporter construct suppression sensitivity on account of transfection method: electroporation was found to be permissive to suppression where lipofection was not. It is unclear to what extent other reports in the literature might be impacted by this problem.

Nevertheless, this molecular mechanism has not escaped the machinations of hCMV in manipulating cellular mechanisms to its advantage. Thus, the hCMV-encoded miR-US25-1 appears to repress cyclin E2 and TRIM28 in 5′-UTR dependent fashion.[320] This claim is better substantiated in comparison to other work, through studies on the impact of miR-US25-1 mutation on target expression. Yet others also reported the existence of transcripts harbouring dual regulatory, suppressive MRE's in both their 5′ and 3′ UTRs involving non-seed (3′ end) and seed sequence miRNA interactions, respectively (so-called miBridge).[321] Later studies associated, in part, these 5′ UTR interactions to upstream start codons (uAUG) as part of the ribosome scanning model of translation.[322] Indeed there seems to be no shortage of putative MREs in 5′ UTR's, at least from a bioinformatics perspective.[323]

7.4 Concluding Remarks: the Impact of the Developing miRNA Knowledge Base on Biotherapeutics Development Efforts

In the 12 years since the description of RNAi in man, our understanding of this important molecular phenomenon has transformed from a simplistic target recognition model to an unexpectedly complicated, multi-dimensional, network-level regulation system. Its ever-evolving complexity features a plethora of potentially RNAi-active miRNA processing variants, mechanisms of miRISC assembly and action, as well as opportunities for integrative gene expression regulation. What was initially believed to be a collection of defined, single molecular entities suitable for complementation therapy or inhibitor targeting have evolved into molecular populations (isomiRs) representing hundreds of potential drugs, drug by-products or drug targets per miRNA precursor. Throughout this chapter, key examples have been put forward of how dogmatic precepts have been continuously subjected to re-evaluation with major impact on our understanding of RNAi and implications in therapy development. Yet despite these lessons, two schools of academic thought continue to dominate the field. On the one hand, the experimental biochemistry approach aims to characterize *in vitro* the molecular components of RNAi, paying basic tribute to the physiological environment where these events take place. On the other, molecular medicine efforts present ever-increasing associations of miRNA expression deregulation in disease, extrapolating conclusions drawn from simplistic experimental tools in understanding miRNA molecular action. That is not to say that findings from experimental setups are incorrect: rather, their therapeutic relevance could be questioned due to commonly held assumptions around tool relevance and specificity, let alone extrapolations between miRNAs. Additional barriers include inadequate analytical methods, subjective statistical approaches and economic challenges in both the financial and data management sense. Collectively, these limitations prevent comprehensive elucidation of the molecular processes interrogated in disease to develop sufficient confidence in the mechanism of action, safety and clinical rationale of potential miRNA-targeted therapeutics. Notably, major strides in our understanding of RNAi have been achieved principally through observational studies carried out under physiological conditions. The value of patient sample-based analysis in the understanding of disease versus animal modelling seems not yet to have translated into the study of molecular phenomena. It would be therefore premature to comment on the survival potential of miRNA biotherapeutic candidates given the context of many of the early findings that have driven drug development efforts.

Yet from a drug discovery and development perspective, ectopic miRNA expression could be considered as the presently most promising approach, where a simple goal is attempted: the complementation of a miRNA under-expressed or entirely deleted in disease. Indeed, after nearly four decades of development, the regulatory approval of antisense drugs is still limited to one topical (fomivirsen) and one liver indication (mipomersen). Whilst the former has not proven commercially successful, the safety of the latter is still under intense scrutiny by European regulators. History indicates this discrepancy might present a considerable approval hurdle, likely to require a personalized medicine approach.[315] This would complicate miRNA inhibitor efforts further beyond their safety/inefficient delivery bottlenecks. Moreover, clinically relevant and safe synthetic RNA duplex delivery systems still remain elusive, whereas synthetic miRNA analogues are faced with considerable chemistry challenges around impact on function.

On the other hand, the natural propensity of viruses to express miRNAs and the recent approval of Glybera® in Europe pave the way for viral miRNA biotherapeutic development. That is not to say that the regulatory challenge of any such efforts will be less complex: the indication Glybera® is marketed against is an inherited disease with a robust companion biomarker assay.[324] Moreover, adenoassociated viruses, such as the Glybera® vector, have yet to be shown to encode vmiRNA. In addition, given the supportive nature of presently available miRNA mechanistic data, these efforts would assume miRNA expression loss to be the causal agent of disease. Arguing, therefore, the administration of a miRNA biotherapeutic would require robust results around vector tropism, miRNA expression intensity and isomiR reconstitution against canonical miRNA levels at the targeted tissue to support safe exposure at the site of action. Target engagement (Argonaute loading and mRNA targeting) would only be partially possible to document (HITS-CLIP). As a conclusive method demonstrating specific miRNA-mediated molecular action on specific mRNA targets is presently lacking, evidence around functional impact would be at best supporting (transcriptomics/proteomics). On the other hand, not all diseases and indications are permissive to biopsy collection, which could be an interesting risk/benefit opportunity, increasing reliance on biomarkers. However, biomarker modulation does not equate to on-target mechanism of action in the absence of such explicit data.

Approvals for the initiation of clinical trials on both simple/defined biologics, such as peptides and antibodies, as well as complex/partially defined ones, such as stem cells, have been repeatedly achieved in many territories. However, the risk-averse nature of regulatory authorities should not be underestimated on the long road to market approval. The journey of Glybera® and some recently marketed live vaccines are pertinent to the efforts of any scientist engaged in realizing the clinical potential of miRNA biotherapeutics. Presently, the tip of the spear aims at the perceived low hanging fruit of oncology indications, as is typical for novel approaches.[325] Whether orphan drug status, niche indications or fast-track routes will yield favourable outcomes remains to be determined. The fascinating roles of these biomolecules and their promise in delivering the next generation of molecular therapies will surely be the source of riveting stories and important lessons for the global biotechnology arena.

References

1. A. Fire, S. Xu, M. K. Montgomery, S. A. Kostas, S. E. Driver and C. C. Mello, *Nature*, 1998, **391**, 806–811.
2. S. M. Elbashir, J. Harborth, W. Lendeckel, A. Yalcin, K. Weber and T. Tuschl, *Nature*, 2001, **411**, 494–498.
3. M. Lagos-Quintana, R. Rauhut, W. Lendeckel and T. Tuschl, *Science*, 2001, **294**, 853–858.
4. C. Matranga, Y. Tomari, C. Shin, D. P. Bartel and P. D. Zamore, *Cell*, 2005, **123**, 607–620.
5. P. J. Leuschner, S. L. Ameres, S. Kueng and J. Martinez, *EMBO reports*, 2006, **7**, 314–320.
6. T. Kawamata and Y. Tomari, *Trends Biochem Sci*, 2010, **35**, 368–376.
7. L. Zekri, D. Kuzuoglu-Ozturk and E. Izaurralde, *The EMBO journal*, 2013, **32**, 1052–1065.
8. C. Y. Chen and A. B. Shyu, *Adv Exp Med Biol*, 2013, **768**, 183–195.
9. O. Matveeva, *Methods Mol Biol*, 2013, **942**, 1–16.
10. P. C. Scacheri, O. Rozenblatt-Rosen, N. J. Caplen, T. G. Wolfsberg, L. Umayam, J. C. Lee, C. M. Hughes, K. S. Shanmugam, A. Bhattacharjee, M. Meyerson and F. S. Collins, *Proc Natl Acad Sci USA*, 2004, **101**, 1892–1897.
11. K. Kariko, P. Bhuyan, J. Capodici and D. Weissman, *J Immunol*, 2004, **172**, 6545–6549.
12. M. E. Kleinman, K. Yamada, A. Takeda, V. Chandrasekaran, M. Nozaki, J. Z. Baffi, R. J. Albuquerque, S. Yamasaki, M. Itaya, Y. Pan, B. Appukuttan, D. Gibbs, Z. Yang, K. Kariko, B. K. Ambati, T. A. Wilgus, L. A. DiPietro, E. Sakurai, K. Zhang, J. R. Smith, E. W. Taylor and J. Ambati, *Nature*, 2008, **452**, 591–597.
13. A. D. Judge, V. Sood, J. R. Shaw, D. Fang, K. McClintock and I. MacLachlan, *Nat Biotechnol*, 2005, **23**, 457–462.
14. S. A. Moschos, M. Frick, B. Taylor, P. Turnpenny, H. Graves, K. G. Spink, K. Brady, D. Lamb, D. Collins, T. D. Rockel, M. Weber, O. Lazari, L. Perez-Tosar, S. A. Fancy, C. Lapthorn, M. X. Green, S. Evans, M. Selby, G. Jones, L. Jones, S. Kearney, H. Mechiche,

D. Gikunju, R. Subramanian, E. Uhlmann, M. Jurk, J. Vollmer, G. Ciaramella and M. Yeadon, *Mol Ther*, 2011, **19**, 2163–2168.
15. H. Yu, Z. Wang, G. Sun and Y. Yu, *Curr Med Chem*, 2012, **19**, 1365–1377.
16. M. A. Behlke, *Oligonucleotides*, 2008, **18**, 305–319.
17. G. A. Calin, C. D. Dumitru, M. Shimizu, R. Bichi, S. Zupo, E. Noch, H. Aldler, S. Rattan, M. Keating, K. Rai, L. Rassenti, T. Kipps, M. Negrini, F. Bullrich and C. M. Croce, *Proc Natl Acad Sci USA*, 2002, **99**, 15524–15529.
18. G. J. Brock, S. Moschos, S. D. Spivack and G. J. Hurteau, *Epigenetics*, 2011, **6**, 268–272.
19. S. A. Moschos, *Int Drug Disc*, 2010, 10–17.
20. S. A. Moschos, A. E. Williams, M. M. Perry, M. A. Birrell, M. G. Belvisi and M. A. Lindsay, *BMC Genomics*, 2007, **8**, 240.
21. J. Krutzfeldt, N. Rajewsky, R. Braich, K. G. Rajeev, T. Tuschl, M. Manoharan and M. Stoffel, *Nature*, 2005, **438**, 685–689.
22. R. Kumar, S. K. Singh, A. A. Koshkin, V. K. Rajwanshi, M. Meldgaard and J. Wengel, *Bioorg Med Chem Lett*, 1998, **8**, 2219–2222.
23. U. A. Orom, S. Kauppinen and A. H. Lund, *Gene*, 2006, **372**, 137–141.
24. R. E. Lanford, E. S. Hildebrandt-Eriksen, A. Petri, R. Persson, M. Lindow, M. E. Munk, S. Kauppinen and H. Orum, *Science*, 2010, **327**, 198–201.
25. C. E. Grueter, E. van Rooij, B. A. Johnson, S. M. DeLeon, L. B. Sutherland, X. Qi, L. Gautron, J. K. Elmquist, R. Bassel-Duby and E. N. Olson, *Cell*, 2012, **149**, 671–683.
26. J. Worm, J. Stenvang, A. Petri, K. S. Frederiksen, S. Obad, J. Elmen, M. Hedtjarn, E. M. Straarup, J. B. Hansen and S. Kauppinen, *Nucleic Acids Res*, 2009, **37**, 5784–5792.
27. A. M. Krieg, *Mol Ther Nucleic Acids*, 2012, **1**.
28. M. Negrini, L. Gramantieri, S. Sabbioni and C. M. Croce, *Anti-Cancer Agents Med Chem*, 2011, **11**, 500–521.
29. M. de Planell-Saguer and M. C. Rodicio, *Clin Biochem*, 2013.
30. M. Miyagishi and K. Taira, *Nat Biotechnol*, 2002, **20**, 497–500.
31. T. R. Brummelkamp, R. Bernards and R. Agami, *Science*, 2002, **296**, 550–553.
32. L. S. Lambeth and C. A. Smith, *Methods Mol Biol*, 2013, **942**, 205–232.
33. D. Boden, O. Pusch, R. Silbermann, F. Lee, L. Tucker and B. Ramratnam, *Nucleic Acids Res*, 2004, **32**, 1154–1158.
34. D. Grimm, K. L. Streetz, C. L. Jopling, T. A. Storm, K. Pandey, C. R. Davis, P. Marion, F. Salazar and M. A. Kay, *Nature*, 2006, **441**, 537–541.
35. N. M. Robas, *Bioanalysis*, 2012, **4**, 2485–2497.
36. S. A. Moschos, M. Frick, B. Taylor, P. Turnpenny, H. Graves, K. G. Spink, K. Brady, D. Lamb, D. Collins, T. D. Rockel, M. Weber, O. Lazari, L. Perez-Tosar, S. A. Fancy, C. Lapthorn, M. X. Green, S. Evans, M. Selby, G. Jones, L. Jones, S. Kearney, H. Mechiche,

D. Gikunju, R. Subramanian, E. Uhlmann, M. Jurk, J. Vollmer, G. Ciaramella and M. Yeadon, *Mol Ther*, 2011, **19**, 2163–2168.
37. Y. Lee, K. Jeon, J. T. Lee, S. Kim and V. N. Kim, *The EMBO journal*, 2002, **21**, 4663–4670.
38. M. Megraw, P. Sethupathy, B. Corda and A. G. Hatzigeorgiou, *Nucleic Acids Res*, 2007, **35**, D149–155.
39. P. Alexiou, T. Vergoulis, M. Gleditzsch, G. Prekas, T. Dalamagas, M. Megraw, I. Grosse, T. Sellis and A. G. Hatzigeorgiou, *Nucleic Acids Res*, 2010, **38**, D137–141.
40. C. H. Chien, Y. M. Sun, W. C. Chang, P. Y. Chiang-Hsieh, T. Y. Lee, W. C. Tsai, J. T. Horng, A. P. Tsou and H. D. Huang, *Nucleic Acids Res*, 2011, **39**, 9345–9356.
41. S. Griffiths-Jones, H. K. Saini, S. van Dongen and A. J. Enright, *Nucleic Acids Res*, 2008, **36**, D154–158.
42. A. Rodriguez, S. Griffiths-Jones, J. L. Ashurst and A. Bradley, *Genome Res*, 2004, **14**, 1902–1910.
43. V. N. Kim, J. Han and M. C. Siomi, *Nat Rev Mol Cell Biol*, 2009, **10**, 126–139.
44. R. C. Lee, R. L. Feinbaum and V. Ambros, *Cell*, 1993, **75**, 843–854.
45. X. Cai, C. H. Hagedorn and B. R. Cullen, *RNA*, 2004, **10**, 1957–1966.
46. Y. Lee, C. Ahn, J. Han, H. Choi, J. Kim, J. Yim, J. Lee, P. Provost, O. Radmark, S. Kim and V. N. Kim, *Nature*, 2003, **425**, 415–419.
47. S. Y. Ying, C. P. Chang and S. L. Lin, *Methods Mol Biol*, 2010, **629**, 205–237.
48. E. Berezikov, W. J. Chung, J. Willis, E. Cuppen and E. C. Lai, *Mol Cell*, 2007, **28**, 328–336.
49. C. R. Sibley, Y. Seow, S. Saayman, K. K. Dijkstra, S. El Andaloussi, M. S. Weinberg and M. J. Wood, *Nucleic Acids Res*, 2012, **40**, 438–448.
50. M. A. Havens, A. A. Reich, D. M. Duelli and M. L. Hastings, *Nucleic Acids Res*, 2012, **40**, 4626–4640.
51. L. W. Barrett, S. Fletcher and S. D. Wilton, *Cell Mol Life Sci*, 2012, **69**, 3613–3634.
52. A. Grundhoff and C. S. Sullivan, *Virology*, 2011, **411**, 325–343.
53. S. Pfeffer, M. Zavolan, F. A. Grasser, M. Chien, J. J. Russo, J. Ju, B. John, A. J. Enright, D. Marks, C. Sander and T. Tuschl, *Science*, 2004, **304**, 734–736.
54. X. Cai, A. Schafer, S. Lu, J. P. Bilello, R. C. Desrosiers, R. Edwards, N. Raab-Traub and B. R. Cullen, *PLoS Pathog*, 2006, **2**, e23.
55. A. Grundhoff, C. S. Sullivan and D. Ganem, *RNA*, 2006, **12**, 733–750.
56. P. Landgraf, M. Rusu, R. Sheridan, A. Sewer, N. Iovino, A. Aravin, S. Pfeffer, A. Rice, A. O. Kamphorst, M. Landthaler, C. Lin, N. D. Socci, L. Hermida, V. Fulci, S. Chiaretti, R. Foa, J. Schliwka, U. Fuchs, A. Novosel, R. U. Muller, B. Schermer, U. Bissels, J. Inman, Q. Phan, M. Chien, D. B. Weir, R. Choksi, G. De Vita, D. Frezzetti, H. I. Trompeter, V. Hornung, G. Teng, G. Hartmann, M. Palkovits, R. Di Lauro, P. Wernet, G. Macino, C. E. Rogler, J. W. Nagle, J. Ju, F. N. Papavasiliou, T. Benzing, P. Lichter, W. Tam, M. J. Brownstein,

A. Bosio, A. Borkhardt, J. J. Russo, C. Sander, M. Zavolan and T. Tuschl, *Cell*, 2007, **129**, 1401–1414.
57. K. Cosmopoulos, M. Pegtel, J. Hawkins, H. Moffett, C. Novina, J. Middeldorp and D. A. Thorley-Lawson, *J Virol*, 2009, **83**, 2357–2367.
58. C. S. Sullivan, C. K. Sung, C. D. Pack, A. Grundhoff, A. E. Lukacher, T. L. Benjamin and D. Ganem, *Virology*, 2009, **387**, 157–167.
59. N. Stern-Ginossar, N. Saleh, M. D. Goldberg, M. Prichard, D. G. Wolf and O. Mandelboim, *J Virol*, 2009, **83**, 10684–10693.
60. T. J. Stark, J. D. Arnold, D. H. Spector and G. W. Yeo, *J Virol*, 2012, **86**, 226–235.
61. K. Dhuruvasan, G. Sivasubramanian and P. E. Pellett, *Virus Res*, 2011, **157**, 180–192.
62. X. Cai, S. Lu, Z. Zhang, C. M. Gonzalez, B. Damania and B. R. Cullen, *Proc Natl Acad Sci USA*, 2005, **102**, 5570–5575.
63. S. Pfeffer, A. Sewer, M. Lagos-Quintana, R. Sheridan, C. Sander, F. A. Grasser, L. F. van Dyk, C. K. Ho, S. Shuman, M. Chien, J. J. Russo, J. Ju, G. Randall, B. D. Lindenbach, C. M. Rice, V. Simon, D. D. Ho, M. Zavolan and T. Tuschl, *Nat Methods*, 2005, **2**, 269–276.
64. M. A. Samols, J. Hu, R. L. Skalsky and R. Renne, *J Virol*, 2005, **79**, 9301–9305.
65. X. Cai and B. R. Cullen, *J Virol*, 2006, **80**, 2234–2242.
66. Y. T. Lin, R. P. Kincaid, D. Arasappan, S. E. Dowd, S. P. Hunicke-Smith and C. S. Sullivan, *RNA*, 2010, **16**, 1540–1558.
67. C. Cui, A. Griffiths, G. Li, L. M. Silva, M. F. Kramer, T. Gaasterland, X. J. Wang and D. M. Coen, *J Virol*, 2006, **80**, 5499–5508.
68. I. Jurak, M. F. Kramer, J. C. Mellor, A. L. van Lint, F. P. Roth, D. M. Knipe and D. M. Coen, *J Virol*, 2010, **84**, 4659–4672.
69. J. L. Umbach, K. Wang, S. Tang, P. R. Krause, E. K. Mont, J. I. Cohen and B. R. Cullen, *J Virol*, 2010, **84**, 1189–1192.
70. M. I. Besecker, M. E. Harden, G. Li, X. J. Wang and A. Griffiths, *J Virol*, 2009, **83**, 3413–3416.
71. M. A. Amen and A. Griffiths, *J Virol*, 2011, **85**, 7296–7311.
72. J. L. Umbach and B. R. Cullen, *J Virol*, 2010, **84**, 695–703.
73. J. L. Umbach, M. F. Kramer, I. Jurak, H. W. Karnowski, D. M. Coen and B. R. Cullen, *Nature*, 2008, **454**, 780–783.
74. S. Lu and B. R. Cullen, *J Virol*, 2004, **78**, 12868–12876.
75. O. Aparicio, N. Razquin, M. Zaratiegui, I. Narvaiza and P. Fortes, *J Virol*, 2006, **80**, 1376–1384.
76. O. Aparicio, E. Carnero, X. Abad, N. Razquin, E. Guruceaga, V. Segura and P. Fortes, *Nucleic Acids Res*, 2010, **38**, 750–763.
77. E. Gottwein, X. Cai and B. R. Cullen, *Cold Spring Harbor Symp Quant Biol*, 2006, **71**, 357–364.
78. R. L. Skalsky and B. R. Cullen, *Annu Rev Microbiol*, 2010, **64**, 123–141.
79. M. Hussain, S. Torres, E. Schnettler, A. Funk, A. Grundhoff, G. P. Pijlman, A. A. Khromykh and S. Asgari, *Nucleic Acids Res*, 2012, **40**, 2210–2223.

80. E. Schnettler, M. G. Sterken, J. Y. Leung, S. W. Metz, C. Geertsema, R. W. Goldbach, J. M. Vlak, A. Kohl, A. A. Khromykh and G. P. Pijlman, *J Virol*, 2012, **86**, 13486–13500.
81. D. L. Ouellet, I. Plante, P. Landry, C. Barat, M. E. Janelle, L. Flamand, M. J. Tremblay and P. Provost, *Nucleic Acids Res*, 2008, **36**, 2353–2365.
82. Y. Bennasser, S. Y. Le, M. L. Yeung and K. T. Jeang, *Retrovirology*, 2004, **1**, 43.
83. V. R. Sanghvi and L. F. Steel, *PloS one*, 2011, **6**, e17246.
84. K. Chiang and A. P. Rice, *Future Virol*, 2011, **6**, 209–221.
85. G. Sun and J. J. Rossi, *Trends Pharmacol Sci*, 2011, **32**, 675–681.
86. B. R. Cullen, *Genes & development*, 2011, **25**, 1881–1894.
87. Z. Klase, P. Kale, R. Winograd, M. V. Gupta, M. Heydarian, R. Berro, T. McCaffrey and F. Kashanchi, *BMC Mol Biol*, 2007, **8**, 63.
88. M. L. Yeung, Y. Bennasser, K. Watashi, S. Y. Le, L. Houzet and K. T. Jeang, *Nucleic Acids Res*, 2009, **37**, 6575–6586.
89. D. A. Suhy, S. C. Kao, T. Mao, L. Whiteley, H. Denise, B. Souberbielle, A. D. Burdick, K. Hayes, J. F. Wright, H. Lavender, P. Roelvink, A. Kolykhalov, K. Brady, S. A. Moschos, B. Hauck, O. Zelenaia, S. Zhou, C. Scribner, K. A. High, S. H. Renison and R. Corbau, *Mol Ther*, 2012, **20**, 1737–1749.
90. M. Imbeault, K. Giguere, M. Ouellet and M. J. Tremblay, *PLoS Pathog*, 2012, **8**, e1002861.
91. C. Riou, V. V. Ganusov, S. Campion, M. Mlotshwa, M. K. Liu, V. E. Whale, N. Goonetilleke, P. Borrow, G. Ferrari, M. R. Betts, B. F. Haynes, A. J. McMichael and C. M. Gray, *J Immunol*, 2012, **188**, 2198–2206.
92. P. Parameswaran, E. Sklan, C. Wilkins, T. Burgon, M. A. Samuel, R. Lu, K. M. Ansel, V. Heissmeyer, S. Einav, W. Jackson, T. Doukas, S. Paranjape, C. Polacek, F. B. dos Santos, R. Jalili, F. Babrzadeh, B. Gharizadeh, D. Grimm, M. Kay, S. Koike, P. Sarnow, M. Ronaghi, S. W. Ding, E. Harris, M. Chow, M. S. Diamond, K. Kirkegaard, J. S. Glenn and A. Z. Fire, *PLoS Pathog*, 2010, **6**, e1000764.
93. N. Xu, S. Gkountela, K. Saeed and G. Akusjarvi, *Nucleic Acids Res*, 2009, **37**, 6950–6959.
94. R. P. Kincaid, J. M. Burke and C. S. Sullivan, *Proc Natl Acad Sci USA*, 2012, **109**, 3077–3082.
95. A. Varble, M. A. Chua, J. T. Perez, B. Manicassamy, A. Garcia-Sastre and B. R. tenOever, *Proc Natl Acad Sci USA*, 2010, **107**, 11519–11524.
96. A. Varble and B. R. Tenoever, *RNA biology*, 2011, **8**, 190–194.
97. H. Rouha, C. Thurner and C. W. Mandl, *Nucleic Acids Res*, 2010, **38**, 8328–8337.
98. J. S. Shapiro, A. Varble, A. M. Pham and B. R. Tenoever, *RNA*, 2010, **16**, 2068–2074.
99. R. A. Langlois, J. S. Shapiro, A. M. Pham and B. R. Tenoever, *Mol Ther*, 2012, **20**, 367–375.
100. R. A. Langlois, J. S. Shapiro, A. M. Pham and B. R. Tenoever, *Mol Ther*, 2012, **20**, 367–375.

101. Y. Lee, M. Kim, J. Han, K. H. Yeom, S. Lee, S. H. Baek and V. N. Kim, *EMBO J*, 2004, **23**, 4051–4060.
102. G. M. Borchert, W. Lanier and B. L. Davidson, *Nat Struct Mol Biol*, 2006, **13**, 1097–1101.
103. A. M. Denli, B. B. Tops, R. H. Plasterk, R. F. Ketting and G. J. Hannon, *Nature*, 2004, **432**, 231–235.
104. R. I. Gregory, K. P. Yan, G. Amuthan, T. Chendrimada, B. Doratotaj, N. Cooch and R. Shiekhattar, *Nature*, 2004, **432**, 235–240.
105. Y. Wang, R. Medvid, C. Melton, R. Jaenisch and R. Blelloch, *Nat Genet*, 2007, **39**, 380–385.
106. J. Han, Y. Lee, K. H. Yeom, J. W. Nam, I. Heo, J. K. Rhee, S. Y. Sohn, Y. Cho, B. T. Zhang and V. N. Kim, *Cell*, 2006, **125**, 887–901.
107. Y. Zeng and B. R. Cullen, *J Biol Chem*, 2005, **280**, 27595–27603.
108. X. Zhang and Y. Zeng, *Nucleic Acids Res*, 2010, **38**, 7689–7697.
109. Y. Zeng and B. R. Cullen, *Nucleic Acids Res*, 2004, **32**, 4776–4785.
110. R. Yi, Y. Qin, I. G. Macara and B. R. Cullen, *Genes Devel*, 2003, **17**, 3011–3016.
111. E. Lund, S. Guttinger, A. Calado, J. E. Dahlberg and U. Kutay, *Science*, 2004, **303**, 95–98.
112. G. Hutvagner, J. McLachlan, A. E. Pasquinelli, E. Balint, T. Tuschl and P. D. Zamore, *Science*, 2001, **293**, 834–838.
113. E. Maniataki and Z. Mourelatos, *Genes Devel*, 2005, **19**, 2979–2990.
114. R. I. Gregory, T. P. Chendrimada, N. Cooch and R. Shiekhattar, *Cell*, 2005, **123**, 631–640.
115. T. P. Chendrimada, R. I. Gregory, E. Kumaraswamy, J. Norman, N. Cooch, K. Nishikura and R. Shiekhattar, *Nature*, 2005, **436**, 740–744.
116. A. D. Haase, L. Jaskiewicz, H. Zhang, S. Laine, R. Sack, A. Gatignol and W. Filipowicz, *EMBO Reports*, 2005, **6**, 961–967.
117. Y. Lee, I. Hur, S. Y. Park, Y. K. Kim, M. R. Suh and V. N. Kim, *EMBO J*, 2006, **25**, 522–532.
118. G. S. Tan, B. G. Garchow, X. Liu, D. Metzler and M. Kiriakidou, *BMC Mol Biol*, 2011, **12**, 19.
119. L. Peters and G. Meister, *Mol Cell*, 2007, **26**, 611–623.
120. G. Meister, M. Landthaler, A. Patkaniowska, Y. Dorsett, G. Teng and T. Tuschl, *Mol Cell*, 2004, **15**, 185–197.
121. J. Liu, M. A. Carmell, F. V. Rivas, C. G. Marsden, J. M. Thomson, J. J. Song, S. M. Hammond, L. Joshua-Tor and G. J. Hannon, *Science*, 2004, **305**, 1437–1441.
122. A. Azuma-Mukai, H. Oguri, T. Mituyama, Z. R. Qian, K. Asai, H. Siomi and M. C. Siomi, *Proc Natl Acad Sci USA*, 2008, **105**, 7964–7969.
123. M. Hafner, M. Landthaler, L. Burger, M. Khorshid, J. Hausser, P. Berninger, A. Rothballer, M. Ascano, Jr., A. C. Jungkamp, M. Munschauer, A. Ulrich, G. S. Wardle, S. Dewell, M. Zavolan and T. Tuschl, *Cell*, 2010, **141**, 129–141.
124. A. M. Burroughs, Y. Ando, M. J. de Hoon, Y. Tomaru, H. Suzuki, Y. Hayashizaki and C. O. Daub, *RNA Biol*, 2011, **8**, 158–177.

125. A. Dueck, C. Ziegler, A. Eichner, E. Berezikov and G. Meister, *Nucleic Acids Res*, 2012, **40**, 9850–9862.
126. H. Su, M. I. Trombly, J. Chen and X. Wang, *Genes Devel*, 2009, **23**, 304–317.
127. S. Gu, L. Jin, F. Zhang, Y. Huang, D. Grimm, J. J. Rossi and M. A. Kay, *Proc Natl Acad Sci USA*, 2011, **108**, 9208–9213.
128. M. M. Perry, S. A. Moschos, A. E. Williams, N. J. Shepherd, H. M. Larner-Svensson and M. A. Lindsay, *J Immunol*, 2008, **180**, 5689–5698.
129. C. L. Noland, E. Ma and J. A. Doudna, *Mol Cell*, 2011, **43**, 110–121.
130. J. G. Betancur and Y. Tomari, *RNA*, 2012, **18**, 24–30.
131. K. Sakurai, M. Amarzguioui, D. H. Kim, J. Alluin, B. Heale, M. S. Song, A. Gatignol, M. A. Behlke and J. J. Rossi, *Nucleic Acids Res*, 2011, **39**, 1510–1525.
132. F. Frank, N. Sonenberg and B. Nagar, *Nature*, 2010, **465**, 818–822.
133. M. Yoda, T. Kawamata, Z. Paroo, X. Ye, S. Iwasaki, Q. Liu and Y. Tomari, *Nat Struct Mol Biol*, 2010, **17**, 17–23.
134. M. Amarzguioui and J. J. Rossi, *Methods Mol Biol*, 2008, **442**, 3–10.
135. M. Tanudji, D. Machalek, G. M. Arndt and L. Rivory, *Oligonucleotides*, 2010, **20**, 27–32.
136. D. Grimm, L. Wang, J. S. Lee, N. Schurmann, S. Gu, K. Borner, T. A. Storm and M. A. Kay, *J Clin Invest*, 2010, **120**, 3106–3119.
137. A. Khvorova, A. Reynolds and S. D. Jayasena, *Cell*, 2003, **115**, 209–216.
138. J. Krol, K. Sobczak, U. Wilczynska, M. Drath, A. Jasinska, D. Kaczynska and W. J. Krzyzosiak, *J Biol Chem*, 2004, **279**, 42230–42239.
139. H. Y. Hu, Z. Yan, Y. Xu, H. Hu, C. Menzel, Y. H. Zhou, W. Chen and P. Khaitovich, *BMC Genomics*, 2009, **10**, 413.
140. S. C. Li, Y. L. Liao, M. R. Ho, K. W. Tsai, C. H. Lai and W. C. Lin, *BMC Benomics*, 2012, **13**(Suppl 1), S13.
141. T. Kawamata, M. Yoda and Y. Tomari, *EMBO Reports*, 2011, **12**, 944–949.
142. Y. Liu, X. Ye, F. Jiang, C. Liang, D. Chen, J. Peng, L. N. Kinch, N. V. Grishin and Q. Liu, *Science*, 2009, **325**, 750–753.
143. X. Ye, N. Huang, Y. Liu, Z. Paroo, C. Huerta, P. Li, S. Chen, Q. Liu and H. Zhang, *Nat Struct Mol Biol*, 2011, **18**, 650–657.
144. T. Kawamata, H. Seitz and Y. Tomari, *Nat Struct Mol Biol*, 2009, **16**, 953–960.
145. D. A. Suhy, S. C. Kao, T. Mao, L. Whiteley, H. Denise, B. Souberbielle, A. D. Burdick, K. Hayes, J. F. Wright, H. Lavender, P. Roelvink, A. Kolykhalov, K. Brady, S. A. Moschos, B. Hauck, O. Zelenaia, S. Zhou, C. Scribner, K. A. High, S. H. Renison and R. Corbau, *Mol Ther*, 2012, **20**, 1737–1749.
146. A. Ulusoy, G. Sahin, T. Bjorklund, P. Aebischer and D. Kirik, *Mol Ther*, 2009, **17**, 1574–1584.
147. L. T. Bish, M. M. Sleeper, C. Reynolds, J. Gazzara, E. Withnall, G. E. Singletary, G. Buchlis, D. Hui, K. A. High, G. Gao, J. M. Wilson and H. L. Sweeney, *Hum Gene Ther*, 2011, **22**, 969–977.

148. D. S. An, F. X. Qin, V. C. Auyeung, S. H. Mao, S. K. Kung, D. Baltimore and I. S. Chen, *Mol Ther*, 2006, **14**, 494–504.
149. A. Calado, N. Treichel, E. C. Muller, A. Otto and U. Kutay, *EMBO J*, 2002, **21**, 6216–6224.
150. Y. Bennasser, C. Chable-Bessia, R. Triboulet, D. Gibbings, C. Gwizdek, C. Dargemont, E. J. Kremer, O. Voinnet and M. Benkirane, *Nat Struct Mol Biol*, 2011, **18**, 323–327.
151. D. Langenberger, M. V. Cakir, S. Hoffmann and P. F. Stadler, *J Exp Zool B Mol Dev Evol*, 2013, **320**, 35–46.
152. C. Gwizdek, B. Ossareh-Nazari, A. M. Brownawell, A. Doglio, E. Bertrand, I. G. Macara and C. Dargemont, *J Biol Chem*, 2003, **278**, 5505–5508.
153. E. M. Ehlert, R. Eggers, S. P. Niclou and J. Verhaagen, *BMC Neurosci*, 2010, **11**, 20.
154. E. Bernstein, A. A. Caudy, S. M. Hammond and G. J. Hannon, *Nature*, 2001, **409**, 363–366.
155. K. Okamura, J. W. Hagen, H. Duan, D. M. Tyler and E. C. Lai, *Cell*, 2007, **130**, 89–100.
156. J. G. Ruby, C. H. Jan and D. P. Bartel, *Nature*, 2007, **448**, 83–86.
157. E. Ladewig, K. Okamura, A. S. Flynt, J. O. Westholm and E. C. Lai, *Genome Res*, 2012, **22**, 1634–1645.
158. M. A. Havens, A. A. Reich, D. M. Duelli and M. L. Hastings, *Nucleic Acids Res*, 2012, **40**, 4626–4640.
159. C. R. Sibley, Y. Seow, S. Saayman, K. K. Dijkstra, S. El Andaloussi, M. S. Weinberg and M. J. Wood, *Nucleic Acids Res*, 2012, **40**, 438–448.
160. H. P. Bogerd, H. W. Karnowski, X. Cai, J. Shin, M. Pohlers and B. R. Cullen, *Mol Cell*, 2010, **37**, 135–142.
161. L. J. Scherer, R. Frank and J. J. Rossi, *Nucleic Acids Res*, 2007, **35**, 2620–2628.
162. M. Ono, M. S. Scott, K. Yamada, F. Avolio, G. J. Barton and A. I. Lamond, *Nucleic Acids Res*, 2011, **39**, 3879–3891.
163. M. S. Scott, F. Avolio, M. Ono, A. I. Lamond and G. J. Barton, *PLoS Comput Biol*, 2009, **5**, e1000507.
164. C. Ender, A. Krek, M. R. Friedlander, M. Beitzinger, L. Weinmann, W. Chen, S. Pfeffer, N. Rajewsky and G. Meister, *Mol Cell*, 2008, **32**, 519–528.
165. R. J. Taft, E. A. Glazov, T. Lassmann, Y. Hayashizaki, P. Carninci and J. S. Mattick, *RNA*, 2009, **15**, 1233–1240.
166. S. Cheloufi, C. O. Dos Santos, M. M. Chong and G. J. Hannon, *Nature*, 2010, **465**, 584–589.
167. J. S. Yang, T. Maurin, N. Robine, K. D. Rasmussen, K. L. Jeffrey, R. Chandwani, E. P. Papapetrou, M. Sadelain, D. O'Carroll and E. C. Lai, *Proc Natl Acad Sci USA*, 2010, **107**, 15163–15168.
168. D. Cifuentes, H. Xue, D. W. Taylor, H. Patnode, Y. Mishima, S. Cheloufi, E. Ma, S. Mane, G. J. Hannon, N. D. Lawson, S. A. Wolfe and A. J. Giraldez, *Science*, 2010, **328**, 1694–1698.

169. J. S. Yang and E. C. Lai, *Cell Cycle*, 2010, **9**, 4455–4460.
170. J. S. Yang, T. Maurin and E. C. Lai, *RNA*, 2012, **18**, 945–957.
171. W. Shi, D. Hendrix, M. Levine and B. Haley, *Nat Struct Mol Biol*, 2009, **16**, 183–189.
172. D. Langenberger, C. Bermudez-Santana, J. Hertel, S. Hoffmann, P. Khaitovich and P. F. Stadler, *Bioinformatics*, 2009, **25**, 2298–2301.
173. S. Bortoluzzi, A. Bisognin, M. Biasiolo, P. Guglielmelli, F. Biamonte, R. Norfo, R. Manfredini and A. M. Vannucchi, *Blood*, 2012, **119**, e120–130.
174. A. M. Burroughs, M. Kawano, Y. Ando, C. O. Daub and Y. Hayashizaki, *Nucleic Acids Res*, 2012, **40**, 1424–1437.
175. H. Zhou, M. L. Arcila, Z. Li, E. J. Lee, C. Henzler, J. Liu, T. M. Rana and K. S. Kosik, *Nucleic Acids Res*, 2012, **40**, 5864–5875.
176. E. Meiri, A. Levy, H. Benjamin, M. Ben-David, L. Cohen, A. Dov, N. Dromi, E. Elyakim, N. Yerushalmi, O. Zion, G. Lithwick-Yanai and E. Sitbon, *Nucleic Acids Res*, 2010, **38**, 6234–6246.
177. J. L. Umbach, L. I. Strelow, S. W. Wong and B. R. Cullen, *Virology*, 2010, **405**, 592–599.
178. J. Song Gao, Y. Zhang, M. Li, L. D. Tucker, J. T. Machan, P. Quesenberry, I. Rigoutsos and B. Ramratnam, *Nucleic Acids Res*, 2010, **38**, 2775–2787.
179. B. Vennstrom, U. Pettersson and L. Philipson, *Nucleic Acids Res*, 1978, **5**, 195–204.
180. N. Xu, B. Segerman, X. Zhou and G. Akusjarvi, *J Virol*, 2007, **81**, 10540–10549.
181. A. E. Williams, M. M. Perry, S. A. Moschos, H. M. Larner-Svensson and M. A. Lindsay, *Biochem Soc Trans*, 2008, **36**, 1211–1215.
182. J. Starega-Roslan, J. Krol, E. Koscianska, P. Kozlowski, W. J. Szlachcic, K. Sobczak and W. J. Krzyzosiak, *Nucleic Acids Res*, 2011, **39**, 257–268.
183. J. M. Cummins, Y. He, R. J. Leary, R. Pagliarini, L. A. Diaz, Jr., T. Sjoblom, O. Barad, Z. Bentwich, A. E. Szafranska, E. Labourier, C. K. Raymond, B. S. Roberts, H. Juhl, K. W. Kinzler, B. Vogelstein and V. E. Velculescu, *Proc Natl Acad Sci USA*, 2006, **103**, 3687–3692.
184. E. Koscianska, J. Starega-Roslan and W. J. Krzyzosiak, *PloS one*, 2011, **6**, e28548.
185. S. C. Li, K. W. Tsai, H. W. Pan, Y. M. Jeng, M. R. Ho and W. H. Li, *BMC Sys Biol*, 2012, **6**(Suppl 2), S14.
186. J. Huang, Z. Liang, B. Yang, H. Tian, J. Ma and H. Zhang, *J Biol Chem*, 2007, **282**, 33632–33640.
187. Z. Peng, Y. Cheng, B. C. Tan, L. Kang, Z. Tian, Y. Zhu, W. Zhang, Y. Liang, X. Hu, X. Tan, J. Guo, Z. Dong, Y. Liang, L. Bao and J. Wang, *Nat Biotechnol*, 2012, **30**, 253–260.
188. A. M. Burroughs, Y. Ando, M. J. de Hoon, Y. Tomaru, T. Nishibu, R. Ukekawa, T. Funakoshi, T. Kurokawa, H. Suzuki, Y. Hayashizaki and C. O. Daub, *Genome Res*, 2010, **20**, 1398–1410.
189. P. K. Juvvuna, P. Khandelia, L. M. Lee and E. V. Makeyev, *Nucleic Acids Res*, 2012, **40**, 6808–6820.

190. H. A. Ebhardt, H. H. Tsang, D. C. Dai, Y. Liu, B. Bostan and R. P. Fahlman, *Nucleic Acids Res*, 2009, **37**, 2461–2470.
191. A. D. Jayaprakash, O. Jabado, B. D. Brown and R. Sachidanandam, *Nucleic Acids Res*, 2011, **39**, e141.
192. S. Alon, F. Vigneault, S. Eminaga, D. C. Christodoulou, J. G. Seidman, G. M. Church and E. Eisenberg, *Genome Res*, 2011, **21**, 1506–1511.
193. M. J. de Hoon, R. J. Taft, T. Hashimoto, M. Kanamori-Katayama, H. Kawaji, M. Kawano, M. Kishima, T. Lassmann, G. J. Faulkner, J. S. Mattick, C. O. Daub, P. Carninci, J. Kawai, H. Suzuki and Y. Hayashizaki, *Genome Res*, 2010, **20**, 257–264.
194. N. Cloonan, S. Wani, Q. Xu, J. Gu, K. Lea, S. Heater, C. Barbacioru, A. L. Steptoe, H. C. Martin, E. Nourbakhsh, K. Krishnan, B. Gardiner, X. Wang, K. Nones, J. A. Steen, N. A. Matigian, D. L. Wood, K. S. Kassahn, N. Waddell, J. Shepherd, C. Lee, J. Ichikawa, K. McKernan, K. Bramlett, S. Kuersten and S. M. Grimmond, *Genome Biol*, 2011, **12**, R126.
195. C. E. Joyce, X. Zhou, J. Xia, C. Ryan, B. Thrash, A. Menter, W. Zhang and A. M. Bowcock, *Hum Mol Genet*, 2011, **20**, 4025–4040.
196. Y. Kawahara, B. Zinshteyn, P. Sethupathy, H. Iizasa, A. G. Hatzigeorgiou and K. Nishikura, *Science*, 2007, **315**, 1137–1140.
197. L. W. Lee, S. Zhang, A. Etheridge, L. Ma, D. Martin, D. Galas and K. Wang, *RNA*, 2010, **16**, 2170–2180.
198. W. Zhang, S. Gao, X. Zhou, J. Xia, P. Chellappan, X. Zhou, X. Zhang and H. Jin, *Genome Biol*, 2010, **11**, R81.
199. E. Marti, L. Pantano, M. Banez-Coronel, F. Llorens, E. Minones-Moyano, S. Porta, L. Sumoy, I. Ferrer and X. Estivill, *Nucleic Acids Res*, 2010, **38**, 7219–7235.
200. H. Wu, J. R. Neilson, P. Kumar, M. Manocha, P. Shankar, P. A. Sharp and N. Manjunath, *PloS One*, 2007, **2**, e1020.
201. H. Zhou, M. L. Arcila, Z. Li, E. J. Lee, C. Henzler, J. Liu, T. M. Rana and K. S. Kosik, *Nucleic Acids Res*, 2012, **40**, 5864–5875.
202. D. T. Humphreys, C. J. Hynes, H. R. Patel, G. H. Wei, L. Cannon, D. Fatkin, C. M. Suter, J. L. Clancy and T. Preiss, *PloS One*, 2012, **7**, e30933.
203. R. D. Morin, M. D. O'Connor, M. Griffith, F. Kuchenbauer, A. Delaney, A. L. Prabhu, Y. Zhao, H. McDonald, T. Zeng, M. Hirst, C. J. Eaves and M. A. Marra, *Genome Res*, 2008, **18**, 610–621.
204. F. Llorens, M. Banez-Coronel, L. Pantano, J. A. Del Rio, I. Ferrer, X. Estivill and E. Marti, *BMC Genomics*, 2013, **14**, 104.
205. S. M. Elbashir, W. Lendeckel and T. Tuschl, *Genes Devel*, 2001, **15**, 188–200.
206. M. Van Stry, T. H. Oguin, 3rd, S. Cheloufi, P. Vogel, M. Watanabe, M. R. Pillai, P. Dash, P. G. Thomas, G. J. Hannon and M. Bix, *J Virol*, 2012, **86**, 4151–4157.
207. R. Zhou and T. M. Rana, *Immunol Rev*, 2013, **253**, 97–111.
208. B. Berkhout and A. T. Das, *Mol Ther Nucleic Acids*, 2012, **1**, e26.

209. S. Takasaki, *Methods Mol Biol*, 2013, **942**, 17–55.
210. N. Nasheri, J. P. Pezacki and S. M. Sagan, *Methods Mol Biol*, 2013, **942**, 69–86.
211. C. Shin, J. W. Nam, K. K. Farh, H. R. Chiang, A. Shkumatava and D. P. Bartel, *Mol Cell*, 2010, **38**, 789–802.
212. Y. Naito and K. Ui-Tei, *Methods Mol Biol*, 2013, **942**, 57–68.
213. B. P. Lewis, I. H. Shih, M. W. Jones-Rhoades, D. P. Bartel and C. B. Burge, *Cell*, 2003, **115**, 787–798.
214. S. W. Chi, G. J. Hannon and R. B. Darnell, *Nat Struct Mol Biol*, 2012, **19**, 321–327.
215. D. P. Bartel, *Cell*, 2009, **136**, 215–233.
216. R. Parker and U. Sheth, *Mol Cell*, 2007, **25**, 635–646.
217. J. Liu, M. A. Valencia-Sanchez, G. J. Hannon and R. Parker, *Nat Cell Biol*, 2005, **7**, 719–723.
218. D. J. Gibbings, C. Ciaudo, M. Erhardt and O. Voinnet, *Nat Cell Biol*, 2009, **11**, 1143–1149.
219. Y. S. Lee, S. Pressman, A. P. Andress, K. Kim, J. L. White, J. J. Cassidy, X. Li, K. Lubell, H. Lim, I. S. do, K. Cho, J. B. Nakahara, P. Preall, E. J. Bellare, Sontheimer and R. W. Carthew, *Nat Cell Biol*, 2009, **11**, 1150–1156.
220. W. Filipowicz, S. N. Bhattacharyya and N. Sonenberg, *Nature reviews. Genetics*, 2008, **9**, 102–114.
221. J. Wen, B. J. Parker, A. Jacobsen and A. Krogh, *RNA*, 2011, **17**, 820–834.
222. K. D. Taganov, M. P. Boldin, K. J. Chang and D. Baltimore, *Proc Natl Acad Sci USA*, 2006, **103**, 12481–12486.
223. D. W. Thomson, C. P. Bracken and G. J. Goodall, *Nucleic Acids Res*, 2011, **39**, 6845–6853.
224. D. W. Thomson, C. P. Bracken, J. M. Szubert and G. J. Goodall, *PLoS One*, 2013, **8**, e55214.
225. J. Zhou, M. S. Song, A. M. Jacobi, M. A. Behlke, X. Wu and J. J. Rossi, *Mol Ther Nucleic Acids*, 2012, **1**, e17.
226. A. Valoczi, C. Hornyik, N. Varga, J. Burgyan, S. Kauppinen and Z. Havelda, *Nucleic Acids Res*, 2004, **32**, e175.
227. I. Naguibneva, M. Ameyar-Zazoua, N. Nonne, A. Polesskaya, S. Ait-Si-Ali, R. Groisman, M. Souidi, L. L. Pritchard and A. Harel-Bellan, *Biomed Pharmacother*, 2006, **60**, 633–638.
228. E. M. Zaghloul, A. S. Madsen, P. M. Moreno, S. Oprea, II, B. El-Andaloussi, P. Bestas, E. B. Gupta, K. E. Pedersen, J. Lundin, Wengel and C. I. Smith, *Nucleic Acids Res*, 2011, **39**, 1142–1154.
229. C. Laxton, K. Brady, S. Moschos, P. Turnpenny, J. Rawal, D. C. Pryde, B. Sidders, R. Corbau, C. Pickford and E. J. Murray, *Antimicrob Agents Chemother*, 2011, **55**, 3105–3114.
230. A. A. Khan, D. Betel, M. L. Miller, C. Sander, C. S. Leslie and D. S. Marks, *Nat Biotechnol*, 2009, **27**, 549–555.
231. G. Sun, H. Li and J. J. Rossi, *Nucleic Acids Res*, 2010, **38**, 239–252.
232. P. Brodersen and O. Voinnet, *Nat Rev Mol Cell Biol*, 2009, **10**, 141–148.

233. G. Sun and J. J. Rossi, *RNA Biol*, 2009, **6**, 406–411.
234. R. Sandberg, J. R. Neilson, A. Sarma, P. A. Sharp and C. B. Burge, *Science*, 2008, **320**, 1643–1647.
235. L. P. Lim, N. C. Lau, P. Garrett-Engele, A. Grimson, J. M. Schelter, J. Castle, D. P. Bartel, P. S. Linsley and J. M. Johnson, *Nature*, 2005, **433**, 769–773.
236. D. Baek, J. Villen, C. Shin, F. D. Camargo, S. P. Gygi and D. P. Bartel, *Nature*, 2008, **455**, 64–71.
237. M. Selbach, B. Schwanhausser, N. Thierfelder, Z. Fang, R. Khanin and N. Rajewsky, *Nature*, 2008, **455**, 58–63.
238. R. B. Darnell, *Wiley Interdiscip Rev RNA*, 2010, **1**, 266–286.
239. M. Hammell, D. Long, L. Zhang, A. Lee, C. S. Carmack, M. Han, Y. Ding and V. Ambros, *Nat Methods*, 2008, **5**, 813–819.
240. S. W. Chi, J. B. Zang, A. Mele and R. B. Darnell, *Nature*, 2009, **460**, 479–486.
241. P. T. Nelson, M. De Planell-Saguer, S. Lamprinaki, M. Kiriakidou, P. Zhang, U. O'Doherty and Z. Mourelatos, *RNA*, 2007, **13**, 1787–1792.
242. Z. Fang and N. Rajewsky, *PLoS One*, 2011, **6**, e18067.
243. Y. Tay, J. Zhang, A. M. Thomson, B. Lim and I. Rigoutsos, *Nature*, 2008, **455**, 1124–1128.
244. W. F. Shen, Y. L. Hu, L. Uttarwar, E. Passegue and C. Largman, *Mol Cell Biol*, 2008, **28**, 4609–4619.
245. Z. Xia, P. Clark, T. Huynh, P. Loher, Y. Zhao, H. W. Chen, I. Rigoutsos and R. Zhou, *Sci Rep*, 2012, **2**, 569.
246. M. Schnall-Levin, O. S. Rissland, W. K. Johnston, N. Perrimon, D. P. Bartel and B. Berger, *Genome Res*, 2011, **21**, 1395–1403.
247. A. Lal, H. H. Kim, K. Abdelmohsen, Y. Kuwano, R. Pullmann, Jr., S. Srikantan, R. Subrahmanyam, J. L. Martindale, X. Yang, F. Ahmed, F. Navarro, D. Dykxhoorn, J. Lieberman and M. Gorospe, *PLoS One*, 2008, **3**, e1864.
248. J. J. Forman, A. Legesse-Miller and H. A. Coller, *Proc Natl Acad Sci USA*, 2008, **105**, 14879–14884.
249. A. M. Duursma, M. Kedde, M. Schrier, C. le Sage and R. Agami, *RNA*, 2008, **14**, 872–877.
250. K. J. Riley, T. A. Yario and J. A. Steitz, *RNA*, 2012, **18**, 1581–1585.
251. C. P. Bracken, J. M. Szubert, T. R. Mercer, M. E. Dinger, D. W. Thomson, J. S. Mattick, M. Z. Michael and G. J. Goodall, *Nucleic Acids Res*, 2011, **39**, 5658–5668.
252. J. H. Yang, J. H. Li, P. Shao, H. Zhou, Y. Q. Chen and L. H. Qu, *Nucleic Acids Res*, 2011, **39**, D202–209.
253. T. Vergoulis, I. S. Vlachos, P. Alexiou, G. Georgakilas, M. Maragkakis, M. Reczko, S. Gerangelos, N. Koziris, T. Dalamagas and A. G. Hatzigeorgiou, *Nucleic Acids Res*, 2012, **40**, D222–229.
254. C. H. Chou, F. M. Lin, M. T. Chou, S. D. Hsu, T. H. Chang, S. L. Weng, S. Shrestha, C. C. Hsiao, J. H. Hung and H. D. Huang, *BMC Genomics*, 2013, **14**(Suppl 1), S2.

255. O. Larsson, B. Tian and N. Sonenberg, *Cold Spring Harbor Perspect Biol*, 2013, **5**, a012302.
256. C. E. Vejnar and E. M. Zdobnov, *Nucleic Acids Res*, 2012, **40**, 11673–11683.
257. J. H. Yang and L. H. Qu, *Adv Exp Med Biol*, 2013, **774**, 251–266.
258. D. C. Ellwanger, F. A. Buttner, H. W. Mewes and V. Stumpflen, *Bioinformatics*, 2011, **27**, 1346–1350.
259. S. Lee and S. Vasudevan, *Adv Exp Med Biol*, 2013, **768**, 97–126.
260. C. Fehr, K. D. Conrad and M. Niepmann, *Cell Cycle*, 2012, **11**, 277–285.
261. S. Vasudevan, Y. Tong and J. A. Steitz, *Science*, 2007, **318**, 1931–1934.
262. S. S. Truesdell, R. D. Mortensen, M. Seo, J. C. Schroeder, J. H. Lee, O. LeTonqueze and S. Vasudevan, *Sci Rep*, 2012, **2**, 842.
263. L. Tserel, T. Runnel, K. Kisand, M. Pihlap, L. Bakhoff, R. Kolde, H. Peterson, J. Vilo, P. Peterson and A. Rebane, *J Biol Chem*, 2011, **286**, 26487–26495.
264. C. C. Lin, L. Z. Liu, J. B. Addison, W. F. Wonderlin, A. V. Ivanov and J. M. Ruppert, *Mol Cell Biol*, 2011, **31**, 2513–2527.
265. T. Ohrt, J. Mutze, W. Staroske, L. Weinmann, J. Hock, K. Crell, G. Meister and P. Schwille, *Nucleic Acids Res*, 2008, **36**, 6439–6449.
266. L. Weinmann, J. Hock, T. Ivacevic, T. Ohrt, J. Mutze, P. Schwille, E. Kremmer, V. Benes, H. Urlaub and G. Meister, *Cell*, 2009, **136**, 496–507.
267. D. Castanotto, R. Lingeman, A. D. Riggs and J. J. Rossi, *Proc Natl Acad Sci USA*, 2009, **106**, 21655–21659.
268. H. W. Hwang, E. A. Wentzel and J. T. Mendell, *Science*, 2007, **315**, 97–100.
269. N. S. Lee, D. H. Kim, J. Alluin, M. Robbins, S. Gu, H. Li, J. Kim, P. M. Salvaterra and J. J. Rossi, *RNA*, 2008, **14**, 1823–1833.
270. J. Y. Liao, L. M. Ma, Y. H. Guo, Y. C. Zhang, H. Zhou, P. Shao, Y. Q. Chen and L. H. Qu, *PLoS One*, 2010, **5**, e10563.
271. B. Chen, B. Zhang, H. Luo, J. Yuan, G. Skogerbo and R. Chen, *International journal of cell biology*, 2012, **2012**, 672462.
272. C. D. Jeffries, H. M. Fried and D. O. Perkins, *The international journal of biochemistry & cell biology*, 2010, **42**, 1236–1242.
273. J. C. Schwartz, S. T. Younger, N. B. Nguyen, D. B. Hardy, B. P. Monia, D. R. Corey and B. A. Janowski, *Nat Struct Mol Biol*, 2008, **15**, 842–848.
274. D. H. Kim, P. Saetrom, O. Snove, Jr. and J. J. Rossi, *Proc Natl Acad Sci USA*, 2008, **105**, 16230–16235.
275. L. C. Li, S. T. Okino, H. Zhao, D. Pookot, R. F. Place, S. Urakami, H. Enokida and R. Dahiya, *Proc Natl Acad Sci USA*, 2006, **103**, 17337–17342.
276. B. A. Janowski, S. T. Younger, D. B. Hardy, R. Ram, K. E. Huffman and D. R. Corey, *Nat Chem Biol*, 2007, **3**, 166–173.
277. R. F. Place, L. C. Li, D. Pookot, E. J. Noonan and R. Dahiya, *Proc Natl Acad Sci USA*, 2008, **105**, 1608–1613.
278. T. Iida, J. Nakayama and D. Moazed, *Mol Cell*, 2008, **31**, 178–189.

279. M. S. Weinberg, L. M. Villeneuve, A. Ehsani, M. Amarzguioui, L. Aagaard, Z. X. Chen, A. D. Riggs, J. J. Rossi and K. V. Morris, *RNA*, 2006, **12**, 256–262.
280. G. Zardo, A. Ciolfi, L. Vian, M. Billi, S. Racanicchi, F. Grignani and C. Nervi, *Cell Cycle*, 2012, **11**, 3543–3549.
281. T. Adilakshmi, I. Sudol and N. Tapinos, *PLoS One*, 2012, **7**, e39674.
282. M. Allo, V. Buggiano, J. P. Fededa, E. Petrillo, I. Schor, M. de la Mata, E. Agirre, M. Plass, E. Eyras, S. A. Elela, R. Klinck, B. Chabot and A. R. Kornblihtt, *Nat Struct Mol Biol*, 2009, **16**, 717–724.
283. J. Liu, J. Hu and D. R. Corey, *Nucleic Acids Res*, 2012, **40**, 1240–1250.
284. K. V. Morris, *Oligonucleotides*, 2009, **19**, 299–306.
285. R. Beane, S. Gabillet, C. Montaillier, K. Arar and D. R. Corey, *Biochemistry*, 2008, **47**, 13147–13149.
286. J. K. Watts, D. Yu, K. Charisse, C. Montaillier, P. Potier, M. Manoharan and D. R. Corey, *Nucleic Acids Res*, 2010, **38**, 5242–5259.
287. C. L. Ahlenstiel, H. G. Lim, D. A. Cooper, T. Ishida, A. D. Kelleher and K. Suzuki, *Nucleic Acids Res*, 2012, **40**, 1579–1595.
288. S. Pitchiaya, J. R. Androsavich and N. G. Walter, *EMBO reports*, 2012, **13**, 709–715.
289. S. Bandiera, S. Ruberg, M. Girard, N. Cagnard, S. Hanein, D. Chretien, A. Munnich, S. Lyonnet and A. Henrion-Caude, *PLoS One*, 2011, **6**, e20746.
290. E. Maniataki and Z. Mourelatos, *RNA*, 2005, **11**, 849–852.
291. L. Sripada, D. Tomar, P. Prajapati, R. Singh, A. K. Singh and R. Singh, *PLoS One*, 2012, **7**, e44873.
292. B. T. Kren, P. Y. Wong, A. Sarver, X. Zhang, Y. Zeng and C. J. Steer, *RNA Biol*, 2009, **6**, 65–72.
293. E. Barrey, G. Saint-Auret, B. Bonnamy, D. Damas, O. Boyer and X. Gidrol, *PLoS One*, 2011, **6**, e20220.
294. K. S. Makarova, N. V. Grishin, S. A. Shabalina, Y. I. Wolf and E. V. Koonin, *Biol Direct*, 2006, **1**, 7.
295. B. Wiedenheft, S. H. Sternberg and J. A. Doudna, *Nature*, 2012, **482**, 331–338.
296. U. A. Orom, F. C. Nielsen and A. H. Lund, *Mol Cell*, 2008, **30**, 460–471.
297. P. A. Thibault, A. Huys, P. Dhillon and J. A. Wilson, *Virology*, 2013, **436**, 179–190.
298. J. A. Wilson, C. Zhang, A. Huys and C. D. Richardson, *J Virol*, 2011, **85**, 2342–2350.
299. T. Fukuhara, H. Kambara, M. Shiokawa, C. Ono, H. Katoh, E. Morita, D. Okuzaki, Y. Maehara, K. Koike and Y. Matsuura, *J Virol*, 2012, **86**, 7918–7933.
300. C. M. Narbus, B. Israelow, M. Sourisseau, M. L. Michta, S. E. Hopcraft, G. M. Zeiner and M. J. Evans, *J Virol*, 2011, **85**, 12087–12092.
301. J. I. Henke, D. Goergen, J. Zheng, Y. Song, C. G. Schuttler, C. Fehr, C. Junemann and M. Niepmann, *EMBO J*, 2008, **27**, 3300–3310.
302. J. Chang, J. T. Guo, D. Jiang, H. Guo, J. M. Taylor and T. M. Block, *J Virol*, 2008, **82**, 8215–8223.

303. Y. Li, T. Masaki, D. Yamane, D. R. McGivern and S. M. Lemon, *Proc Natl Acad Sci USA*, 2013, **110**, 1881–1886.
304. T. Shimakami, D. Yamane, R. K. Jangra and B. J. Kempf, C. Spaniel, D. J. Barton and S. M. Lemon, *Proc Natl Acad Sci USA*, 2012, **109**, 941–946.
305. K. D. Conrad, F. Giering, C. Erfurth, A. Neumann, C. Fehr, G. Meister and M. Niepmann, *PLoS One*, 2013, **8**, e56272.
306. A. P. Roberts, A. P. Lewis and C. L. Jopling, *Nucleic Acids Res*, 2011, **39**, 7716–7729.
307. C. T. Pager, S. Schutz, T. M. Abraham, G. Luo and P. Sarnow, *Virology*, 2013, **435**, 472–484.
308. C. Zhang, A. Huys, P. A. Thibault and J. A. Wilson, *Virology*, 2012, **433**, 479–488.
309. P. S. Pang, E. A. Pham, M. Elazar, S. G. Patel, M. R. Eckart and J. S. Glenn, *J Virol*, 2012, **86**, 1250–1254.
310. D. Goergen and M. Niepmann, *Virus Res*, 2012, **167**, 343–352.
311. S. A. Mortimer and J. A. Doudna, *Nucleic Acids Res*, 2013.
312. Y. Ariumi, M. Kuroki, Y. Kushima, K. Osugi, M. Hijikata, M. Maki, M. Ikeda and N. Kato, *J Virol*, 2011, **85**, 6882–6892.
313. R. T. Marquez, S. Bandyopadhyay, E. B. Wendlandt, K. Keck, B. A. Hoffer, M. S. Icardi, R. N. Christensen, W. N. Schmidt and A. P. McCaffrey, *Lab Invest J Tech Methods Pathol*, 2010, **90**, 1727–1736.
314. M. Sarasin-Filipowicz, J. Krol, I. Markiewicz, M. H. Heim and W. Filipowicz, *Nat Med*, 2009, **15**, 31–33.
315. S. A. Moschos, *Bioanalysis*, 2012, **4**, 2499–2511.
316. W. C. Tsai, S. D. Hsu, C. S. Hsu, T. C. Lai, S. J. Chen, R. Shen, Y. Huang, H. C. Chen, C. H. Lee, T. F. Tsai, M. T. Hsu, J. C. Wu, H. D. Huang, M. S. Shiao, M. Hsiao and A. P. Tsou, *J Clin Invest*, 2012, **122**, 2884–2897.
317. X. Ding, J. Ding, J. Ning, F. Yi, J. Chen, D. Zhao, J. Zheng, Z. Liang, Z. Hu and Q. Du, *Mol Med Rep*, 2012, **5**, 1428–1432.
318. S. H. Hsu, B. Wang, J. Kota, J. Yu, S. Costinean, H. Kutay, L. Yu, S. Bai, K. La Perle, R. R. Chivukula, H. Mao, M. Wei, K. R. Clark, J. R. Mendell, M. A. Caligiuri, S. T. Jacob, J. T. Mendell and K. Ghoshal, *J Clin Invest*, 2012, **122**, 2871–2883.
319. J. R. Lytle, T. A. Yario and J. A. Steitz, *Proc Natl Acad Sci USA*, 2007, **104**, 9667–9672.
320. F. Grey, R. Tirabassi, H. Meyers, G. Wu, S. McWeeney, L. Hook and J. A. Nelson, *PLoS Pathog*, 2010, **6**, e1000967.
321. I. Lee, S. S. Ajay, J. I. Yook, H. S. Kim, S. H. Hong, N. H. Kim, S. M. Dhanasekaran, A. M. Chinnaiyan and B. D. Athey, *Genome Res*, 2009, **19**, 1175–1183.
322. S. S. Ajay, B. D. Athey and I. Lee, *BMC Genomics*, 2010, **11**, 155.
323. X. Zhou, X. Duan, J. Qian and F. Li, *Genetica*, 2009, **137**, 159–164.
324. D. Gaudet, J. de Wal, K. Tremblay, S. Dery, S. van Deventer, A. Freidig, D. Brisson and J. Methot, *Atheroscler Suppl*, 2010, **11**, 55–60.
325. A. G. Bader, *Front Genet*, 2012, **3**, 120.

CHAPTER 8

Novel Therapeutic Agents from Bacterial Toxins

JOHN A. CHADDOCK

Syntaxin Ltd., Units 4–10 The Quadrant, Barton Lane, Abingdon, OXON, OX14 3YS, UK
Email: john.chaddock@syntaxin.com

8.1 Introduction

Although it is difficult to arise at a consensus for the number of bacterial species on earth, it is well understood it represents one of the most diverse groups of living organisms and, from this diversity and the pressure to adapt to a variety of different environments, it is almost inevitable that evolutionary pressure has led to proteins with interesting and potentially useful properties. Bacterial protein toxins[1] are one such class of molecules with unique biology that can be both harmful and yet also useful.

A toxin is defined as a poisonous substance, such as a protein, that is capable of causing disease when introduced into the body. The ability of bacteria to produce toxins is termed toxigenesis. There are two main types of bacterial toxins in lipopolysaccharides and proteins, and this chapter will focus on protein toxins; specifically extracellular difusable toxins (exotoxins) that are produced from a subset of bacterial species. Released into the host by secretion or bacterial cell lysis, exotoxins act enzymatically within a host cell or by direct action with host cells leading to a stimulation of host cell responses. This chapter will focus on bacterial protein toxins that act intracellularly within a host cell, leading to a modification of host cell biology.

RSC Drug Discovery Series No. 36
Biotherapeutics: Recent Developments using Chemical and Molecular Biology
Edited by Lyn H. Jones and Andrew J. McKnight

Published by the Royal Society of Chemistry, www.rsc.org

This modification can be lethal to the intoxicated cell, leading to local cellular destruction and, if the intoxication is widespread, subequent host death. Alternatively, the host cellular modification can be non-lethal and result in a modified interaction with the environment. A good example of this is botulinum neurotoxin, produced from *Clostridium botulinum* species, in which the mechanism of action leads to inhibition of secretion from the target cell (a motor neuron). The motor neuron is not destroyed by the toxin, and indeed will recover functionality after the toxic effects have worn off. Do not be lulled into a false sense of security however; botulism outbreaks (a result of botulinum toxin intoxication) are frequent and can have fatal consequences. In these severe cases, the impact of toxin action is to inihibit muscle contraction, manifest as a descending paralysis, which leads to an inability to maintain respiration.

In terms of relative toxicity, it is often not understood that bacterial protein toxins are the most powerful human poisons known and are effective at very low concentration. For example, the toxicity of the *C. botulinum* and *Clostridium tetanus* producted toxins (botulinum neurotoxin (BoNT) and tetanus neurotoxin (TeNT), respectively) can be measured in the low picogram range in mice. BoNT is the most toxic protein known, with approximately 2×10^8 LD_{50} units per milligram of material. Estimates for doses that result in human lethality vary depending on the route of administration, but it has been reported to be in the range of 0.09–0.15 μg intravenously or intramuscularly, 0.70–0.90 μg inhalationally, and 70 μg orally.[2] This means that BoNT is a million fold more toxic than arsenic.

Bacterial toxins damage the host at the site of bacterial infection or distant from the site. Bacterial toxins can be single proteins or oligomeric protein complexes that are organized with distinct AB structure-function properties. The A domain encodes a catalytic activity. ADP ribosylation of host proteins is the earliest post-translational modification determined to be performed by bacterial toxins; other modifications include glucosylation and proteolysis. Bacterial toxins also catalyse the non-covalent modification of host protein function or can modify host cell properties through direct protein–protein interactions. The B domain includes two functional domains: a receptor-binding domain, which defines the tropism of a toxin for a cell, and a translocation domain that delivers the A domain across a lipid bilayer, either on the plasma membrane or the endosome. In some cases, for example in Shiga toxin (from *Shigella dysenteriae* and a toxigenic group of *Escherichia coli,* which includes serotypes O157:H7 and other enterohemorrhagic *E. coli*) the B domain comprises five B subunits in a classical AB_5 architecture. Interestingly, the A domain of Shiga toxin utilises the same mechanism of action, N-glycosidase cleavage of a specific adenine from the 28S RNA of the 60S subunit of the ribosome (consequently inhibiting protein synthesis), as the major class of plant ribosome inactivating toxins (RIPs). Though this article focuses on the application of bacterial toxins for therapeutic use, there is also a rich parallel literature for the application of RIPs in the design of novel biologics.

8.2 Therapeutic Opportunities from Cytotoxic Bacterial Toxins

8.2.1 Pre-Clinical Studies with Cytotoxic Bacterial Toxin Domains

Since the early 1980's, researchers have been developing the knowledge base that has enabled an investigation of the potential for cytotoxic bacterial toxin domains to be used in a theraueptic setting. The two main lines of investigation have been focused on the application of Diphtheria toxin and Pseudomonas exotoxin A. Table 8.1 summarises the relevant main similarities and differences.

8.2.1.1 Diphtheria Toxin-Based Opportunities

Diphtheria toxin (DT) is a bacterial exotoxin of approximately 60 kDa comprising two major functional units: the A subunit, which is an ADP-ribosylating enzyme, and the B subunit, which is responsible for binding to the membrane of a susceptible host cell. In common with many of the examples later in this chapter, the toxin incorporates a membrane translocation function into the binding domain which facilitates translocation of the enzymatic component into the cytoplasm of the target cell. This multi-domain, multi-functional architecture is a potential template for protein engineering and for isolation of 'useful' domains to apply to new purposes. In the case of DT, the majority of research has focused on the utilisation of truncated forms of the toxin in which the natural binding domain has been deleted. Such truncations

Table 8.1 Main characteristics of two cytotoxic bacterial toxins.

	Diphtheria Toxin	*Pseudomonas Exotoxin A*
Natural source	*Corynebacterium diphtheriae*	*Pseudomonas aeruginosa*
Structure	Multi-domain disulphide-bonded heterodimer originating from an Enzyme domain-Translocation domain-Binding domain single-chain expressed polypeptide	Multi-domain disulphide-bonded heterodimer originating from a Binding domain-Translocation domain-Enzyme domain single-chain expressed polypeptide
Size	535 amino acids	613 amino acids
Mechanism of action	Modification of Elongation Factor 2 by ADP-ribosylation	Modification of Elongation Factor 2 by ADP-ribosylation
Receptor	Complex of heparin-binding epidermal growth factor-like growth factor precursor and CD9	$\alpha 2$ macroglobulin receptor
Internalisation pathway	Endosome internalisation followed by low-pH triggered translocation of the enzymatic into the cytosol, probably from early endosomes	Endosome internalisation to transreticular Golgi to endoplasmic reticulum

(most frequently termed DAB389 due to truncation from position 389 onwards) retain the catalytic A subunit enzyme portion of the toxin attached to the translocation function of the B subunit. Such a molecule is capable of translocation and ADP-ribosylation but is not capable of binding selectively to a target cell.

To enable DAB389 to have therapeutic utility, researchers have prepared engineered versions that incorporate additional proteins that provide a receptor binding function. The hypothesis, now extensively tested, is that the new domain will provide the cellular selectivity and cellular interaction function necessary to target the 'warhead' (the A subunit) to a cell that requires destruction. Such a concept, that of the immunotoxin, has been widely postulated as a novel approach for the treatment of cancer for example. In the case of DT, one of the most reported binding ligands in the construction of immunotoxins is interleukin 2 (IL2). IL2 binds with high affinity to the interleukin 2 receptor (IL2R); a complex receptor that is found in B- and T-cells cells that are fundamental to progression of cutaneous T-cell lymphoma, Hodgkin disease, adult T-cell leukaemia and other leukaemias and lymphomas.

Construction of immunotoxins based on DT has benefitted from an advanced understanding of the structure of DT and the contribution that parts of the polypeptide make to biological activity. Indeed, some of the early DT truncates used to construct immunotoxins utilised DAB486,[3] a longer version of DAB389 that was less tolerated in animal studies and had a reduced half-life. Understanding that conventional chemical conjugation of proteins (e.g. a toxin truncate to a binding ligand) to create a homogenous protein suitable for therapeutic use is difficult and costly, significant effort was expended to establish recombinant toxin expression and putification methods. Often using *E. coli* as the host organism for expression, many examples of recombinant DT-ligand proteins have been prepared. The clinical progress of some of these materials is described below.

8.2.1.2 *Pseudomonas* Exotoxin-Based Opportunities

With many similarities to DT, *Pseudomonas* exotoxin (PE) is a bacterial exotoxin of approximately 60 kDa comprising three major structural domains; domain Ia is the cell binding domain, domain II contains a furin site that is necessary to release domain III from the cell binding domain, and domain III comprises the ADP-ribosylation activity that inactivates elongation factor 2 in a manner akin to that of DT subunit A. In contrast to DT, PE utilises an alternative method of cellular entry and intracellular routing; hijacking the KDEL intracellular sorting receptor which transports the 37 kDa C-terminal fragment from the trans-reticular Golgi apparatus to the endoplasmic reticulum (ER), from which translocation of the ADP-ribosylating enzyme component can occur.

In common with DT, truncated versions of PE have been prepared for new molecule creation. The most common form of truncated PE is PE38; a protein that comprises the translocation and ADP-ribosylation functions of the holotoxin.

Expression of single chain recombinant PE38 incorporating a range of cell-binding ligand variants has also been achieved.[4] It is worth noting that the architecture of a recombinant PE-ligand construct differs from a DT equivalent in the position of the new ligand binding domain: in the case of PE the new ligand is positioned at the amino terminus of PE whereas it is positioned at the carboxy terminus of DT.

8.2.2 Clinical Application of Cytotoxic Bacterial Toxin Domains

From the description of the functional domains of PE and DT it should be clear that both toxins possess a valuable enzymatic function: that of ADP-ribosylation of elongation factor 2. ADP-ribosylation of EF2 leads to inactivation of EF2 and an inability of the intoxicated cell to achieve protein synthesis. In the absence of protein synthesis the cell cannot survive and death of the cell results. By virtue of the cell-killing aspect of the mechanism of action, PE and DT truncates have been explored for their utility in combatting diseases that result from proliferation. A wide range of pre-clinical research and clinical investigation has been achieved with such engineered proteins, henceforth referred to as targeted toxins. Table 8.2 illustrates just some of the ligands that have been used in their construction.

Possibly the most commonly referred to targeted toxin is a protein termed denileukin diftitox, a recombinant DAB389-IL2 protein[5] originally assessed in patients with cutaneous T-cell lymphoma (CTCL). Denileukin diftitox (also referred to as ONTAK®) received initial approval in the United States in 1999 and is established as a marketed product for the treatment of patients with persistent or recurrent cutaneous T-cell lymphoma whose malignant cells express the CD25 component of the IL-2 receptor. At the time of preparing this report, there were a number of active clinical trials of denileukin diftitox in alternative conditions, for example melanoma (Stage IIIC & Stage IV) and ovarian cancer. Though successful, denileukin diftitox treated clinical study

Table 8.2 Examples of ligands used in the construction of targeted toxins.

Examples of ligands used in the construction of targeted toxins
Interleukin 2
Interleukin 13
Transferrin
Granulocyte-macrophage colony-stimulating factor
Anti-CD3
Anti-PSMA
Anti-CD25
Anti-CD22
Anti-folate receptor β
Anti-ADAM17
Bispecific anti-EGFR + anti-uPAR

patients have experienced a range of side effects; the most common serious adverse reactions were capillary leak syndrome (11.1%), infusion reactions (8.1%), and visual changes including loss of visual acuity (4%). Indeed denileukin diftitox was discontinued in 28.2% of patients due to adverse reactions. Capillary leak syndrome continues to be an area of intense exploratory activity and optimisation of treatment regimen is an important strategic consequence. In June 2011, Eisai Inc. released a statement that reported an interruption in the supply of ONTAK due to a trend in certain product characteristics, including out-of-specification results, that required an in-depth process review prior to releasing any further lots. A compassionate use program was put in place to maintain an option to supply to patients already receiving the drug, but it is a sharp reminder that the complexities of preparing complex recombinant biologics must not be underestimated.

8.3 Therapeutic Opportunities from Non-Cytotoxic Bacterial Toxins

Although fragments of bacterial cytotoxins such as DT and PE have proved to be valuable tools in the design of therapeutic agents, and some success has been achieved through this route, they are not suitable for use as therapeutic tools in their natural holotoxin state. They are, after all, cytotoxins that are associated with the clinical symptoms of two human pathogens.

By contrast, the development of therapeutics based on another bacterial toxin, BoNT, has been led by the use of the holotoxin and has only latterly been considered for expansion through protein engineering routes. This section of the chapter focuses on the opportunities that exist for development of therapeutics based on BoNT and fragments of BoNT.

8.3.1 Clinical Application of Natural Toxin Products

To many, the application of the world's most potent toxin to the treatment of human disease will seem counter-intuitive. However, to practitioners in the field, the elegant simplicity of application of vanishingly small quantities of protein to a patient to result in long-term clinical benefit with minimal side effects is evident in multiple therapeutic arena.

In 1897 Professor Emile-Pierre-Marie van Ermengem identified an anaerobic bacterium that he termed *Bacillus botulinus* (subsequently classified as *C. botulinum*) that released a potent toxin that caused the symptoms associated with food-borne botulism. In the latter half of the 20th century, research into the structure, mechanism of action and clinical opportunities resulted in a greatly enhanced knowledge base from which the innovations of the last 10 or so years have developed.

The fundamentals of BoNT are illustrated in Table 8.3 and make an interesting comparison to the properties of PE and DT described in Table 8.1. Differing in many respects to PE and DT, BoNT is fundamentally

Table 8.3 Key properties of botulinum neurotoxin (BoNT).

Natural source	*Clostridium botulinum*
Structure	Multi-domain disulphide-bonded heterodimer originating from an Enzyme domain-Translocation domain-Binding domain single-chain expressed polypeptide
Size	1296 amino acids for BoNT/A
Mechanism of action	Protease cleavage of SNARE proteins
Receptor	Luminal domain of synaptic vesicle protein SV2 (for BoNT/A)
Internalisation pathway	Endosome internalisation followed by low-pH triggered translocation of the enzymatic into the cytosol

differentiated from its bacterial stablemates by its non-cytotoxic mechanism of action. BoNT does not kill its target cell, the motor neuron; rather it inhibits neurotransmitter release from the neuron. BoNTs inhibit ACh release from the pre-synaptic terminals of peripheral cholinergic neurons, particularly the motor-end plates of motor neurons. The human and animal lethality of BoNT intoxication is a result of the impact of inhibition of neurotransmission rather than the impact on the target cell itself. A BoNT intoxicated neuron is unable to initiate muscle contraction and the muscle weakness of botulism, characteristically starting as a descending paralysis, initially manifest as double vision, loss of facial expression, swallowing problems, leading to weakness in the arms, legs and eventually the respiratory muscles. In severe botulism, the patient is unable to move the diaphragm effectively and consequently has imparied respiration. Artifical ventilation can be used to maintain the patient's respiration needs until the effect of BoNT poisoning wears off (which many take many weeks).

There are seven 'families' or classifications of BoNTs (termed serotypes and referred to as BoNT/A, B, C, D, E, F or G) and multiple sub-types within a given serotype, such that there are well over 20 different BoNTs known. All are multi-domain heterodimers comprising an enzyme, translocation and binding domain and, though the precise substrates and receptors differ between serotypes, the end result of inhibition of neurotransmission is the same. Clinical products are currently available for the A and B serotypes.

The first clinical use of BoNT was achieved in the 1970s by Dr Alan Scott.[6] Dr Scott investigated the effects of BoNT complex on hyperactive muscles in the condition of strabismus and established that small doses of material could be used to very effectively, and temporarily, weaken the muscles sufficient to achieve a therapeutic outcome. Since this seminal work, the clinical use of BoNT products has increased the market size three-fold since the start of the century (to over $1.6 billion in 2010). This growth has been achieved using product preparations that are a complex mixture of active BoNT and associated proteins (with onabotulinumtoxinA (BOTOX®) and abobotulinumtoxinA (Dysport®) being the major two products), and only very recently have products come onto the marketplace that comprise the active BoNT in the absence of associated proteins (such as incobotulinumtoxinA; marketed as XEOMIN®).

Following Scott's pioneering studies, BoNT/A was shown to relieve the muscle spasm involved in various focal dystonias and then later, during treatment of blepharospasm, it was noted that vertical glabellar creases (frown lines) disappeared and the cosmetic use of Botox® came of age. In addition to dystonia, BoNT injections can treat other hyperkinetic movement disorders, including tremors of the hands and head. Furthermore, BoNT has been used to treat muscle tone disorders, including spasticity associated with cerebral palsy and multiple sclerosis for example. The treatment of leg spasticity in children with cerebral palsy with BoNT has had a particular impact in negating the need for alternative surgical techniques and also prevention of musculoskeletal deformities and other orthopaedic problems later in life.

8.3.2 The Domain Structure of BoNT as a Template for New Product Engineering

Despite the early knowledge on the biological impact of the toxin, the structural understanding only gained pace in 1946 when a pure crystalline botulinum toxin of serotype A (BoNT/A) was obtained. Over the subsequent 20 years the level of understanding of the multi-protein complex nature of BoNT increased and in 1998 publication of the first X-ray structure of BoNT/A[7] was achieved, twenty years after the first clinical use of BoNT.

Lacy's seminal work established the current view of BoNT structural organisation and provided a platform to study the primary, secondary, tertiary and quaternary structure of the botulinum neurotoxin family of proteins. With respect to our understanding of BoNT molecular architecture, the gross three dimensional arrangements of BoNT/A and BoNT/B were determined in 1998[7] and 2000,[8] respectively, and were demonstrated to be essentially similar: a three domain protein in which the domains are arranged in a linear fashion with the translocation domain in the middle flanked by the binding and catalytic domains (Figure 8.1). In 2008[9] the tertiary structure of BoNT/E was determined and, though the individual domains showed a high degree of structural similarity with the A and B cases, in serotype E the catalytic and binding domains are arranged on the same side of the translocation domain and consequently share interactions that would be absent in A or B.

Structural data from crystallographic studies are now available for (i) the individual catalytic domains (LC) of serotypes A, B, C, D, E, F and G; (ii) the binding domain (HC) of A, B, C, D, F and G; (iii) the di-chain LH_N species comprising the LC and the H_N domain of serotypes A and B; and (iv) the holotoxins BoNT/A, BoNT/B and BoNT/E. Understanding the molecular architecture of the individual domains and their inter-relationships within the macromolecule has provided opportunities for a number of paths for application of protein engineering strategies.[10] Opportunities to utilise the individual domain components of BoNT (the LC, the H_N, the HC domains or subcomponents/combinations of each) are discussed below.

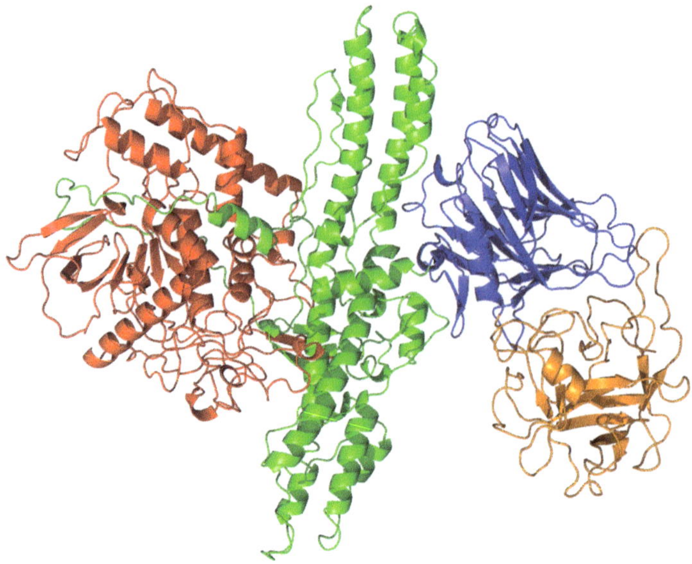

Figure 8.1 Tertiary structure of BoNT/A holotoxin. LC is depicted in red, HN domain in green, H_{CN} in blue and H_{CC} in orange.
By permission of Springer-Verlag GmbH.

8.3.3 Utilising Toxin Domains to Deliver Therapeutic Cargo

The HC domain has evolved to selectively bind to neurons, although not all BoNTs bind to the target neuron using precisely the same mechanism. In a fascinating outcome of evolution, BoNTs have developed a binding mechanism that involves gangliosides specificity, exposed protein receptor specificity, or both. For example, BoNT/B and G use gangliosides and the intraluminal domain of synaptotagmin as dual receptors, whilst BoNT/A, D, E uses gangliosides and the intraluminal domain of synaptic vesicle glycoprotein 2 (SV2). A protein binding domain has however not yet been identified in BoNT/C, suggesting the BoNT/C uses ganglioside only, or an as yet unidentified proteinaceous component.[11] Recent research has led to a greater understanding of the molecular basis of this important binding interaction and is leading to the development of proteins with improved properties, for example enhanced potency BoNTs. The HC domain therefore acts as a delivery vehicle to locate the SNARE-cleaving LC on the outside surface of its target cell.

Therefore, BoNTs represent a unique opportunity to develop for delivery of therapeutic molecules, particularly biological molecules, into nerve cells via their HC binding and the beneficial properties of the in-built membrane translocation function. For these reasons, full-length BoNTs lacking a functional endopeptidase have been proposed as preferred delivery tools.[12,13] Such constructs retain the evolved connectivity and functionality of BoNTs whilst being non-toxic by virtue of an inactivated endopeptidase domain. Considering what applications there may be for using BoNT heavy chains, or fragments thereof, enzyme replacement studies, such as to treat lysosomal storage diseases and treatment of oxidative injury, have been considered as have the delivery of DNA and viral vectors. Whilst an attractive opportunity supported by model system studies, there are not yet any clinical applications of drug delivery using a BoNT-based vehicle and more research is required to understand the rate of uptake of BoNT HCs and the kinetics and mechanism of the translocation step to fully exploit BoNT HCs for delivery purposes.

8.3.4 Harnessing the Properties of Clostridial Neurotoxin Domains in Engineered Hybrids

As described earlier, the various BoNT serotype families comprise the same overall domain architecture and the same domain functionality. Therefore, like a box of coloured model bricks that can be connected together to create a new product, there is interest in exploring whether such variations in domain can be combined to create novel proteins of therapeutic benefit. In fact, a number of the natural subtypes would appear to have originated through natural combination: for example BoNT/A2 subtype comprises domains that are similar to the LC of BoNT/A1 and the HC of BoNT/A3. Using recombinant DNA techniques to isolate individual BoNT domains or to express multi-domain proteins in laboratory expression hosts, the creation of hybrids has indeed been achieved. For example, Wang and colleagues[13] have prepared a chimera of BoNT/A and /E in which the HC domain of one serotype was expressed recombinantly fused to the LC and HN (LHN) of the other serotype. When assessed for key functional characteristics, it was noted that the translocation and neuronal specificity properties of the hybrid were clearly differentiated: translocation reflected the behaviour of the parent LHN domains, whilst neuronal specificity was influenced by the identity of the HC. In the situation where the HC/A is replaced with the HC/B, the resultant hybrid induced a more prolonged neuromuscular paralysis in mice than the parent BoNT/A. In the complementary switch where the HC/A is inserted into the BoNT/B structure, the resultant molecule retained neuronal SNARE cleavage ability but also enabled cleavage of VAMP within fibroblast-like synoviocytes that lack SNAP-25 and would ordinarily be resistant to the SNARE-cleavage ability of BoNT/A.

8.3.4.1 Creation of Improved Product Opportunities

Such approaches to engineering of the native BoNT structure through domain switching has been further described by Dolly as the basis for the construction of 'BoTIMs' (full-length BoNTs incorporating catalytic-inactive LC/A).[14] By recombinantly combining BoTIMs incorporating LC/E domains, a hybrid protein was constructed that utilised components within the LC/A element to extend the intracellular persistence of the LC/E and therefore the duration of action of LC/E-induced SNAP-25 cleavage.[13] Dolly proposed that the

LC/E-induced cleavage of SNAP-25 would be advantageous for specific conditions, for example in the treatment of various pain states, including chronic pain.

8.3.5 Targeted Secretion Inhibitors

8.3.5.1 Concept and Potential

The multi-domain, multi-functional organisation of BoNTs has already been described to be a useful platform from which to engineer a new class of neuronal delivery vehicles for therapeutic use. However, the most significant progress in engineering novel agents using BoNT domains has so far been achieved not by using the binding domain but by harnessing the power of the light chain endopeptidase. Acknowledging that SNAREs are integral components to a universal mechanism of secretion from cells of neuronal and non-neuronal origin, the potential to utilise the exquisite selectivity of BoNT light chains for the SNARE substrate is clear. Such an approach would also expand the exploitation of BoNTs (through the use of BoNT fragments) to non-neuronal cells not targeted naturally by CNTs.

Cleavage of SNAREs leads to inhibition of secretion and, if this activity could be targeted to cells that are critical to a hypersecretory event then one could envisage the concept of targeted inhibition of secretion from cells. This concept has been realised in the Targeted Secretion Inhibitors (TSI) platform developed by Syntaxin Ltd., and is occasionally referred to elsewhere in the literature as targeted exocytosis modulators (TEMs). By creating novel recombinant proteins, it has proved possible to deliver the LC into cell types not sensitive to native CNTs, cleave the relevant SNARE protein and thereby inhibit secretion. In addition to the LC endopeptidase domain, the proteins incorporate the HN translocation domain to ensure cytosolic delivery, together with a binding domain that binds to a cell surface receptor on a proposed target cell. The binding domain can be a peptide or a protein that mimics a natural protein binding to cell surface receptors. It can also be derived from a binding library of, for example, antibodies or scaffold proteins. Targeting membrane receptors or cell surface proteins that internalise and enter an acidifying endosomal compartment ensures that the HN domain undergoes the necessary conformational change and forms a translocation pore in the endosomal membrane to enable entry of the LC into the cytosol. This approach also enables creation of neuronally targeted therapeutics with improved neuronal selectivity and an enhanced therapeutic window than native BoNTs.

8.3.5.2 Implementing the Concept at the Research Phase

The first description of modifying a clostridial neurotoxin cell-binding domain was actually achieved using a fragment of tetanus neurotoxin (TeNT), a closely related multi-domain multi-functional protein to BoNT, to study the biology to TeNT. Such studies demonstrated that clostridial neurotoxin domains could be coupled chemically to other proteins that exhibited binding activity (in this case the lectin domains of wheat germ agglutinin or ricin). The first report that a BoNT light chain could be delivered to a target cell through a non-BoNT binding domain was described in 2000.[15] In this study a conjugate of nerve growth factor and a fragment of BoNT/A lacking the HC domain (LHN/A fragment) was demonstrated to cleave SNAP-25 and inhibit noradrenaline release from PC12 cells.

Subsequently, a chemical conjugate of a lectin (wheat germ agglutinin) and the LH_N fragment of BoNT/A was found to deliver the light chain endopeptidase into both neuronal and non-neuronal cell types, to cleave SNAP-25 and consequently inhibit secretion.[16] In one arm of this study, concentration-dependent inhibition of stimulated insulin release from a hamster pancreatic β cell, HIT-T15, was observed. Furthermore, the inhibition of secretion was found to correlate with cleavage of SNAP-25. HIT-T15 cells are resistant to the effects of BoNT/A, so this result confirmed the hypothesis that it is possible to internalise the endopeptidase into the cytosol of a cell normally resistant to the effect of BoNT. As a further impact of this work, it was also confirmed that the HN domain was able to function in the new target cell following binding and endocytosis and deliver the endopeptidase.

Following these early proofs of concept, a therapeutically relevant application of this approach was initiated through the development of a conjugate of *Erythrina crista-galli* lectin and LHN/A (ECL-LHN/A). Such a conjugate was designed to use the signature of surface carbohydrate expression to target a light chain endopeptidase to nociceptive afferents. Such a conjugate resulted in inhibition of both substance P and glutamate release from embryonic dorsal root ganglion neurons in culture for an extended period following a single treatment.[17] Developing this concept further, intrathecally administered ECL-LHN/A significantly reduced the nociceptive inputs to convergent dorsal horn neurons by primary sensory afferents of the C-fibre and Aδ types with little or no effect on sensory inputs from Aβ-fibres. Furthermore, intrathecal ECL-LHN/A also resulted in a prolonged withdrawal latency in a 'hotplate' model of acute thermal pain that was sustained for more than 30 days post-single administration of the conjugate.[18] Such early studies into the TSI concept demonstrated that the retargeted endopeptidase proteins retain the prolonged duration of action that is the hallmark of BoNTs and are thus particularly suitable for treating chronic diseases.

8.3.5.3 Development of Recombinant TSI

Though successful for early proof-of-concept, preparation of novel agents based on chemical conjugation of fragments derived from clostridial sources is not ideal. Additionally, preparation of the LH_N fragment from BoNTs other than BoNT/A by limited proteolytic cleavage is very difficult due to the susceptibility of the L and HN domains to proteolytic degradation. To overcome such challenges, recombinant expression of a catalytically active, stable LH_N fragments was required. This was first reported for the LH_N fragment of BoNT/A in

2002.[19] Subsequently, the expression and purification of catalytically active LHN/B and LHN/C from *E. coli* was also reported.[20] In addition to ease of manufacture and supply, such recombinant fragments were shown to be superior to material derived from native sources with respect to low toxicity. The LH_N fragment is effectively non-toxic because it lacks the necessary HC domain with which to bind to acceptors on the neuronal surface. Importantly, recent studies have shown that recombinant LH_N of BoNT/A and BoNT/B retains the crystal structure of the equivalent domains in the intact BoNT/A protein.

There are several drawbacks to using chemical conjugation to produce therapeutic proteins, foremost of which is the inevitable heterogeneous mixture of species that is created using this process. This inherent heterogeneity of chemical conjugates makes developing a regulatory-compliant process based upon them difficult. A fully recombinant expression platform is the preferred route for the development of a therapeutic protein. Given the size and complexity of such a multi-domain, multi-functional fusion protein, developing a fully recombinant chimera protein was a challenging task. It has, however, been achieved and a number of examples are now in the public domain. Firstly, a fully recombinant fusion protein consisting of the LH_N-fragment of BoNT/C1 and epidermal growth factor (EGF) was reported in 2006.[21] Such a recombinant material was expressed as a single chain precursor comprising the light chain and translocation domain of BoNT/C genetically fused to the N-terminus of EGF, with the EGF domain taking the functional and structural position ordinarily occupied by the native BoNT H_C.

The potential of the TSI technology was therefore established and it is clear that creation of fully recombinant proteins that target and deliver LC into a specified cell represents a tremendous opportunity to develop therapeutic proteins that inhibit secretion from cells involved in a wide variety of diseases. For example, in conditions where the release of multiple inflammatory mediators is a feature of allergic and inflammatory diseases such as rheumatoid arthritis, the application of a TSI that could inhibit such secretions for an extended period of time is of great interest. In oncology, the potential to inhibit over-secretion of growth factors and modify the tumour microenvironment could have a significant impact on the progression of disease. In the next two sections, two examples of TSI technology that has progressed beyond initial research are described.

8.3.5.4 Pre-Clinical Evidence of TSI Development

One area of focus for the development of TSI is in the endocrine therapeutic space. Here, disease progression is often a result of over-secretion of one or more factors, for example the release of hormones and growth factors is a core feature of orphan diseases caused by benign endocrine tumours, such as in acromegaly and Cushing's Disease. Furthermore, various metabolic diseases are caused by the inappropriate release of hormones.

Acromegaly is a debilitating human condition that can result from hyper-secretion of growth hormone (GH) from pituitary adenomas, often of a benign nature. It is currently treated by surgery (transsphenoidal resection), radiation therapy or pharmacological agents, such as somatostatin analogues, dopamine agonists or, recently, GH receptor antagonists. The aim of these treatments is to reduce the impact of circulating GH (either by removal or blocking) on peripheral receptors, for example the GH receptor on the liver that leads to the production of circulating Insulin-like Growth Factor 1 (IGF1). Reported in a recent publication by Somm and colleagues,[22] a TSI comprising the LH_N fragment of BoNT/D and a modified growth hormone-releasing hormone (GHRH) peptide produced a dose-dependent inhibition of GH synthesis, storage and secretion in a juvenile rat model. Consequently, hepatic IGF1 production and resultant levels of circulating IGF1 were reduced. Such a reduction in circulating IGF1 led to decreases in body weight, body length, organ weight and bone mass acquisition. These exciting data illustrated the potential of a botulinum fragment-derived TSI to inhibit secretion from a target tissue that is a key factor in the development and maintenance of disease.

The data reported by Somm *et al.* are significant in a number of aspects beyond the clear therapeutic potential described previously. Firstly, the TSI protein used in this study was a fully recombinant protein constructed with the GHRH receptor targeting domain located to the N-terminus of the translocation domain rather than to the C-terminus as described for the EGF construct in the section previously. This novel orientation of binding domain was demonstrated to successfully facilitate activity of the protein in *in vitro* and *in vivo* models and illustrates the robustness of the recombinant platform to modification. Secondly, the TSI was injected into the animal intravenously rather than specifically into the target organ and the GHRH targeting domain successfully localised the TSI to the GHRH-receptor on the pituitary adenoma. This illustrates that a suitably selective targeting domain can achieve organ-specific targeting of the LH_N fragment and in so doing minimise off target effects. Finally, it is noticeable that the significant long-term effects of the TSI were achieved after a single application of the protein. The extended duration of action is a result of the SNARE-cleavage mechanism of action and extended intracellular lifetime of the LC rather than longevity of the material in circulation. In fact, elimination of the material from the circulation was relatively fast (approximately 12 hours).

8.3.5.5 *Targeted Secretion Inhibitors in the Clinic*

In terms of progression of TSI into the clinic, the most advanced example is that of AGN-214868, a TSI designed to treat chronic pain by targeting C fibres and inhibiting release of nociceptive neurotransmitters.

It is well known that the medical applications of BoNTs is expanding widely and the number of approved indications are increasing; of particular relevance here are the emerging opportunities in certain types of migraine. The use of BoNT is naturally limited by its mechanism of cell targeting and, though the TSI platform has the potential to significantly expand such opportunities, clinical proof of the use of TSI has not been available. In the first quarter of

2011, Syntaxin Ltd. announced that its partner Allergan had initiated two Phase II trials to evaluate the safety and efficacy of its re-targeted endopeptidase drug candidate AGN-214868. The Phase II trials are focused on patients with post herpetic neuralgia (PHN; ClinicalTrials.gov Identifier NCT01129531) administrating 3 and 16 μg of AGN-214868 and Idiopathic Overactive Bladder and Urinary Incontinence (ClinicalTrials.gov Identifier NCT01157377) injecting 0.5 to 60 μg of AGN-214868. The PHN trial is reported to be complete and Allergan provided early insight into some of the data at their R&D day in March 2012. AGN-214868 was reported to be well tolerated and did not exhibit dose-limiting toxicity in the earlier Phase I study, also in which proof-of-principle was established in a capsaicin model of pain. In the Phase II study, AGN-214868 was reported to be safe and well tolerated. Furthermore, AGN-214868 resulted in an apparent reduction in perceived pain intensity that was maintained at least to 12 weeks post treatment. With the initiation of Phase II trials and the early illustration of clinical efficacy, the TSI technology platform has reached a significant point of development.

8.4 Conclusions

Bacterial toxins are versatile, powerful and innovative platforms from which to base the development of novel therapies. This chapter has focused on two main classes of toxin that have been considered for this use: the cytotoxic proteins (as exemplified by DT and PE) and the non-cytotoxic proteins (as exemplified by BoNT). Both classes of toxin exhibit the common feature of a complex multi-domain multifunctional structure comprising a cell binding activity, a protein translocating activity that enables a 'warhead' to traverse an intracellular membrane and, finally, a 'warhead' that catalytically modifies one or more components that are essential to a cellular function. The catalytic functional domains of bacterial toxins are fundamental to the engineering opportunity: the need to internalise a relatively small molar amount of warhead to achieve a significant biological effect is a major advantage over other protein:protein interactions that occur with 1 : 1 stoichiometry. Compared to a small molecule receptor antagonist, these 50–150 kDa bacterial toxins are indeed complex beasts.

Where these two example classes of toxin differ is in their impact on the cell. In the case of the cytotoxins DT and PE, their mode of action leads to inhibition of protein synthesis and consequently cell death. Such a mechanism is potentially indiscriminate and therefore a powerful tool to harness in pursuit of anti-proliferative treatments, particularly therefore for cancer applications. Of course, with such a broad ranging mechanism of action comes the danger of off-target effects where small, but effective amounts of the toxin inadvertently gain access to tissues neighbouring the target cell and result in unwanted cell killing. To minimise this event, researchers have utilised functional toxin domains of minimal size (to remove inherent cell binding capability) and optimised the specific cell targeting ligand to focus the warhead to the target cell. The example of denileukin diftitox demonstrates that the principles of genetic engineering can be applied to creation of an effective novel multi-domain agent. In this case, the product is engineered to display an interleukin 2 binding domain that interacts with over-expressed receptors on certain cancer cells. The relationship between receptor expression, ligand affinity and ligand selectivity are critical components in the design of an effective therapeutic.

The understanding of cytotoxic bacterial toxin biology has definitely enabled significant progress to be made in the pursuit of new therapies for a range of human diseases and they continue to be a field of active research and development. Against a background where the natural toxin has such a harmful impact on human health, such innovations are both scientifically interesting and potentially highly impactful. In parallel, the opportunity to use non-cytotoxic bacterial protein toxins has developed significantly, particularly in the last 10 years. In this case, the background situation for the toxin is quite different. As noted earlier, BoNT proteins are established clinical products and have already made a major impact in the treatment of a number of debilitating human neurological and neuromuscular conditions. However, a really exciting opportunity now exists, with the latest understanding of the structure of BoNT, to develop a new range of therapeutics that utilise BoNT domains.

The therapeutic success of the BoNTs results from two structure-function derived outputs: (i) the specific and potent inhibition of neurotransmitter release from peripheral cholinergic neurons that results from the neuronal binding event and the unique substrate cleavage activity; and (ii) the LC-derived duration of action that can often be measured in months. The clinical utility of the neurotoxins is, however, severely constrained, both by their limited range of target cells and narrow therapeutic window. Overcoming such constraints, for example through the creation of novel hybrids or through the development of the TSI platform, has the potential to open up many new opportunities. The initial Phase II study success reported for senrebotase demonstrates that the concept of retargeting fragments of BoNT is sound and can have significant advantages over using the parent holotoxin. The TSI approach is already progressing in the pre-clinical setting for a range of indications not treatable with neurotoxin products. For example, the evidence presented for the impact of TSI targeted to the pituitary indicate broad potential beyond just the neuronal cell.

Hopefully, further clinical and pre-clinical progress with bacterial toxins will inspire further effort onto maximising the potential of this exciting class of proteins.

References

1. J. S. Henkel, M. R. Baldwin and J. T. Barbieri, *EXS*, 2010, **100**, 1–29.
2. S. S. Arnon, R. Schechter, T. V. Inglesby, D. A. Henderson, J. G. Bartlett, M. S. Ascher, E. Eitzen, A. D. Fine, J. Hauer, M. Layton, S. Lillibridge, M. T. Osterholm, T. O'Toole, G. Parker, T. M. Perl, P. K. Russell, D. L. Swerdlow and K. Tonat, *J Am Med Assoc*, 2001, **285**, 1059–1070.
3. D. P. Williams, C. E. Snider, T. B. Strom and J. R. Murphy, *J Biol Chem*, 1990, **265**, 11885–11889.

4. D. Fitzgerald and I. Pastan, *Ann N Y Acad Sci*, 1993, **685**, 740–745.
5. T. Kiyokawa, D. P. Williams, C. E. Snider, T. B. Strom and J. R. Murphy, *Protein Eng*, 1991, **4**, 463–468.
6. A. B. Scott, *J Pediatr Ophthalmol Strabismus*, 1980, **17**, 21–25.
7. D. B. Lacy, W. Tepp, A. C. Cohen, B. R. DasGupta and R. C. Stevens, *Nat Struct Biol*, 1998, **5**, 898–902.
8. S. Swaminathan and S. Eswaramoorthy, *Acta Crystallogr D Biol Crystallogr*, 2000, **56**, 1024–1026.
9. D Kumaran, S. Eswaramoorthy, W. Furey, J. Navaza, M. Sax and S. Swaminathan, *J Mol Biol*, 2009, **386**, 233–245.
10. S. Swaminathan S, *FEBS J*, 2011, **278**, 4467–4485.
11. J. Strotmeier, S. Gu, S. Jutzi, S. Mahrhold, J. Zhou, A. Pich, T. Eichner, H. Bigalke, A. Rummel, R. Jin and T. Binz, *Mol Microbiol*, 2011, **81**, 143–156.
12. S. Bade, A. Rummel, C. Reisinger, T. Karnath, G. Ahnert-Hilger, H. Bigalke and T. Binz, *J Neurochem*, 2004, **91**, 1461–1472.
13. J. Wang, J. Meng, G. W. Lawrence, T. H. Zurawski, A. Sasse, M. O. Bodeker, M. A. Gilmore, E. Fernández-Salas, J. Francis, L. E. Steward, K. R. Aoki and J. O. Dolly, *J Biol Chem*, 2008, **283**, 16993–17002.
14. J. O. Dolly, J. Wang, T. H. Zurawski and J. Meng, *FEBS J*, 2011, **278**, 4454–4466.
15. J. A. Chaddock, J. R. Purkiss, M. J. Duggan, C. P. Quinn, C. C. Shone and K. A. Foster, *Growth Factors*, 2000, **18**, 147–155.
16. J. A. Chaddock, J. R. Purkiss, L. M. Friis, J. D. Broadbridge, M. J. Duggan, S. J. Fooks, C. C. Shone, C. P. Quinn and K. A. Foster, *Infect Immun*, 2000, **68**, 2587–2593.
17. M. J. Duggan, C. P. Quinn, J. A. Chaddock, J. R. Purkiss, F. C. Alexander, S. Doward, S. J. Fooks, L. M. Friis, Y. H. Hall, E. R. Kirby, N. Leeds, H. J. Moulsdale, A. Dickenson, G. M. Green, W. Rahman, R. Suzuki, C. C. Shone and K. A. Foster, *J Biol Chem*, 2002, **277**, 34846–34852.
18. J. A. Chaddock, J. R. Purkiss, F. C. Alexander, S. Doward, S. J. Fooks, L. M. Friis, Y. H. Hall, E. R. Kirby, N. Leeds, H. J. Moulsdale, A. Dickenson, G. M. Green, W. Rahman, R. Suzuki, M. J. Duggan, C. P. Quinn, C. C. Shone and K. A. Foster, *Mov Disord.*, 2004, **19**, S42–S47.
19. J. A. Chaddock, M. H. Herbert, R. J. Ling, F. C. Alexander, S. J. Fooks, D. F. Revell, C. P. Quinn, C. C. Shone and K. A. Foster, *Protein Expression Purif*, 2002, **25**, 219–228.
20. J. M. Sutton, J. Wayne, A. Scott-Tucker, S. M. O'Brien, P. M. Marks, F. C. Alexander, C. C. Shone and J. A. Chaddock, *Protein Expression Purif*, 2005, **40**, 31–41.
21. K. A. Foster, E. J. Adams, L. Durose, C. J. Cruttwell, E. Marks, C. C. Shone, J. A. Chaddock, C. L. Cox, C. Heaton, J. M. Sutton, J. Wayne, F. C. Alexander and D. F. Rogers, *Neurotox Res*, 2006, **9**, 101–107.
22. E. Somm, N. Bonnet, A. Martinez, P. M. Marks, V. A. Cadd, M. Elliott, A. Toulotte, S. L. Ferrari, R. Rizzoli, P. S. Hüppi, E. Harper, S. Melmed, R. Jones and M. L. Aubert, *J Clin Invest*, 2012, **122**, 3295–3306.

CHAPTER 9

Phage Selection of Mono- and Bicyclic Peptide Ligands

SHIYU CHEN AND CHRISTIAN HEINIS*

Institute of Chemical Sciences and Engineering, Ecole Polytechnique Fédérale de Lausanne, CH-1015 Lausanne, Switzerland
*Email: christian.heinis@epfl.ch

9.1 Introduction

Cyclic peptides are a fascinating class of molecules that have found application as therapeutics and as research tools.[1,2] In comparison to linear peptides, the backbone of cyclic peptides is conformationally less flexible, endowing these ring-shaped structures with a number of favorable properties.[3,4] Firstly, constrained peptides show often a smaller loss in entropy on target binding and thus higher binding affinities. Secondly, the constraint allows the cyclic peptides to adopt a reduced set of conformations, rendering them more selective. And thirdly, the lower flexibility makes the peptides fitting less well to the active sites of peptidases and hence metabolically more stable than linear analogues.

With the advent of phage display, it has become possible to develop cyclic peptides binding to virtually any protein target.[5,6] In phage display, polypeptides are genetically fused to bacteriophage coat proteins, leading to display of the polypeptide on the surface. DNA encoding for the phage proteins and the displayed polypeptides is encapsulated in the phage particle. Large libraries containing more than a billion bacteriophage, all displaying different polypeptides, are simultaneously subjected to affinity selection with a target of interest. The physical link between polypeptide and DNA allows the iterative amplification and enrichment of peptide binders and eventually the

RSC Drug Discovery Series No. 36
Biotherapeutics: Recent Developments using Chemical and Molecular Biology
Edited by Lyn H. Jones and Andrew J. McKnight

Published by the Royal Society of Chemistry, www.rsc.org

identification of peptide binders by sequencing of the phage DNA. Peptides and antibodies specifically binding to immobilized targets were isolated from phage display libraries in the early 1990s.[7–10] Phage display was subsequently used to develop binders based on many different protein scaffolds and peptide formats, including the cyclic peptides published first in 1992[11] and the bicyclic peptides presented in 2009.[12] Several articles[6,13] and book chapters[14–16] broadly review phage display technology. An article about recent progress in phage display, discussing technology improvements of the last years, was published by Bratkovic.[17]

In this chapter, we review phage display strategies to generate monocyclic and bicyclic peptide ligands. In the first section, phage technologies and peptide libraries used to generate monocyclic peptide ligands are discussed. While cyclic peptide phage libraries were also applied for the mapping of antibody epitopes,[18,19] we focus in this chapter on work that was performed to generate ligands. In the second section, impressive work to generate monocyclic peptide ligands to two targets, vascular endothelial growth factor and erythropoietin receptor are reviewed. In the following section, we discuss a more recently developed phage display-based strategy to generate bicyclic peptide ligands. Finally, we present examples of bicyclic peptides that were evolved with this latter technology.

9.2 Phage Selection of Monocyclic Peptides

The first peptide libraries screened by phage display were based on linear peptides containing 6[8,10] or 15[9] randomized amino acids. Several peptides isolated from the latter 15-mer library contained a pair of cysteines, which suggested that they were captured as disulfide-linked, cyclic structures.[8] To generate phage libraries rich in cyclic peptides, random peptide sequences flanked by a pair of cysteines were subsequently displayed on phage and oxidized prior to affinity selection (Figure 9.1a). O'Neil *et al.* constructed a library of random hexapeptides flanked by cysteines (CX_6C) and isolated high-affinity ligands of platelet glycoprotein IIb/IIIa (integrin $\alpha_{IIb}\beta_3$).[11] McLafferty *et al.* cloned a library of the format XCX_4CX-phage and isolated cyclic peptide binders to streptavidin and anti-β-endorphin.[20] Subsequently, many libraries of disulfide-cyclized peptides were cloned and cyclic peptides to numerous targets isolated (Table 9.1). These libraries, the applied methods, as well as some of the isolated cyclic peptides are discussed in this chapter. Some early work about phage-selected cyclic peptides has been reviewed already.[3,21] Several reviews about phage display discuss other polypeptide formats in addition to cyclic peptides.[22–29]

9.2.1 Phage Display Systems

Most disulfide-cyclized peptides were isolated using the filamentous phage strains M13, fd or f1 as display systems. Lytic bacteriophage such as lambda, T4 and T7 phage were applied rarely to screen cyclic peptide libraries. The latter T7 display system was reported to show less sequence bias in random

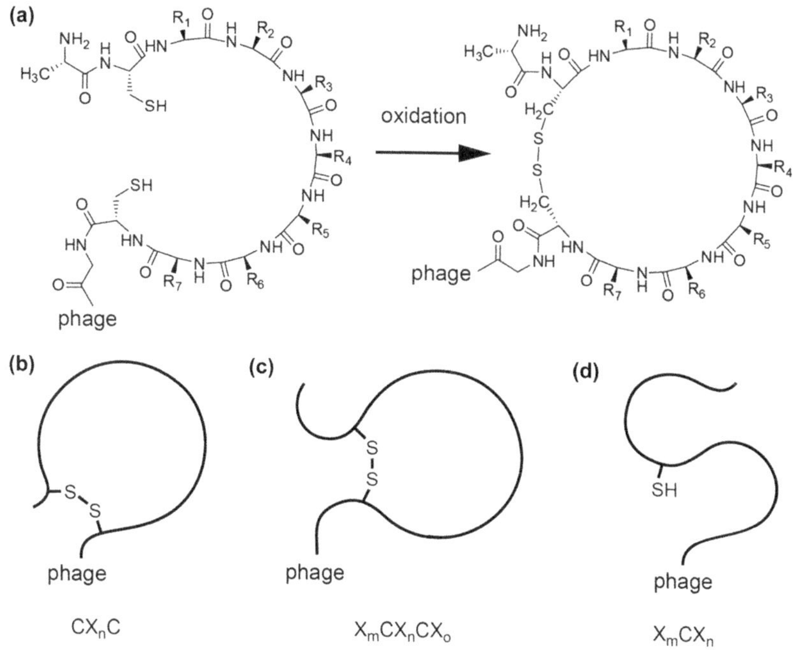

Figure 9.1 Phage selection of disulfide-cyclized peptides. (a) Oxidative cyclization of peptides on phage. The commercially available cyclic peptide phage library termed Ph.D.™-C7C is shown. The library contains seven random amino acids bridged by two cysteines. All peptides of this library contain a constant alanine residue at the *N*-terminus and are linked via a Gly–Gly–Gly–Ser; linker to the *N*-terminus of phage pIII. (b)–(d) Peptide library formats applied to generate cyclic peptide ligands. (b) Basic cyclic peptide library format. The number of random amino acids *n* ranges from 4 to 10. (c) In some cyclic peptide libraries, amino acids flanking the cyclic ring are also randomized. (d) In some peptide libraries, only one cysteine residue is placed in a fixed position. A second cysteine appearing in one of the $m+n$ randomized positions can pair with the fixed cysteine.

peptide libraries than filamentous phage-displayed libraries.[30] This observation was explained by the different export mechanisms, as in the case of the lytic phage where the peptides do not interact with exporting proteins. Cyclic peptides were also isolated by other display systems, such as mRNA display, CIS display, yeast display, bacterial display or ribosome display. Application of the latter systems to peptide libraries was recently reviewed.[31] As the vast majority of cyclic peptide ligands were isolated by filamentous phage display, we herein introduce this display system.

The wild-type filamentous phage is a rod of around 1 μm in length and 6–7 nm in diameter. Its detailed structure and biology is described in book chapters[32] and review articles.[6] In most phage libraries, the polypeptides are fused to the *N*-termini of the pIII or pVIII phage coat proteins.[6,15] This leads to a display of five copies per phage in the case of pIII fusion proteins and thousands of copies in the case of pVIII fusion proteins. The smaller valence

Table 9.1 Examples of disulfide-cyclized peptides isolated by phage display. Either dissociation constants K_ds or IC_{50} values are indicated. The IC_{50} values depend largely on the assay format and may not directly be compared.

Target[a]	*Peptide*	*Affinity*	*Ref.*
Antibody IgG Fc	DCAWHLGELVWCT-NH2 (Fc-III)	25 nM (K_i)	105
$\alpha_5\beta_1$ integrin	GACRGDCLGA	20 nM (IC_{50})	85
	GACRRETAWACGA	8 nM (IC_{50})	39
α-chymotrypsin	Ac-CCFSWRCRC	1 μM (IC_{50})	86
$\alpha_v\beta_3/\alpha_v\beta_5$ integrins	CDCRGDCFC, (RGD4C)	200 nM (IC_{50})	40
	DGARYCRGDCFDG, (RGD10)	200 nM (IC_{50})	87
EPOR	GGTYSCHFGPLTWVCKPQGG-NH_2 (EMP1)	200 nM (IC_{50})	56
	DREGCRRGWVGQCKAWFN (ERB1-7)	45 nM (IC_{50})	63
Factor VIIa	Ac-ALCDDPRVDRWYCQFVEG-NH2 (peptide E-76)	8.5 nM (K_d)	106
Fibrin	WECPYGLCWIQ-NH_2 (Tn6-2a)	2.9 μM (K_d)	88
gp41	(D)-Ac-KKGACELLGWEWAWLCAA-NH_2 (D10-p5-2k)	3.65 μM (IC_{50})	89
GPIbα	Ac-ACTERMALHNLCGG-NH_2 (os-1)	0.74 μM (K_d)	90
Grb2 SH2 domain	CELYENVGMYC (G1)	10 μM (IC_{50})	91
Grb7	CPFSGYDNQFLC-NH_2 (08-6046)	18.1 μM (K_d)	92
HBcAg	CWSFFSNIC (A1)	12 μM (K_d)	93
HSA/RSA	Ac-RLIEDICLPRWGCLWEDD-NH_2 (SA21)	467 nM/ 266 nM (K_d)	94
IGFBP-1	SEVGCRAGPLQWLCEKYFG (p1-02)	50 nM (IC_{50})	107
Integrin CD11c/ CD18	CGRWSGWPADLC (N1)	56 μM (K_d)	95
LEDGF	CVMGHPLWC (CP63)	19.37 μM (IC_{50})	96
MMP-9	CRRHWGFEFC	10 μM (IC_{50})	97
Ndst1	CNMQALSMPVTC	140 μM (K_i)	98
Neutravidin	CDRATPYCG-NH_2	12.5 μM (K_D)	99
Pancreatic lipase	CQPHPGQTC-NH_2 (D23)	16 μM (K_i)	100
PKA	CTFRVFGC-NH_2 (BS6-3)	57 μM (IC_{50})	99
PSA	CVAYCIEHHCWTC (C-4)	2.9 μM (K_d)	101
SH3 domain	(D)-RCLSGLRLGLVPCA-OH (Pep-D1)	63 μM (K_d)	102
Streptavidin	Ac-AECHPQGPPCIEGRK-NH_2	230 nM (K_d)	38
	Ac-AECHPQFCIEGRK-NH_2	23 nM (K_d)	103
Thrombin	CNRPFIPTC (TIP)	497 μM(K_i)	104
Trypsin-3	GIPCSWFRTGCA (TRY3-X10-1 #1)	0.2 μM (IC_{50})	101
Urokinase	CSWRGLENHRMC (upain-1)	6.7 μM (K_i)	82
VEGF	VEPNCDIHVMWEWECFERL-NH_2 (v114)	220 nM (IC_{50})	42

[a]*Note:* Targets: SH3 domain (Src homology 3 domain), Grb2 SH2 domain (growth factor receptor-bound protein 2 Src homology 2 domain), VEGF (vascular endothelial growth factor), EPOR (erythropoietin receptor), MMP-9 (matrix metalloproteinase 9), PSA (prostate-specific antigen), HSA (human serum albumin), HBcAg (hepatitis B core antigen), PKA (protein kinase A), GPIbα (glycoprotein Ibα), Ndst1 (*N*-deacetylase-*N*-sulfotransferase 1), Grb7 (growth-receptor-bound protein 7), LEDGF (lens epithelium-derived growth factor/p75), IGFBP-1 (insulin-like growth factor-binding protein 1).

resulting for pIII display allows better discrimination between peptides with different affinities due to the smaller impact of avidity effects compared to pVIII display. With the so-called phagemid system, the number of displayed peptides per phage virion can be reduced.[33,34] Phagemid systems displaying peptides on pIII allow presenting of a single peptide per phage. Phagemid systems combined with pVIII display leads to the display of a few hundred peptides. The choice of display valence depends on the protein target and the type of peptide library. In a selection of a naïve peptide library against a new protein target, weak affinities are expected and a polyvalent display is more suitable. For the affinity maturation of cyclic peptide leads having affinities in the nanomolar range, phage displaying a single peptide proved efficient. A comprehensive overview of phage vectors and modes of display can be found in a book written by Petrenko and Smith.[35]

9.2.2 Disulfide-Cyclized Peptide Phage Libraries

Many different disulfide-cyclized peptide phage libraries were constructed and subjected to affinity selections.[6,23,36,37] An overview of the different types of cyclic peptide phage libraries is shown in Figure 9.1. In most of the libraries, a region of 4–10 amino acids is randomized and flanked by two cysteine residues on both sides (CX_nC, $n=4$–10; Figure 9.1b). Often, multiple libraries were subjected to selections in parallel or as a mixture. Giebel *et al.* cloned and panned three libraries of the format CX_nC ($n=5$, 6, 7) against streptavidin. Conformationally constrained cyclic peptide ligands isolated in this work bound streptavidin up to three orders of magnitude better than linear peptides identified in a previous screen.[38] Koivunen *et al.* developed and screened libraries of the form CX_nC ($n=6$, 7, 8) and isolated cyclic peptide ligands to integrins.[39,40] A peptide isolated from one of these libraries actually contained two disulfide bridges. This peptide was 20-fold more potent in inhibiting the attachment of αvβ3 or αvβ5 integrin-expressing cells to vitronectin than a corresponding peptide with one disulfide bridge, and 200-fold more potent than linear peptides.[40] In some cyclic peptide phage libraries, amino acids of the exocyclic regions were additionally randomized (Figure 9.1c). Bonnycastle *et al.* developed large libraries of the format XCX_4CX ($n=4$, 6, 8) and screened them against a panel of monoclonal antibodies.[41] Fairbrother *et al.* developed peptide libraries of the format $X_lCX_mCX_n$ ($m=4$, 5, 6, 7, 8, 9, 10; $l+m+n=18$) and panned them against several monoclonal antibodies.[42] This latter work, and other studies, showed that each target had a different, unpredictable preference for certain peptide formats. It is therefore advisable to screen simultaneously multiple libraries containing cyclic peptides with different ring sizes and formats. Towards this end, phage peptide libraries were designed to contain a single fixed cysteine residue flanked by regions of random amino acids (Figure 9.1d).[40,41,43] This residue could then bridge with a cysteine appearing in the randomized regions. McConnell *et al.* developed a peptide phage library of the format $X_{20}GCGX_{20}$, having as many as 40 randomized amino acid positions.[43] Other phage peptide libraries with a single constant cysteine have the formats CX_9,[40] or X_8CX_8, $X_{15}CX$ and XCX_{15}.[41] As anticipated, the vast majority of the isolated peptides contained an even number of cysteines.[40,41,43] One peptide isolated from such a library format contained as many as four cysteines that most likely formed two disulfide bridges.[43] In several cyclic peptide libraries, invariable sequences were inserted to either introduce a specific peptide conformation or to introduce a specific peptide motif that already provided a basal affinity for a protein target. For example, a Gly–Pro motif was inserted to induce a β-turn conformation ($X_4CX_2GPX_4CX_4$).[42] A commercial disulfide-cyclized peptide phage library is offered by New England Biolabs. The library has the format CX_7C and is provided in the form of a kit that includes appropriate *Escherichia coli* strains, control reagents, sequencing primers and protocols. A starting vector is also provided for construction of custom peptide libraries.[44]

Most phage display peptide libraries were produced by ligating polymerase chain reaction (PCR) products into double-stranded circular phage DNA.[44,45] The random regions in the PCR product are generated with DNA primers that contain random oligonucleotide regions. Another popular library cloning strategy is based on the principle of Kunkel mutagenesis in which a DNA oligonucleotide with a randomized region is annealed to a single stranded phage DNA and the complementary strand is enzymatically synthesized.[46] The random regions in DNA oligonucleotides are preferentially encoded by the degenerate codons NNK or NNB (N = any of the four nucleotides, K = guanosine and thymidine, B = cytosine, guanosine and thymidine). In contrast to NNN codons, the NNK degenerate codons minimize the over-representation of the commonest amino acids. The reduced codon set NNB has the smallest probability of coding for a stop codon. The library size is limited by the number of bacterial cells that can be transformed with a DNA plasmid. Most of the above described phage display libraries comprise between 10^7 and 10^9 different peptides. The phage coat proteins pIII and pVIII contain signal sequences that direct them to the periplasm of the bacteria. In this oxidizing environment, the cysteines in the peptides fused to the phage coat protein are readily oxidized. If not oxidized at this stage, the cysteines may get oxidized in the medium once the phage particles are extruded from the cells.

9.2.3 Phage Panning of Disulfide-Cyclized Peptide Libraries

Phage panning, also termed 'biopanning', is the procedure by which phage expressing the desired binding specificity are enriched from a library of phage displaying a pool of variant polypeptides. This process involves immobilization of the target, exposing the phage library to the exposed target, washing away unbound phage and elution of the bound phage. The eluted phage are amplified by infecting bacteria and again panned against the target. The process of phage production and biopanning is typically repeated 2–3 times to enrich specifically binding phage. The biopanning procedures applied to cyclic peptide libraries are the same as those applied in phage selection of other polypeptides formats, such as antibodies. Phage panning strategies and protocols are described in various review articles[47] and books.[14,48–50]

Most often, one or several families of related peptides are found in phage selections with a given target. The finding of a consensus sequence is indicative of a successful selection. In many selections with cyclic peptide phage libraries, a consensus sequence was already found after 1–2 rounds of panning. Identified consensus motifs may be matched to a region of a protein that is the natural binding partner of the target protein. In fact, often the binding sites of the phage-selected peptides coincide with natural ligand-binding sites.[24] If the binding affinity of isolated peptides is not sufficiently strong, consensus motifs are built into secondary libraries to generate higher-affinity peptides.

9.3 Examples of Phage-Selected Monocyclic Peptides

Cyclic peptide phage display libraries were panned against more than one hundred different protein targets. Table 9.1 shows examples of phage-selected cyclic peptides and their binding affinities. This list is by far not complete but should give an idea of protein targets to which cyclic peptide ligands could be generated, and it should provide an idea of typical binding affinities that can be achieved. The binding affinities vary significantly with the median dissociation constant being around 1 μm. Most of the ligands have good binding selectivity, particularly those with high affinity. To illustrate the strategies applied to evolve cyclic peptide ligands, we reviewed two prime examples. In the first example, a series of model experiments are described that led to the isolation of a nanomolar antagonist of the rather challenging target vascular endothelial growth factor (VEGF). In the second example, the generation of cyclic peptides mimicking erythropoietin are reviewed. This work was also highly challenging because the ligands not only needed to bind to the erythropoietin receptor but also required triggering receptor dimerization and signaling. This goal was achieved and an erythropoietin mimetic was approved by the FDA in 2012, but unfortunately, in early 2013 the product had to be retracted from the market due to serious hypersensitivity reactions in some patients.

9.3.1 Ligands of Vascular Endothelial Growth Factor

VEGF is a primary modulator of vascular neogenesis and angiogenesis.[51] The cytokine contains two identical polypeptide chains that form a dimer and mediate signaling by dimerizing tyrosine kinase receptors. In addition to its normal physiological role, VEGF is associated with numerous pathogenic states, including cancer, rheumatoid arthritis, diabetic retinopathy and psoriasis. VEGF-blocking agents based on antibodies, an RNA aptamer or a soluble receptor, are in clinical use.

Aiming at the generation of VEGF antagonists with smaller sizes, scientists at Genentech have screened seven random disulfide-constrained peptide libraries of the form $X_mCX_nCX_l$ with n ranging from 4 to 10 and $m+n+l=18$.[42] Additionally, they screened a library of the form $X_4CX_2GPX_4CX_4$, containing a Gly–Pro motif that can form a type I β-turn. The peptides were displayed on the phage pVIII for multivalent display, allowing the capture of weak binders. After four selection rounds, a total of three VEGF-specific peptides were isolated, all having relatively low affinities. The peptides were affinity matured by applying two different mutagenesis strategies. In a first approach, the peptides were "soft" randomized by spiking the nucleotides in the DNA synthesis with a small fraction of the other three nucleotides (2.5% each). The peptide libraries were displayed on phage pIII using a phagemid system to achieve a monovalent display and allow discrimination among the best binders. In a second affinity maturation approach, a "tailored" randomization strategy was applied. In this approach, peptide variants were displayed and screened that combine mutations identified in the first affinity maturation. The best binders isolated are v107 (GGNEC-DIARMWEWECFERL-NH_2) and v114 (VEPNCDIHVMWEWECFERL-NH_2), blocking the binding of VEGF to the receptor with IC_{50} values of 0.7 and 0.22 μM, respectively (Figure 9.2a). The peptides inhibited VEGF1-165-induced proliferation of HUVE primary cell cultures. Structural analysis of the cyclic peptides by nuclear magnetic resonance (NMR) indicated that in solution, the ligands adopted multiple conformations rather than a defined structure. In complex with VEGF, the peptides were found to adopt a defined structure.[52,53] The structure of a weakly binding VEGF ligand, v108 ($IC_{50} = 2.2$ μM) was solved in complex with VEGF by X-ray crystallography (PDB entry 1VPP).[53] The structure of the previously mentioned cyclic peptide v107, having a higher affinity, was determined in complex with VEGF by NMR (PDB entry 1KAT, Figure 9.2b).[52] Peptides binding to VEGF were also isolated by Daugherty and colleagues using bacterial display.[54] Screening of an X_{15} peptide library yielded clones that contained pairs of cysteines, of which some contained the consensus motif WE/DWE/D present in the ligands v107 and v114 described earlier. Affinity maturation of those cyclic peptides yielded VEGF antagonists with nanomolar affinities (the best $K_d = 0.47$ μM).

9.3.2 Cyclic Peptide Erythropoietin Receptor Agonists

Erythropoietin (EPO) is the primary regulator of the growth and survival of erythroid progenitors, which mature into red blood cells.[55] It acts by binding to a receptor on the surface of progenitor cells. The EPO receptor (EPOR) possesses an extracellular ligand-binding domain, a single-pass transmembrane region and an intracellular signaling domain. EPO activates the receptor by binding and homodimerization of two receptor monomers in a 2:1 receptor: hormone assembly. EPO and various derivatives thereof are clinically applied to treat severe anemia associated with acute and chronic disease and offer a great benefit to these patients. The EPO-based products must be administered by intravenous or subcutaneous injection.

A long-standing goal of scientists has been the development of smaller agonists that can be applied more conveniently. Towards the generation of lead compounds, researchers from Affymax and Johnson Pharmaceutical have panned a large combinatorial phage library of the format CX_8C displayed on phage pVIII against the extracellular part of EPOR.[56] A peptide termed AF11154, binding weakly to EPOR (around 10 μM), could be isolated

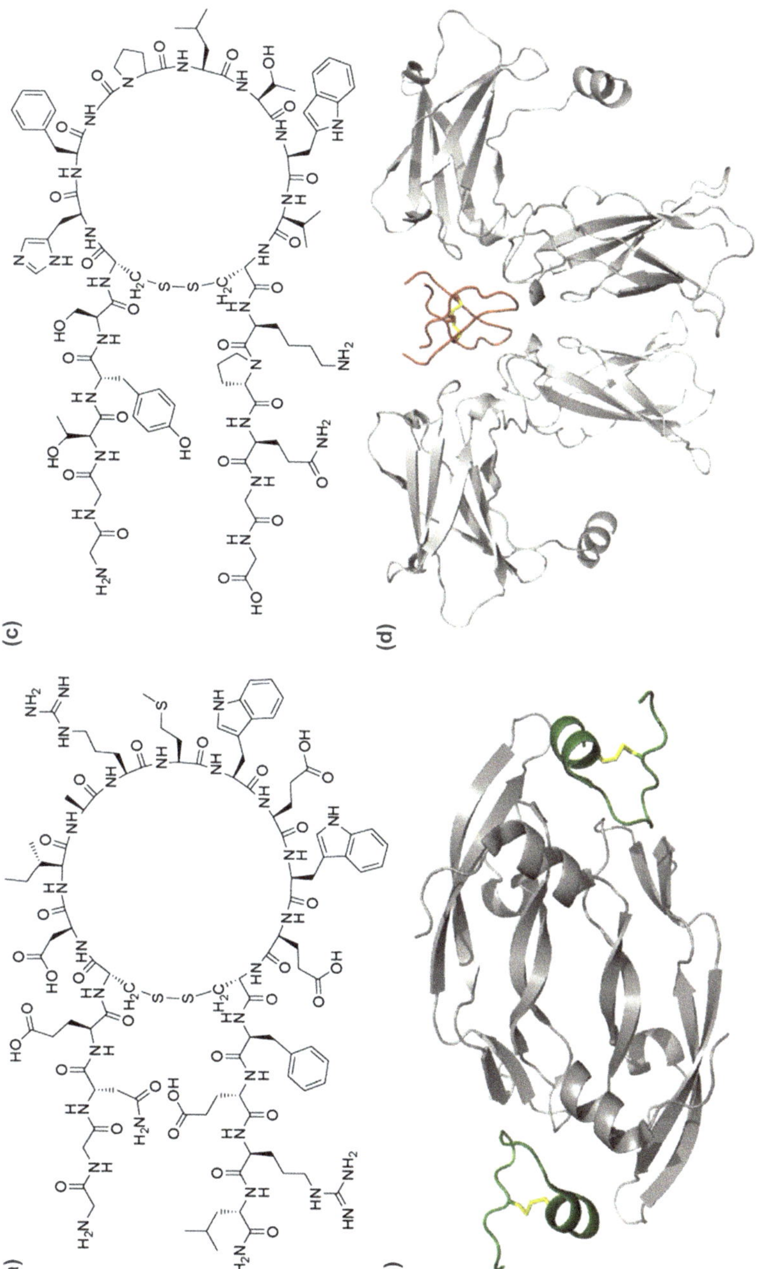

Figure 9.2 Examples of phage-selected monocyclic peptides. (a) Chemical structure of the disulfide-cyclized peptide ligand of VEGF v107. (b) NMR structure of two cyclic peptides v107 (green) bound to a VEGF dimer (grey; PDB entry 1KAT). The disulfide bridges are shown in yellow. (c) Chemical structure of the disulfide-cyclized peptide ligand of EPOR EMP1. (d) Co-crystal structure of EMP1 (red) and EPOR (two fibronectin III domains; grey) (PDB entry 1EBP). The EMP1 dimer assembly is the primary interaction that mediates receptor dimerization, with minimal contact occurring between the receptor monomers. Each peptide monomer contacts both receptor molecules. The disulfide bridges are shown in yellow.

(CRIGPITWVC). Based on this peptide, secondary libraries were generated and displayed on pIII using a phagemid vector. In the first library, the amino acids of peptide AF11154 were "soft" randomized to mutate each amino acid with a frequency of 50%. A second library of the same format was cloned but three random amino acids were appended on each side of the cyclic peptide. From the second library, cyclic peptides with 10 to 50-fold higher affinity could be isolated. The best EPOR binder, EMP1 (GGTYSCHFGPLTWVCKPQGG) showed an IC_{50} of 200 nM in an ELISA-based assay (Figure 9.2c). The phage-selected peptides sharing the consensus sequence XYXCXXGPXTWXCXPXX did not have any homology with EPO. X-ray structure analysis of the cyclic peptide bound to the receptor (EMP1-EPOR) showed a complex of two peptides bound to two receptor monomers in a 2:2 ratio (PDB entry 1EBP, Figure 9.2d).[57] Further studies with chemically linked EPO mimetic peptides revealed a regulatory process for EPOR activation that is dependent on the actual configuration of the receptor–ligand dimer assembly.[58–62] Cyclic peptide EPOR agonists were also isolated from phage libraries by other research groups.[63] Chemical dimerization of an EMP1-related peptide through covalent linkage of polyethylene glycol (PEG) generated the di-peptide with the generic name peginesatide (formerly named hematide). Dimerization increased the affinity and potency, and pegylation significantly prolonged the circulation time in the blood stream. Peginesatide was launched for the treatment of anemia associated with chronic kidney disease[64] and distributed by Affymax and Takeda. Unfortunately, peginesatide showed serious hypersensitivity reactions in some patients. Only introduced in 2012, the drug was voluntarily retracted from the market in early 2013.

9.4 Phage Selection of Bicyclic Peptides

A strategy to generate bicyclic peptide ligands by phage display was recently developed.[12] Combinatorial libraries of bicyclic peptides are generated by chemical cyclization of phage-displayed linear peptides (Figure 9.3a). In a first proof-of-concept work published in 2009, bicyclic peptide inhibitors of two human proteases were developed.[12] In subsequent phage selections, bicyclic peptides were isolated against other targets.[66] An evident characteristic of the isolated bicyclic peptides is the high binding affinities, typically in the nanomolar range.[12,66–69] Selections performed against the human protease plasma kallikrein recently yielded binders with picomolar affinities.[69] The following four sections deal with the molecule format of bicyclic peptides (Section 9.4.1), the phage selection strategy (Section 9.4.2), bicyclic peptide phage libraries (Section 9.4.3), and two examples of phage-selected bicyclic peptides (Section 9.4.4).

9.4.1 Bicyclic Peptides

Bicyclic peptides contain two macrocyclic rings that are generated by cyclization of linear peptides or by connecting two peptide rings.[70] Many

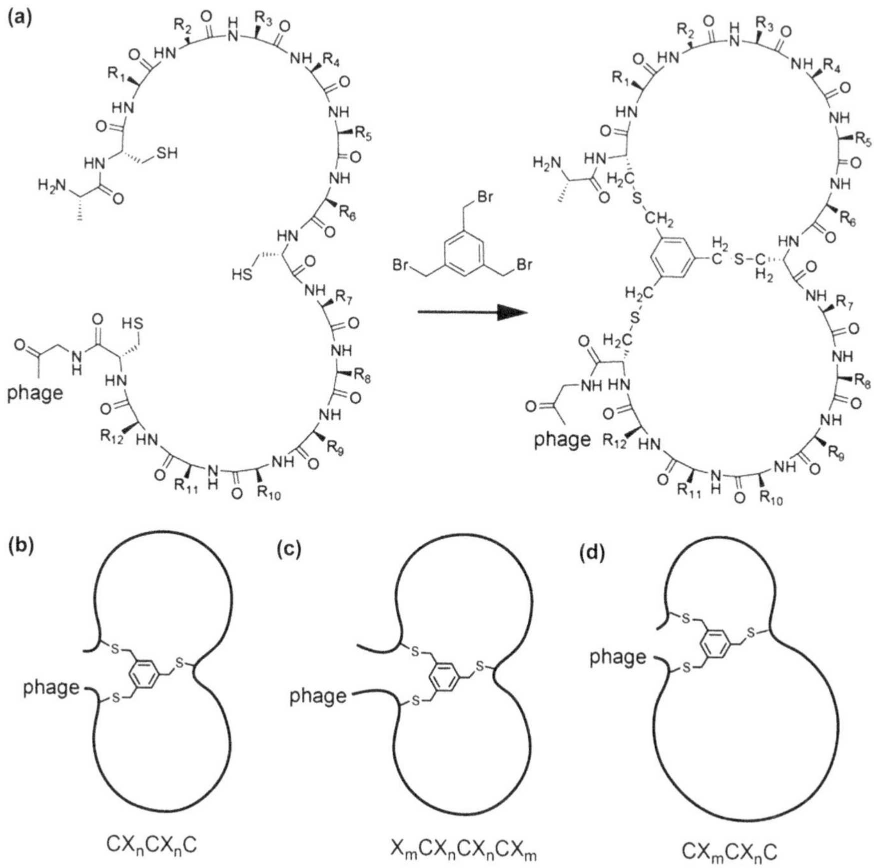

Figure 9.3 Phage selection of bicyclic peptides. (a) Cyclization of cysteine-rich peptides with the alkylating reagent TBMB. The 6×6 bicyclic peptide library is shown. All peptides of this library contain a constant alanine residue at the *N*-terminus and are linked via a Gly–Gly–Ser–Gly linker to the *N*-terminus of disulfide-free phage pIII. (b)–(d) Formats of existing bicyclic peptide phage libraries: (b) bicyclic peptide with two rings of equal size with *n* random amino acids per ring; (c) amino acids flanking the two rings were additionally randomized; and (d) bicyclic peptides having differently sized rings. Existing bicyclic peptide libraries are shown in Table 9.2.

organisms such as bacteria, fungi, plants and animals are producing natural product bicyclic peptides with highly diverse biological activities. Two of these naturally occurring bicyclic peptides, actinomycin D and romidepsin, are used as drugs in cancer therapies.[70] Other bicyclic peptide natural products are used as reagents in research. For example the death cap toxin phalloidin binding to actin is used as a cell imaging reagent.[71] Due to the second macrocyclic ring, the bicyclic peptides tend to be conformationally more rigid than monocyclic peptides wherein this can vary strongly from case to case. Key qualities of monocyclic peptides such as binding affinity, selectivity or proteolytic stability are therefore even more pronounced in many bicyclic peptides.

9.4.2 Bicyclization of Peptides on Phage

Bicyclic peptide phage libraries are generated by display of linear peptides on phage coat proteins and subsequent chemical cyclization (Figure 9.3a).[12,72] The chemical reaction must be efficient in order to convert the linear precursors quantitatively into bicyclic peptides. The reaction ideally yields a single product rather than several product isomers. The reaction conditions must not impair the functionality of the phage because it carries the genetic information required for phage propagation and decoding. Alkylation of sulfhydryl groups in cysteines with bromomethyl functional groups conjugated to benzene rings has been proven suitable. Peptides containing three cysteines were efficiently and selectively cyclized with the reagent 1,3,5-tris(bromomethyl)benzene (TBMB) in aqueous solution and at room temperature.[12,72] This reaction condition is well compatible with filamentous phage. In model experiments, a TBMB concentration as low as 10 μM was sufficient to quantitatively cyclize a peptide fused to the phage pIII coat protein.[12] The phage remained essentially functional under these conditions. Before reaction with TBMB, the phage peptides need to be reduced. A suitable reducing agent is tris-(carboxyethyl)phosphine (TCEP).[73] Cysteines in peptides displayed on the phage were quantitatively reduced after incubation with 1 mM TCEP at 42 °C for 1 h.[12] As TBMB can react with TCEP, the concentration of the reducing agent needs to be lowered before treating the phage with TBMB. This can be achieved by filtering the phage in spin tubes. A 100-fold reduced TCEP concentration (i.e. 10 μM) was found not to impair the alkylation of phage peptide with 10 μM TBMB.[12]

The phage coat protein pIII contains eight cysteines that can potentially interfere with the peptide cyclization reaction. Structure determination of the *N*-terminal two domains of pIII showed that the six cysteines in these domains form three disulfide bridges.[74] One of the cysteines is found close to the *N*-terminus of the mature pIII (Cys7) and there is a high risk that, once reduced, participates in the cyclization reaction, leading to product mixtures. To omit such problems, a disulfide-free phage mutant developed by Kather *et al.*[75] can be applied. In this phage, the six cysteines were mutated and many additional mutations were introduced to rescue the functionality of the pIII and the ability of phage to infect bacteria. The mutated phage has a significantly lower infectivity than wild-type phage but proved to be efficient in the phage selection of bicyclic peptides.[12]

9.4.3 Bicyclic Peptide Phage Libraries

A number of different bicyclic peptide phage libraries were cloned and subjected to affinity selections. In all libraries reported to date, the bicyclic peptides were displayed as a fusion of the minor coat protein pIII wherein phage vectors were applied to display five copies of the peptides per phage. The first bicyclic peptide phage library (library 1, also termed 6×6 library) was designed to contain three cysteines in fixed positions spaced by regions of six random amino acids (CX_6CX_6C, library 1 [6×6 library]) (Figure 9.3a,b, Table 9.2).[12] In the randomized amino acid positions, all 20 natural amino acids were allowed. The library contains more than 4×10^9 different clones of the theoretically 4.1×10^{15} possible peptides. The 6×6 bicyclic peptide library was panned against several protein targets and yielded binders with low nanomolar affinities.[12,66,68] Subsequently, bicyclic peptides having 3, 4 and 5 random amino acids in each of the two rings were cloned and binders isolated (CX_nCX_nC, $n=3$, 4, 5) (Figure 9.3b, Table 9.2).[69] In some of the libraries, amino acid positions flanking the rings were additionally randomized (library formats: XCX_3CX_3CX and XCX_4CX_4CX) (Figure 9.3c, Table 9.2).[67,69] Although the latter bicyclic peptides have smaller rings and fewer amino acids that can potentially interact, phage selections with these libraries also yielded tight binders.[67,69] Bicyclic peptides isolated from the 5×5 library against the plasma kallikrein inhibited the serine protease with even picomolar K_is.[69] Variation of the bicyclic peptide ring size turned out to be a suitable strategy to modulate the target specificity (please see Section 9.5.1).[69] Other phage libraries developed display bicyclic peptides with two differently sized rings of 3, 4, 5 or 6 random amino acids (CX_nCX_mC, n, $m=3$–6) (Figure 9.3d, Table 9.2).[68] Phage selections with these libraries yielded a larger number of different consensus sequences. For a given target, certain ring size combinations were preferentially enriched.[68]

The bicyclic peptide phage display libraries were cloned by ligating DNA encoding the peptide diversity into a phage vector. A detailed protocol for the library construction can be found in Chapter 2 of Rentero Rebollo *et al.*[72] The random DNA sequences were appended in a PCR reaction to the gene of the first two domains of pIII using degenerate primer. The randomized amino acid positions were encoded by NNK codons. The PCR product was inserted into a phage vector between a leader sequence (for periplasmic export) and the third domain of pIII. As mentioned earlier, a mutant of pIII that did not contain the six cysteines in the first two domains was used to construct the phage libraries.[75] The DNA was transformed into *E. coli* bacterial cells by electroporation. The

Table 9.2 Bicyclic peptide phage libraries. The peptide format showing the number and position of randomized amino acids are indicated. All 20 amino acids were allowed in the randomized positions. In these libraries, the peptides are displayed on phage pIII.

Library name	*Bicyclic peptide format*	*Library size*	*Reference*
Library 6×6	ACX_6CX_6CG	4×10^9	12
Library 4×4	XCX_4CX_4CX	7.3×10^8	67
Library 3×3	XCX_3CX_3CX	5.6×10^8	69
Library A	ACX_mCX_nCG; $m+n=7$ or 8	5×10^8	68
Library B	ACX_mCX_nCG; $m+n=9$	1×10^7	68
Library C	ACX_mCX_nCG; $m+n=10$ or 11	6×10^7	68

theoretical and effective diversities of the different libraries are indicated in Table 9.2.

All bicyclic peptides isolated to date are cyclized by the previously described alkylating reagent TBMB. The diversity of bicyclic peptide libraries could potentially be increased by applying additional cyclization reagents. Unlike the peptide, the cyclization reagent is not encoded by the phage DNA and selections with different cyclization linkers would need to be performed in parallel to retain the identity of the cyclization linker. Two new reagents containing three thiol-reactive groups were recently developed.[76] The cyclization linkers termed TATA and TBAB, with either bromoacetamide or acrylamide functional groups, proved suitable to efficiently and selectively cyclize peptides containing three cysteine residues. The two linkers provide a different structural environment to the peptides that are linked to them. This was demonstrated in an experiment in which the TBMB-based linker of the bicyclic peptide PK15, a potent inhibitor of plasma kallikrein ($K_i = 2.9$ nM), was replaced by the new linkers.[76] The inhibitory activity of the bicyclic peptides cyclized with TATA or TBAB were more than 1000-fold weaker than the one cyclized with TBMB. Structural analysis by solution-state NMR revealed different nuclear Overhauser effect (NOE) constraints in the three bicyclic peptides, indicating that the relatively small linkers at the center of bicyclic peptides significantly influence the conformations of the peptides.[76] These results demonstrate the prominent structural role of linkers in peptide macrocycles and suggest that applying different cyclization linkers in a combinatorial fashion could be an attractive strategy for generating topologically diverse macrocycle libraries. Additionally, the two new cyclization linkers contain multiple hydrogen acceptor and donor groups that could potentially interact with amino acids in the peptides to form more structured macrocycles.

9.4.4 Phage Panning of Bicyclic Peptide Phage Libraries

Bicyclic peptide ligands are isolated from phage libraries in affinity selections similarly as disulfide-cyclized peptides (please see Section 9.2.3).[72] A major difference in the biopanning of bicyclic peptides is the production of phage in larger culture volumes. This is necessary because the mutant phage with disulfide-free pIII has a significantly lower infectivity than wild-type phage. One per cent or less of the mutant phage are infective and only around 10^{12} infective phage can be obtained from a 1 L bacterial culture.[75] To represent each clone of the 6×6 bicyclic peptide library (complexity: $4*10^9$) in average around 100 times, phage is typically produced in a 0.5 L culture. A detailed protocol for the biopanning of bicyclic peptide phage libraries can be found in Chapter 4 of Rentero Rebollo *et al.*[72]

9.5 Examples of Phage-Selected Bicyclic Peptides

For the development of the bicyclic peptide phage display technology, the serine proteases plasma kallikrein and cathepsin G were used as targets.[12]

Following the successful isolation of inhibitors with nanomolar inhibitory constants for these proteases, bicyclic peptides were also isolated against other trypsin-like serine proteases, such as urokinase-type plasminogen activator[66,68] and protease of other classes (unpublished data). More recently, the technology was applied to other targets than proteases, such as receptors or a cytokine (unpublished data). In the following two sections, bicyclic peptides isolated against the two targets plasma kallikrein and urokinase-type plasminogen activator will be presented. In the first example involving plasma kallikrein as target, procedures are illustrated to screen a naïve bicyclic peptide library, to affinity mature bicyclic peptide ligands and to modulate the specificity of bicyclic peptide ligands. In the second example, bicyclic peptide inhibitors of urokinase-type plasminogen activator are described. In this section, characteristics of bicyclic peptides important for their therapeutic application, such as stability and pharmacokinetic properties, will be discussed in detail.

9.5.1 Inhibitors of Plasma Kallikrein

The serine protease plasma kallikrein (PK) is implicated in different diseases including hereditary angioedema (HAE), a life-threatening disease characterized by local swelling in subcutaneous tissue.[77] Most HAE patients have reduced levels of the endogenous C1 protease, leading to insufficient control of PK activity. PK cleaves kininogen to release bradykinin, which mediates vasodilation and leads to edema. HAE therapy was significantly improved recently with the development of three new drugs: a recombinant C1 inhibitor, an engineered protein PK inhibitor (ecallantide) and a bradykinin receptor antagonist (icatibant).[78] Further improvement of HAE therapy might be achieved by a PK inhibitor with a long half-life that can be applied prophylactically. The ideal would be a molecule that can be administered orally.

Bicyclic peptides binding to human PK were isolated from the 6×6 phage library described in Section 9.4.3.[12] Most of the peptides contained one of three consensus motifs that were identified by comparing the sequences. The tested bicyclic peptide inhibited PK with inhibitory constants in the medium nanomolar to high nanomolar range. The finding that all isolated peptides were inhibitors was surprising as the phage panning procedure selects just for binders. The three consensus motifs were found either in the first or the second random peptide region, indicating that mainly one of the two rings in the bicyclic peptides interacts with PK. The motifs were inserted into secondary libraries containing one randomized ring to generate bicyclic peptides with even higher affinity.[12] The best bicyclic peptide inhibitor isolated in this affinity maturation, PK15, had a K_i of 2.9 nM (Figure 9.4a). Specificity profiling of the inhibitor PK15 revealed high target specificity. The inhibitor did not block factor XIa, a homologous human serine protease that shares the highest sequence identity with human PK (69% sequence identity) or any of the other human serine proteases tested. Such a high selectivity was also later found for bicyclic peptides isolated against other proteins.[66] Phage-selected bicyclic peptides appear to have high selectivity although no selection pressure was applied to enforce the enrichment of selective binders in any of the biopanning experiments.

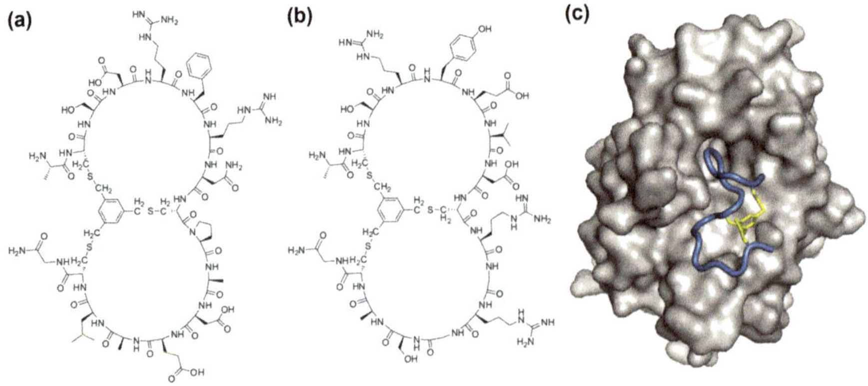

Figure 9.4 Examples of phage-selected bicyclic peptides. (a) Bicyclic peptide inhibitor of PK (PK15). (b) Bicyclic peptide inhibitor of uPA (UK18). (c) Co-crystal structure of bicyclic peptide UK18 (peptide backbone in blue and cyclization linker in yellow) with uPA (grey) (PDB entry 3QN7).

The selective inhibitor PK15 did not efficiently inhibit the orthologous murine PK despite its high sequence identity with the human protease (81%). This prevented testing of its therapeutic effect in small laboratory animals. Comparison of the human and rat PK structures identified a small region around the active site of PK that is identical in the two orthologous proteases. From phage display libraries containing smaller bicyclic peptides (3×3 and 5×5 libraries described earlier), inhibitors binding to both the human and mouse protease could be isolated.[69] A 5×5 bicyclic peptide (PK133) showed the desired specificity profile: it inhibited human and rat PK with K_is of 2.9 and 6.6 nM, respectively, and did not affect any of the paralogous proteases.[69]

9.5.2 Inhibitors of Urokinase-Type Plasminogen Activator

The serine protease urokinase-type plasminogen activator (uPA) participates in the turnover of extracellular matrix proteins and is implicated in tumor growth and invasion.[79] Small molecule inhibitors of uPA, such as the clinical candidate WX-UK1 or the tripeptide derivative CJ-463 (benzylsulfonyl-D-Ser-Ser-4-amidinobenzylamide), were found to inhibit tumor growth in various xenograft tumor mouse models;[80,81] however, these small molecules also inhibit, to some extent, other human serine proteases at concentrations required to fully block uPA, which could potentially lead to toxicity in a clinical setting.

Bicyclic peptide inhibitors of uPA were isolated from the above described 6×6 bicyclic peptide phage library.[66] The selections yielded two families of bicyclic peptides, one having a consensus sequence in the first peptide ring and the second one having a consensus sequence covering the end of the first ring and almost the whole second ring. Peptides of both families inhibited uPA by binding to the active site. A peptide of the second family, UK18, was the most potent inhibitor ($K_i = 53$ nM) (Figure 9.4b). The bicyclic peptide inhibited uPA with up to 200-fold higher affinities than the disulfide-cyclized monocyclic peptide upain-1 isolated previously to the same target by phage display.[82] Structurally related trypsin-like serine proteases were inhibited only at more than 1000-fold higher K_is, indicating high target specificity. Linear and monocyclic analogues of UK18 were synthesized and found to inhibit uPA with 320 and 17-fold weaker inhibitory constants, showing that the bicyclic configuration is important for the activity. The structure of UK18 in complex with uPA was determined by X-ray crystallography (PDB entry 3QN7) (Figure 9.4c).[66] This structure revealed a large binding interface of around 700 $Å^2$ and showed that both peptide rings interact directly with the target.

In relation to the testing of the therapeutic effect of UK18 in tumor-bearing mice, its metabolic stability and renal clearance rate were assessed. The bicyclic peptide UK18 showed a good stability in plasma *ex vivo* and resisted proteases in human plasma significantly better than its linear or monocyclic analogues described earlier.[83] Furthermore, the bicyclic peptide was more stable than the two peptide rings synthesized individually. This indicated that the two macrocyclic rings of the bicyclic peptide mutually protect each other from proteolytic activity. By conjugating UK18 to an IgG Fc fragment[84] or an albumin-binding peptide,[83] fast renal clearance could be prevented and a plasma half-life in mice of 1 day or longer achieved. Conjugation to the albumin-binding peptide, and thus non-covalent tethering to albumin, not only delayed the renal clearance but also additionally stabilized the proteolytic stability of the bicyclic peptide further. The good metabolic stability and the long circulation time suggested that bicyclic peptides are a suitable format for the generation of long-acting peptide therapeutics.

9.6 Conclusions

The power of phage display to screen billions of peptide variants in short time and its simple procedures have led to its widespread application. Cyclic peptides were isolated to numerous protein targets. Some of these ligands have found application as research reagents and several are in pre-clinical and clinical development. An essential quality of cyclic peptide ligands is a good binding affinity. For most clinical applications, affinities in the low nanomolar or subnanomolar range are required. Unfortunately, many of the phage-selected cyclic peptides have, even after affinity maturation, only micromolar or high nanomolar affinities. A promising strategy toward the generation of high-affinity peptide macrocycles to a broader range of targets is the development of cyclic peptide repertoires with more rigid structures, such as the bicyclic peptide reviewed herein. By combining novel peptide library formats with new chemical cyclization reactions, phage libraries of highly diverse conformationally constrained polycyclic peptides might be generated and high-affinity binders to a broad range of targets isolated.

Acknowledgments

The financial contribution from the Swiss National Science Foundation (SNSF Professorship PP00P3_123524/1 to C.H.) is gratefully acknowledged.

References

1. E. M. Driggers, S. P. Hale, J. Lee and N. K. Terrett, *Nat Rev Drug Discovery*, 2008, **7**, 608–624.
2. R. M. J. Liskamp, D. T. S. Rijkers and S. E. Bakker, *Bioactive Macrocyclic Peptidews and Peptide Mimics*, in *Modern Supramolecular Chemistry: Strategies for Macrocycle Synthesis*, ed. F. Diderich, P. J. Stang and R. R. Tykwinski, Wiley: Weinheim, 2008, pp. 1–27.
3. R. C. Ladner, *Trends Biotechnol*, 1995, **13**, 426–430.
4. J. Rizo and L. M. Gierasch, *Annu Rev Biochem*, 1992, **61**, 387–418.
5. G. P. Smith, *Science*, 1985, **228**, 1315–1317.
6. G. P. Smith and V. A. Petrenko, *Chem Rev (Washington, DC, United States)*, 1997, **97**, 391–410.
7. T. Clackson, H. R. Hoogenboom, A. D. Griffiths and G. Winter, *Nature*, 1991, **352**, 624–628.
8. S. E. Cwirla, E. A. Peters, R. W. Barrett and W. J. Dower, *Proc Natl Acad Sci USA*, 1990, **87**, 6378–6382.
9. J. J. Devlin, L. C. Panganiban and P. E. Devlin, *Science*, 1990, **249**, 404–406.
10. J. K. Scott and G. P. Smith, *Science*, 1990, **249**, 386–390.
11. K. T. O'Neil, R. H. Hoess, S. A. Jackson, N. S. Ramachandran, S. A. Mousa and W. F. DeGrado, *Proteins*, 1992, **14**, 509–515.
12. C. Heinis, T. Rutherford, S. Freund and G. Winter, *Nat Chem Biol*, 2009, **5**, 502–507.
13. J. W. Kehoe and B. K. Kay, *Chem Rev (Washington, DC, United States)*, 2005, **105**, 4056–4072.
14. B. K. Kay and R. H. Hoess, *Principles and Applications of Phage Display*. Academic Press, San Diego, CA, 1996.
15. M. Russel, H. B. Lowman and T. Clackson, *Introduction to Phage Biology and Phage Display*, in *Phage Display – A Practical Approach*, ed. H. B. Lowman and T. Clackson, Oxford University Press, UK, 2004, pp. 1–26.
16. R. Webster, *Filamentous Phage Biology*, in *Phage Display: A Laboratory Manual*, ed. C. F. Barbas, D. R. Burton 3rd, J. K. Scott and G. J. Silverman, Cold Spring Harbor Laboratory Press, Woodbury, NY, 2001, pp. 1.1–1.37.
17. T. Bratkovic, *Cell Mol Life Sci*, 2010, **67**, 749–767.
18. M. J. Rowley, K. O'Connor and L. Wijeyewickrema, *Biotechnol Annu Rev*, 2004, **10**, 151–188.
19. S. S. Sidhu, W. J. Fairbrother and K. Deshayes, *ChemBioChem*, 2003, **4**, 14–25.
20. M. A. McLafferty, R. B. Kent, R. C. Ladner and W. Markland, *Gene*, 1993, **128**, 29–36.

21. R. C. Ladner, A. K. Sato, J. Gorzelany and M. de Souza, *Drug Discovery Today*, 2004, **9**, 525–529.
22. T. Clackson and J. A. Wells, *Trends Biotechnol*, 1994, **12**, 173–184.
23. R. Cortese, P. Monaci, A. Luzzago, C. Santini, F. Bartoli, I. Cortese, P. Fortugno, G. Galfre, A. Nicosia and F. Felici, *Curr Opin Biotechnol*, 1996, **7**, 616–621.
24. B. A. Katz, *Annu Rev Biophys Biomol Struct*, 1997, **26**, 27–45.
25. H. B. Lowman, *Annu Rev Biophys Biomol Struct*, 1997, **26**, 401–424.
26. P. Molek, B. Strukelj and T. Bratkovic, *Molecules*, 2011, **16**, 857–887.
27. S. S. Sidhu, H. B. Lowman, B. C. Cunningham and J. A. Wells, *Methods Enzymol*, 2000, **328**, 333–363.
28. M. Szardenings, *J Recept Signal Transduction Res*, 2003, **23**, 307–349.
29. F. Uchiyama, Y. Tanaka, Y. Minari and N. Tokui, *J Biosci Bioeng*, 2005, **99**, 448–456.
30. L. R. Krumpe, A. J. Atkinson, G. W. Smythers, A. Kandel, K. M. Schumacher, J. B. McMahon, L. Makowski and T. Mori, *Proteomics*, 2006, **6**, 4210–4222.
31. C. G. Ullman, L. Frigotto and R. N. Cooley, *Briefings Funct Genomics*, 2011, **10**, 125–134.
32. D. J. Rodi, S. Mandava and L. Makowski, *Filamentous Bacteriophage Structure and Biology*, in *Phage Display in Biotechnology and Drug Discovery*, ed. S. S. Sidhu, Taylor & Francis, London, UK, 2005 pp. 1–62.
33. S. Bass, R. Greene and J. A. Wells, *Proteins*, 1990, **8**, 309–314.
34. H. Qi, H. Lu, H. J. Qiu, V. Petrenko and A. Liu, *J Mol Biol*, 2012, **417**, 129–143.
35. V. Petrenko and G. P. Smith, *Vectors and Modes of Display*, in *Phage Display in Biotechnology and Drug Discovery*, ed. S. S. Sidhu, Taylor & Francis, London, UK, 2005, pp. 63–110.
36. J. B. Burritt, C. W. Bond, K. W. Doss and A. J. Jesaitis, *Anal Biochem*, 1996, **238**, 1–13.
37. M. B. Zwick, J. Shen and J. K. Scott, *Curr Opin Biotechnol*, 1998, **9**, 427–436.
38. L. B. Giebel, R. T. Cass, D. L. Milligan, D. C. Young, R. Arze and C. R. Johnson, *Biochemistry*, 1995, **34**, 15430–15435.
39. E. Koivunen, B. Wang and E. Ruoslahti, *J Cell Biol*, 1994, **124**, 373–380.
40. E. Koivunen, B. Wang and E. Ruoslahti, *Bio/Technology*, 1995, **13**, 265–270.
41. L. L. Bonnycastle, J. S. Mehroke, M. Rashed, X. Gong and J. K. Scott, *J Mol Biol*, 1996, **258**, 747–762.
42. W. J. Fairbrother, H. W. Christinger, A. G. Cochran, G. Fuh, C. J. Keenan, C. Quan, S. K. Shriver, J. Y. Tom, J. A. Wells and B. C. Cunningham, *Biochemistry*, 1998, **37**, 17754–17764.
43. S. J. McConnell, A. J. Uveges, D. M. Fowlkes and D. G. Spinella, *Mol Diversity*, 1996, **1**, 165–176.
44. K. A. Noren and C. J. Noren, *Methods*, 2001, **23**, 169–178.
45. F. A. Fellouse and G. Pal, *Methods for the Construction of Phage-Displayed Libraries*, in *Phage Display in Biotechnoloy and Drug Discovery*, ed. S. S. Sidhu, Taylor & Francis, London, UK, 2005, pp. 111–142.

46. R. Huang, P. Fang and B. K. Kay, *Methods*, 2012, **58**, 10–17.
47. B. K. Kay, J. Kasanov and M. Yamabhai, *Methods*, 2001, **24**, 240–246.
48. S. S. Sidhu, *Phage Display in Biotechnology and Drug Discovery*, Taylor & Francis, London UK, 2005.
49. C. F. Barbas, D. R. Burton 3rd, J. K. Scott and G. J. Silverman, *Phage Display: A Laboratory Manual*. Cold Spring Harbor Laboratory Press, Woodbury, NY, 2001.
50. H. B. Lowman and T. Clackson, *Phage Display – A Practical Approach*. Oxford University Press, Oxford, UK, 2004.
51. N. Ferrara, *Am J Physiol*, 2001, **280**, C1358–1366.
52. B. Pan, B. Li, S. J. Russell, J. Y. Tom, A. G. Cochran and W. J. Fairbrother, *J Mol Biol*, 2002, **316**, 769–787.
53. C. Wiesmann, H. W. Christinger, A. G. Cochran, B. C. Cunningham, W. J. Fairbrother, C. J. Keenan, G. Meng and A. M. de Vos, *Biochemistry*, 1998, **37**, 17765–17772.
54. S. A. Kenrick and P. S. Daugherty, *Protein Eng Des Sel*, 2010, **23**, 9–17.
55. S. B. Krantz, *Blood*, 1991, **77**, 419–434.
56. N. C. Wrighton, F. X. Farrell, R. Chang, A. K. Kashyap, F. P. Barbone, L. S. Mulcahy, D. L. Johnson, R. W. Barrett, L. K. Jolliffe and W. J. Dower, *Science*, 1996, **273**, 458–464.
57. O. Livnah, E. A. Stura, D. L. Johnson, S. A. Middleton, L. S. Mulcahy, N. C. Wrighton, W. J. Dower, L. K. Jolliffe and I. A. Wilson, *Science*, 1996, **273**, 464–471.
58. D. L. Johnson, F. X. Farrell, F. P. Barbone, F. J. McMahon, J. Tullai, K. Hoey, O. Livnah, N. C. Wrighton, S. A. Middleton, D. A. Loughney, E. A. Stura, W. J. Dower, L. S. Mulcahy, I. A. Wilson and L. K. Jolliffe, *Biochemistry*, 1998, **37**, 3699–3710.
59. D. L. Johnson, F. X. Farrell, F. P. Barbone, F. J. McMahon, J. Tullai, D. Kroon, J. Freedy, R. A. Zivin, L. S. Mulcahy and L. K. Jolliffe, *Chem Biol*, 1997, **4**, 939–950.
60. O. Livnah, D. L. Johnson, E. A. Stura, F. X. Farrell, F. P. Barbone, Y. You, K. D. Liu, M. A. Goldsmith, W. He, C. D. Krause, S. Pestka, L. K. Jolliffe and I. A. Wilson, *Nat Struct Biol*, 1998, **5**, 993–1004.
61. I. A. Wilson and L. K. Jolliffe, *Curr Opin Struct Biol*, 1999, **9**, 696–704.
62. N. C. Wrighton, P. Balasubramanian, F. P. Barbone, A. K. Kashyap, F. X. Farrell, L. K. Jolliffe, R. W. Barrett and W. J. Dower, *Nat Biotechnol*, 1997, **15**, 1261–1265.
63. S. J. McConnell, T. Dinh, M. H. Le, S. J. Brown, K. Becherer, K. Blumeyer, C. Kautzer, F. Axelrod and D. G. Spinella, *Biol Chem*, 1998, **379**, 1279–1286.
64. I. C. Macdougall, J. Rossert, N. Casadevall, R. B. Stead, A. M. Duliege, M. Froissart and K. U. Eckardt, *N Engl J Med*, 2009, **361**, 1848–1855.
65. J. M. McKoy, R. E. Stonecash, D. Cournoyer, J. Rossert, A. R. Nissenson, D. W. Raisch, N. Casadevall and C. L. Bennett, *Transfusion*, 2008, **48**, 1754–1762.
66. A. Angelini, L. Cendron, S. Chen, J. Touati, G. Winter, G. Zanotti and C. Heinis, *ACS Chem Biol*, 2012, **7**, 817–821.

67. V. Baeriswyl and C. Heinis, *Protein Eng Des Sel*, 2013, **26**, 81–89.
68. I. Rentero Rebollo, A. Angelini and C. Heinis, *MedChemComm*, 2012, **4**, 145–150.
69. V. Baeriswyl, H. Rapley, L. Pollaro, C. Stace, D. Teufel, E. Walker, S. Chen, G. Winter, J. Tit and C. Heinis, *ChemMedChem*, 2012, **7**, 1173–1176.
70. V. Baeriswyl and C. Heinis, *ChemMedChem*, 2013.
71. S. Auinger and J. V. Small, *Methods Cell Biol*, 2008, **88**, 257–272.
72. I. Rentero Rebollo and C. Heinis, *Methods*, 2013.
73. E. B. Getz, M. Xiao, T. Chakrabarty, R. Cooke and P. R. Selvin, *Anal Biochem*, 1999, **273**, 73–80.
74. P. Holliger, L. Riechmann and R. L. Williams, *J Mol Biol*, 1999, **288**, 649–657.
75. I. Kather, C. A. Bippes and F. X. Schmid, *J Mol Biol*, 2005, **354**, 666–678.
76. S. Chen, J. Morales-Sanfrutos, A. Angelini, B. Cutting and C. Heinis, *ChemBioChem*, 2012, **13**, 1032–1038.
77. H. Longhurst and M. Cicardi, *Lancet*, 2012, **379**, 474–481.
78. L. Bouillet, *La Revue de medecine interne/fondee ... par la Societe nationale francaise de medecine interne*, 2012, **33**, 150–154.
79. P. A. Andreasen, R. Egelund and H. H. Petersen, *Cell Mol Life Sci*, 2000, **57**, 25–40.
80. I. Henneke, S. Greschus, R. Savai, M. Korfei, P. Markart, P. Mahavadi, R. T. Schermuly, M. Wygrecka, J. Sturzebecher, W. Seeger, A. Gunther and C. Ruppert, *Am J Respir Crit Care Med*, 2010, **181**, 611–619.
81. A. H. Mekkawy, D. L. Morris and M. H. Pourgholami, *Future Oncol*, 2009, **5**, 1487–1499.
82. M. Hansen, T. Wind, G. E. Blouse, A. Christensen, H. H. Petersen, S. Kjelgaard, L. Mathiasen, T. L. Holtet and P. A. Andreasen, *J Biol Chem*, 2005, **280**, 38424–38437.
83. A. Angelini, J. Morales-Sanfrutos, P. Diderich, S. Chen and C. Heinis, *J Med Chem*, 2012, **55**, 10187–10197.
84. A. Angelini, P. Diderich, J. Morales-Sanfrutos, S. Thurnheer, D. Hacker, L. Menin and C. Heinis, *Bioconjugate Chem*, 2012, **23**, 1856–1863.
85. E. Koivunen, D. A. Gay and E. Ruoslahti, *J Biol Chem*, 1993, **268**, 20205–20210.
86. M. Krook, C. Lindbladh, J. A. Eriksen and K. Mosbach, *Mol Diversity*, 1997, **3**, 149–159.
87. P. Holig, M. Bach, T. Volkel, T. Nahde, S. Hoffmann, R. Muller and R. E. Kontermann, *Protein Eng Des Sel*, 2004, **17**, 433–441.
88. A. F. Kolodziej, S. A. Nair, P. Graham, T. J. McMurry, R. C. Ladner, C. Wescott, D. J. Sexton and P. Caravan, *Bioconjugate Chem*, 2012, **23**, 548–556.
89. D. M. Eckert, V. N. Malashkevich, L. H. Hong, P. A. Carr and P. S. Kim, *Cell*, 1999, **99**, 103–115.
90. S. A. Benard, T. M. Smith, K. Cunningham, J. Jacob, T. DeSilva, L. Lin, G. D. Shaw, R. Kriz and K. S. Kelleher, *Biochemistry*, 2008, **47**, 4674–4682.

91. L. Oligino, F. D. Lung, L. Sastry, J. Bigelow, T. Cao, M. Curran, T. R. Burke, S. Wang Jr, D. Krag, P. P. Roller and C. R. King, *J Biol Chem*, 1997, **272**, 29046–29052.
92. N. D. Ambaye, S. C. Pero, M. J. Gunzburg, M. Yap, D. J. Clayton, M. P. Del Borgo, P. Perlmutter, M.-I. Aguilar, G. S. Shukla, E. Peletskaya, M. M. Cookson, D. N. Krag, M. C. J. Wilce and J. A. Wilce, *J Mol Biol*, 2011, **412**, 397–411.
93. K. L. Ho, K. Yusoff, H. F. Seow and W. S. Tan, *J Med Virol*, 2003, **69**, 27–32.
94. M. S. Dennis, M. Zhang, Y. G. Meng, M. Kadkhodayan, D. Kirchhofer, D. Combs and L. A. Damico, *J Biol Chem*, 2002, **277**, 35035–35043.
95. C. Frick, A. Odermatt, K. Zen, K. J. Mandell, H. Edens, R. Portmann, L. Mazzucchelli, D. L. Jaye and C. A. Parkos, *Eur J Immunol*, 2005, **35**, 3610–3621.
96. B. A. Desimmie, M. Humbert, E. Lescrinier, J. Hendrix, S. Vets, R. Gijsbers, R. M. Ruprecht, U. Dietrich, Z. Debyser and F. Christ, *Mol Ther*, 2012, **20**, 2064–2075.
97. E. Koivunen, W. Arap, H. Valtanen, A. Rainisalo, O. P. Medina, P. Heikkila, C. Kantor, C. G. Gahmberg, T. Salo, Y. T. Konttinen, T. Sorsa, E. Ruoslahti and R. Pasqualini, *Nat Biotechnol*, 1999, **17**, 768–774.
98. T. F. Gesteira, V. J. Coulson-Thomas, A. Taunay-Rodrigues, V. Oliveira, B. E. Thacker, M. A. Juliano, R. Pasqualini, W. Arap, I. L. S. Tersariol, H. B. Nader, J. D. Esko and M. A. S. Pinhal, *J Biol Chem*, 2011, **286**, 5338–5346.
99. S. C. Meyer, T. Gaj and I. Ghosh, *Chem Biol Drug Des*, 2006, **68**, 3–10.
100. M. Lunder, T. Bratkovic, S. Kreft and B. Strukelj, *J Lipid Res*, 2005, **46**, 1512–1516.
101. P. Wu, J. Leinonen, E. Koivunen, H. Lankinen and U.-H. Stenman, *Eur J Biochem*, 2000, **267**, 6212–6220.
102. T. N. Schumacher, L. M. Mayr, D. L. Minor, M. A. Milhollen Jr, M. W. Burgess and P. S. Kim, *Science*, 1996, **271**, 1854–1857.
103. B. A. Katz and R. T. Cass, *J Biol Chem*, 1997, **272**, 13220–13228.
104. M. S. Meiring, D. Litthauer, J. Hársfalvi, V. van Wyk, P. N. Badenhorst and H. F. Kotzé, *Thromb Res*, 2002, **107**, 365–371.
105. W. L. DeLano, M. H. Ultsch, A. M. de Vos and J. A. Wells, *Science*, 2000, **287**, 1279–1283.
106. M. S. Dennis, C. Eigenbrot, N. J. Skelton, M. H. Ultsch, L. Santell, M. A. Dwyer[1], M. P. O'Connell[1] and R. A. Lazarus, *Nature*, 2000, **404**, 465–470.
107. H. B. Lowman, Y. M. Chen, N. J. Skelton, D. L. Mortensen, E. E. Tomlinson, M. D. Sadick, I. C. A. Robinson and R. G. Clark, *Biochemistry*, 1998, **37**, 8870–8878.

CHAPTER 10

Chemical Approaches for Localization, Characterization and Mimicry of Peptide Epitopes

PAUL R. WERKHOVEN[a] AND ROB M.J. LISKAMP*[a,b]

[a] Medicinal Chemistry and Chemical Biology, Utrecht University, Universiteitsweg 99, 3584 CG Utrecht, 3508 TB Utrecht, The Netherlands;
[b] School of Chemistry, Joseph Black Building, University of Glasgow, University Avenue, Glasgow, G12 8QQ, UK
*Email: robert.liskamp@glasgow.ac.uk

10.1 Introduction, Definitions and Different Types of Epitopes

The term "epitope" originates from the area of immunology where it is mainly used as a synonym of an "antigenic determinant", which is the part of an antigen that is recognized by molecules of the immune system. Different types of epitopes can be distinguished and the most frequently occurring are B- and T-cell epitopes, immuno-dominant epitopes and neutralizing epitopes.[1,2] These names refer to occurrence and/or role. For example, immuno-dominant epitopes indicate epitopes that can diminish the role of other epitopes present in the antigen. B-cell epitopes are the part of an antigen that binds an antibody, either in its free state or membrane-bound to a B-cell, while T-cell sepitopes are peptides that are the product of proteolytic cleavage of the antigen, which are presented by T-cells.

RSC Drug Discovery Series No. 36
Biotherapeutics: Recent Developments using Chemical and Molecular Biology
Edited by Lyn H. Jones and Andrew J. McKnight

Published by the Royal Society of Chemistry, www.rsc.org

More specifically, the epitope of a (protein) antigen molecule is a defined region of its structure where an antibody binds.[3,4] Nowadays, the term "epitope" is increasingly used outside immunology. Since a (peptide/protein) antigen–antibody interaction is essentially a protein–protein interaction, all protein–protein interactions can be considered as interactions between two epitopes. In the antigen–antibody interaction example, this involves interaction between the epitope on the antigen and the epitope of the antibody, also referred to as "paratope". It is hard to overestimate the importance of protein–protein interactions because it is estimated that there are about 650,000 protein–protein interactions in the human interactome.[5]

Epitopes can be divided into two categories: continuous epitopes, sometimes denoted as linear epitopes, and discontinuous epitopes (Figure 10.1).[4,6] Although the latter category is by far the largest, so far approaches to express or mimic discontinuous epitopes have been very limited and therefore this is a challenging area for further research. This chapter will discuss available approaches.

A continuous epitope consists of a single contiguous stretch of amino acids; however, it must be noted that not necessarily each residue in this peptide segment plays a crucial role in binding to the antibody. Most continuous epitopes contain amino acids whose side chain can be replaced without significant loss of binding affinity. These residues are present either for spacing between essential binding residues or their backbone atoms play a role in binding.[7] The collection of residues that cannot be replaced without a significant loss in binding affinity is sometimes referred to as the "functional" epitope or, in the case of non-antibody–antigen interactions, as the hot spot.[4,8]

In contrast to continuous epitopes, discontinuous epitopes are composed of multiple peptide segments that are remote in the protein sequence, but are brought into spatial proximity by folding of the protein into its tertiary structure.[9] Thus, in addition to possible other functions, the remainder of the protein will act as a scaffold presenting the (discontinuous) epitope.

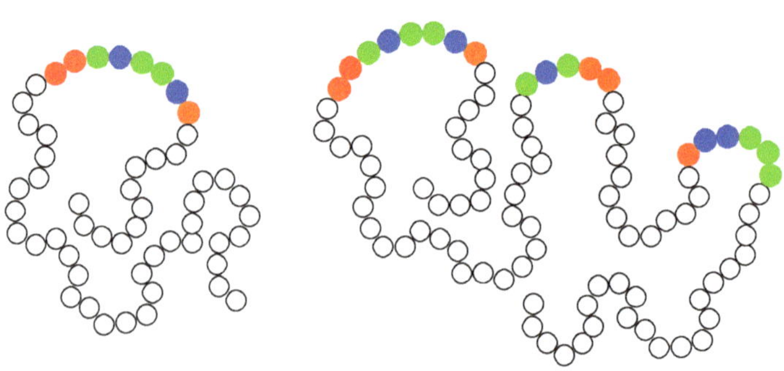

Figure 10.1 Left: continuous, sequential or linear epitope; right: discontinuous epitope.

In the discontinuous epitope, the positioning and shape of the peptide segments with respect to each other is essential for a high affinity and selectivity with an interacting partner. For example, antibodies, which bind to discontinuous epitopes present in the antigen, often do not bind to the denatured antigen in which the spatial characteristics of the epitope have been lost.[2] These features of discontinuous epitopes contribute to the challenging nature of research in this area, especially with respect to structural investigations and mimicry.

The term mimotope was introduced by Geysen *et al.* (1986) and can be defined as (part of) a molecule that mimics the essential features of a continuous or discontinuous epitope, and is therefore able to bind to the same antibody as the parent epitope.[10] Although close structural resemblance of the epitope can be of great significance, it is not an absolute requirement for a mimotope. Examples are known of mimotopes capable of binding an antibody, without possessing any apparent sequence similarity with the epitope.[11] This suggests that interaction between antibody and antigen is not completely dependent on entire amino acid residues, but more so on the correct spatial positioning of certain atoms in either the side chain and/or the backbone.[4]

Mimicry of epitopes present in (antigenic) proteins of pathogens allows induction of antibody production, and therefore possible candidates for synthetic vaccines can be generated. Synthetic vaccines are, in principle, attractive alternatives to the classical vaccines, which are mostly based on the original pathogen. Some contain a killed or attenuated version of the original pathogen, while others contain an inactivated version of the pathogen's toxin. The production of these types of vaccines is very laborious and potentially hazardous because they all have the live pathogen as a starting point. For this reason, considerable research efforts are spent on the development of synthetic vaccines, for example by epitope mimicry.[12,13]

In addition to synthetic vaccines, other possible applications of epitope mimics include inhibitors of protein–protein interactions and even ultimately synthetic antibodies. These will be discussed next.

In antigen–antibody interactions, the epitope of the antigen binds to complementary sequences present in the antibody, which is referred to as the "paratope". The paratope may consist of one to six Complementary Determining Regions (CDRs), which are present as three loops on the light chain and three loops on the heavy chain of an antibody (Figure 10.2).[14,15] These are highly variable regions of this molecule and therefore a paratope can be generated against virtually any antigenic molecule.

Similar to a discontinuous epitope, the CDR-loops are spatially brought together by the framework of an antibody molecule, enabling interaction of several CDR-loops with the antigenic molecule, thereby increasing affinity and selectivity of binding. In this way, a *multivalent* interaction is achieved.[16]

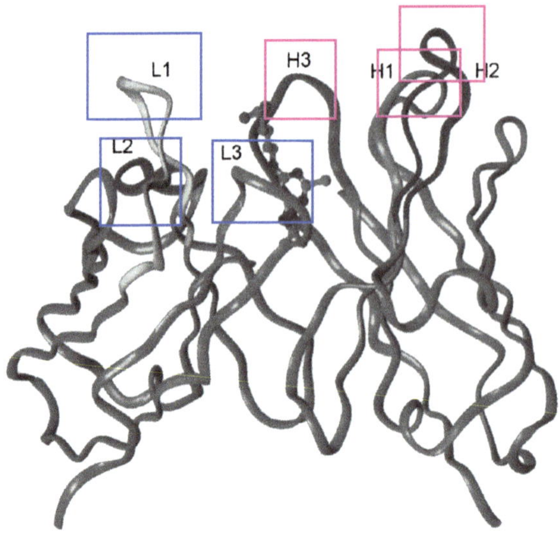

Figure 10.2 Architecture of CDR-loops in both heavy (H1–H3) and light chains (L1–H3) in the anti-atrazine antibody.[15] The spatial arrangement of the CDR loops of the heavy chain (H1–H3) and light chain (L1–L3) of the antibody is determined by the framework of the remainder of the antibody. A combination of residues in several CDR loops composes the antigen-binding paratope.

10.2 Shape and Structure of Epitopes

For an antibody to recognize an epitope in a protein, that particular epitope has to be accessible for binding by relatively large antibody molecules. Consequently, when an epitope is more accessible, it will be easier to generate antibodies binding to this site. It may be even so that this easily accessible epitope obscures the presence of other (more hidden) potential epitopes, and therefore this accessible epitope may become an immuno-dominant epitope (Figure 10.3).[17] Accessible epitope sites of a protein are mostly present on its surface. They are often encountered as loops,[18–21] although helices[22–27] and beta chains/sheets are not uncommon.[28–31] A beta hairpin conformation is also found of which the turn of the hairpin resembles a loop-like epitope, albeit more constrained and ordered, because of the additional beta-strand structure.[32,33]

The amino acid composition of an epitope is clearly of vital importance. Analysis of multiple protein–protein complexes, including antigen–antibody complexes, and alanine-scan studies have provided important insights into amino acid compositions of the interaction sites. These studies have not only identified some amino acids as being more prevalent in interaction sites than others, but have also shown the role of backbone atoms.[7,8,34,35] Furthermore, the exclusion of water molecules between the two proteins from the hydrophobic contact surfaces has also been shown to be a beneficial factor for

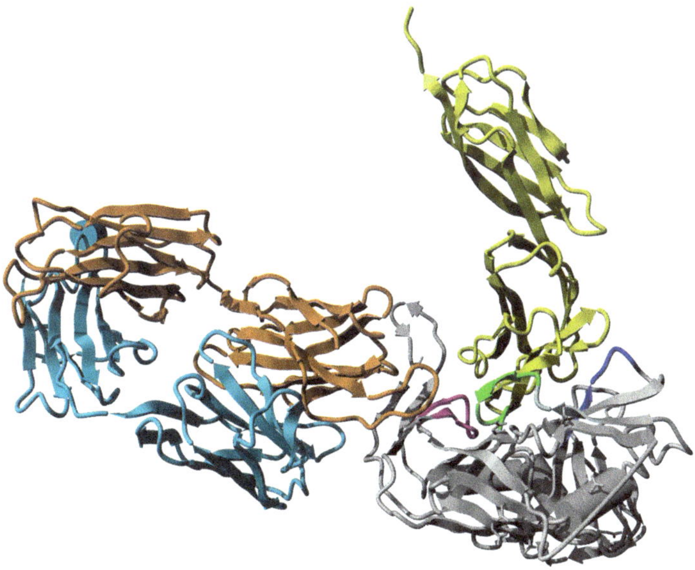

Figure 10.3 X-Ray structure of a gp120-CD4-antibody complex. CD4 (yellow) binds to gp120 (gray) through interactions with three conserved loops (blue, green, and purple). The antibody (cyan and orange) on the other hand binds solely to the variable, immuno-dominant V3 loop, which due to its relative accessibility is the preferred site for binding of antibodies.[21]

binding. After initial creation of the antibody, the immune system enhances the binding affinity by maturation of the antibody. This is achieved by increasing the buried surface of the antibody–antigen interaction site, leading to an increase in the number of excluded water molecules and resulting hydrophobic interactions.[36]

Despite the abundant presence of epitope sequences mainly in loops on the surface of proteins, mimicry of these epitopes, that is the identification and preparation of mimotopes, is still mainly achieved by the preparation of a linear peptide sequence, either by chemical synthesis or by, for example, phage display.[37–40] Although this has led to successful mimicry of epitopes present in antigenic proteins in a number of cases,[37,38,41] it is expected that the success rate can be increased by a closer mimicry of the conformation or secondary structure of the epitope by, for example, a better resemblance of the loop-like structure. Various methods have been reported, capable of constraining peptides in order to control their secondary structure. For example, a loop-like structure can be adequately mimicked by a cyclic peptide, so this is an attractive strategy for obtaining improved epitope mimics.[42,43] However, despite the widespread use of cyclic peptides in every conceivable branch of chemistry and biology, their synthesis and purification, irrespective of their size and amino acid composition, is still often a major hurdle.[44,45] For this reason, linear peptides are still being used in approaches to determine the epitope sequences and ensuing preparations of epitope mimics. Improving access to cyclic peptides by improving their synthetic approaches should broaden the possibilities of obtaining meaningful and adequate epitope mimics. Increased attention for the shape of an epitope should also enable mimicry of shapes of helix and beta-chain character containing epitopes. Approaches to constraining peptides will be highlighted in the following section.

10.3 Determination of the Primary, Secondary or Tertiary Structure of Epitopes

Similar to a protein, the primary structure of a proteinogenic epitope refers to its amino acid sequence, the secondary structure refers to its loop, helix or sheet character, and the tertiary structure relates to the relative positioning of the epitopes in space, for example to form a discontinuous epitope.

Clearly, the most comprehensive information with respect to the identity of the involved peptide segment(s) and its (spatial) structure is supplied by X-ray and/or nuclear magnetic resonance (NMR) structural analysis. This structural biological information is of crucial importance for the identification of epitopes. Using X-ray crystallography, structures of antibody–antigen complexes have been elucidated. These provide a direct visualization and detailed structural insight into which residues of the antigen are in close proximity to the antibody and therefore can be considered as an epitope. Several examples of X-ray crystallography-guided epitope localization of antigens have been published, for example, in the area of influenza and HIV.[21,46,47]

NMR spectroscopy can also provide valuable insights into the structure of an antibody–antigen complex. As an integral part of the structure elucidation using NMR, the binding residues of an epitope can be determined. An advantage of NMR over X-ray crystallography is the ability to measure the complex in solution. As a result, it is possible to determine differences between the free antigen and the antigen bound in the antigen–antibody complex. Binding an antibody restricts the mobility of residues present in the antigen, thus changing their relaxation time. Using different NMR techniques, such as dynamic filtering, heteronuclear nuclear Overhauser effect (NOE) and heteronuclear single quantum coherence (HSQC), these changes can be measured and quantified, resulting in the identification of the residues important for binding.[48–52]

Another increasingly important technique for identification of epitope sites is mass spectrometry (MS). From deuterium exchange and/or acetylation MS experiments, it is, in principle, possible to determine the interaction sites of an antigen–antibody complex. From this, the epitope site can be derived. In addition, mild proteolytic digestion of the antigen–antibody complex and subsequent analysis using MS can provide insight into the peptides capable of binding to the paratope of the antibody. Although this MS method does provide information on the parts of the antigen that are in close proximity to the antibody, it does not supply information on the individual amino acid residues or atoms important for binding.[53–61]

Classical approaches of determining the primary sequence of an epitope include synthesis of overlapping peptides of an antigenic protein and evaluation of the binding properties with an antibody.[62,63] The most potent antibody-binding peptides are believed to include the peptide sequences of the epitope(s). Understandably, this method is best suitable for the identification of continuous epitopes.

A disadvantage of the method is that only short stretches of linear peptides are synthesized, which lack the often-required secondary structure of the peptide segments, because they are present in the context of the antigenic protein. Although this method could in principle lead to the identification and localization of discontinuous epitopes by finding more than one peptide sequence capable of binding to the antibody, multiple binding can also result from binding of antibodies to completely different protein sites. Nevertheless, this approach of using peptide arrays in combination with X-ray structural information on the antigenic protein could shed light on the identity and localization of (discontinuous) epitopes in a protein.[64]

In general, peptide arrays can be used to explore protein–protein interactions to dissect the interacting epitopes;[65] however, most of the current techniques involve the preparation and screening of arrays of linear peptides, while arrays of cyclic peptides would be far more desirable because they mimic loop-like structures more closely.

Another, more "brute force", classical method for localization of epitopes is the generation of a phage display peptide library. This usually involves the generation of a large number (typically $>10^9$) of peptides, followed by establishing which peptides are capable of binding to antibodies, for example.[66] In principle, phage display allows the generation of cyclic peptides by incorporation of two cysteine residues, which subsequently can be oxidized leading to disulfide-bridge-containing cyclic peptides.[67] However, a disulfide bridge is inherently labile and, therefore, approaches towards more stable cyclic peptides are required. The recent approach of alkylating the cysteine thiol moieties of peptides displayed on phage particles with electrophiles, such as benzylbisbromide, which leads to bicyclic peptides represents a very important step for obtaining libraries of (bi)cyclic peptides.[68] The phage display method will of course yield peptides corresponding to the primary structure of the protein of interest,[39] but in addition peptides will be produced with an amino acid sequence not (clearly) related to this sequence. However, as long as these peptides are capable of binding the protein partner, they can be used in the production of mimotopes.[67,69]

Complementary to using the previously discussed crystallographic, MS or NMR spectroscopic structural information, bio-informatics approaches, including modeling, also allow the prediction of peptide segments, which may be accessible for binding by a large molecule such as an antibody. From this, epitope sequences can be derived as suitable candidates for mimicry. The availability of structural data of the epitope-bearing protein is of great value, adding, for example, to the accuracy of modeling.[70–72]

10.4 Applications of Peptide Epitope Mimics

Information on the localization and nature of epitopes is especially valuable in attempts to mimic these epitopes as part of relatively small molecular constructs. This may lead to a number of potential important applications related to prevention of or interference with undesired protein–protein interactions. Moreover, they may serve as small(er) molecular equivalents of a large(r) protein and therefore might be used as synthetic vaccines or even synthetic antibodies. The legitimate questions in all these applications are how accurate epitope mimicry should be and which approaches are required for adequate mimicry.

10.4.1 Epitope Mimics as Synthetic Vaccines

In our opinion a promising, extremely attractive area for the application of epitope information of antigenic proteins is the development of (synthetic) vaccines.

Vaccination is one of the most effective methods in the battle against infectious diseases.[12,13,73,74] Traditionally, (modified) proteins, or other constituents of a pathogen or even complete (attenuated) pathogens, have been used for vaccination purposes. Although these methods have proven to be very effective, they are associated with various risks, such as genetic instability, the use of potentially harmful components, and residual virulence.[13,73] Furthermore, during preparation of the vaccine, (parts of) the original pathogen are used, which in principle creates a bio-hazardous situation. Considering this, the advantages of possible synthetic vaccines become very evident and because a synthetic vaccine is accessible through chemical synthesis, there is no risk of exposure to and contamination by (parts of) the pathogen. A chemical process for its preparation may also be easier to scale up than a corresponding biotechnological process for production of a vaccine involving pathogens.

Nevertheless, development of molecular constructs based on the incorporation of discontinuous epitope mimics may take considerable efforts, especially with respect to adequate molecular mimicry of (pathogenic) protein from which they are derived. Altogether potential applications of such vaccines may be very attractive for tackling diseases like cancer, influenza, HIV, and malaria.[75,76] Using more detailed molecular information about the sites in proteins of disease-causing organisms that are responsible for binding to antibodies or to crucial proteins of the infected host, one can develop better, more selective vaccines for diseases for which there is currently no good or existing vaccine.

Several peptides have entered clinical trials to investigate their synthetic vaccine prospects. In recent reviews, it was stated that of the 125 peptides that enrolled in clinical trial phase I, only 30 survived to enter phase II and none went on to phase III.[3,77] Most of these were single, linear peptides, thus mimicking either a continuous epitope or only one of the peptide segments of a discontinuous epitope. The failure to proceed to phase III trials showed that there are significant hurdles to overcome when designing a synthetic peptide vaccine. The reliance on continuous epitopes and their mimics has so far not yielded any good vaccine candidates. Research is therefore increasingly focused on mimicry of discontinuous epitopes; however, the major hurdle with respect to this is the considerable effort required for adequate mimicry of a discontinuous epitope. Approaches to overcome this hurdle are described in the next section. Other important issues to consider include the (not always justified) confidence in the specificity of antibodies and the underestimation of the difference between antigenicity and immunogenicity.[3] Especially intriguing is the necessity of coupling of peptide epitope mimics to carrier proteins, such as the often-used keyhole limpet hemocyanin (KLH) and tetanus toxoid, to induce immunogenic memory and therefore protection.[78–80]

10.4.2 Epitope Mimics as Inhibitors (Antagonists) of Protein–Protein Interactions

Vaccination is a prophylactic approach and, as such, may represent the ultimate protection strategy against diseases or toxic proteins originating from pathogens by removing them through pathways of the immune system. Alternatively, the therapeutic approach involving molecular constructs, which can inhibit undesired protein–protein interactions, may also be an attractive avenue, perhaps even for diseases like Parkinson's or Alzheimer's.

There is a growing interest in the development of especially "intermediate size" molecular constructs for interfering with undesired protein–protein interactions.[81,82] Many of these constructs are peptide-based and the importance of the conformation of these constructs is often emphasized. Prominent examples include the development of stapled peptides[83–86] and other peptides mimicking a continuous interaction site[26,27] for inhibition of protein–protein interactions.[6,87]

However, development of molecular constructs based on discontinuous epitope sequences for protein–protein interaction inhibition is still in its infancy. A few promising examples[45,88,89] have already appeared in the literature hinting at the potential of these molecular constructs. Efforts towards uncovering the optimal molecular constructs and their efficient and reliable syntheses may greatly facilitate important future developments.

10.4.3 Paratope Mimics as Synthetic Antibodies

As described previously, epitope mimics can be used as synthetic vaccines by mimicking the structure of the antigenic determinant. However, the paratope, the part of the antibody that binds the antigen, can also be mimicked. As mentioned in the introduction, the paratope has a discontinuous epitope-like structure consisting of multiple loops (the complementary-determining region, Figure 10.2). By employing the same strategy as for other discontinuous epitope mimics, these paratope-mimics may ultimately approach the specificity and binding affinity of antibodies. Although no mimics with the same affinity as antibodies have been obtained so far, important progress has been made towards the realization of this goal.[90,91]

10.5 Mimicry of Epitopes by Peptides

In order to adequately mimic the epitope site(s) of a parent protein, one has to establish first whether the epitope is of continuous or discontinuous nature. A continuous epitope can be in principle mimicked by a single peptide, while a discontinuous epitope will usually require several peptides for adequate mimicry. However, even for mimicry of a continuous epitope, it is important to look at its *secondary* structure and try to mimic this as closely as possible, for example mimicry of loops by cyclization or mimicry of helices by "stapling". For mimicry of discontinuous epitopes, information about its *tertiary* structure has to be included, leading to proper arrangement of the peptides in space as well as with respect to each other. The development of such molecular constructs is a great challenge, involving advanced synthetic methodology.

10.5.1 Single Linear Peptides as Continuous Epitope Mimics

As stated in previously, a continuous epitope, of which the amino acid sequence and conformational properties have been determined, might be mimicked by a linear or a constrained (mostly cyclic, see following text) peptide.

Despite the absence of any significant secondary structure, relatively small linear (15-mer) peptides led to the first peptide vaccine providing projection against parvovirus.[92] Linear (12-mer) peptide epitopes of snake venom displayed on phages were also immunogenic in mice.[93] The peptide, which was used in vaccination against influenza virus and provided protection in mice, was linear but probably sufficiently long (>60 residues) to assume a high percentage of its original secondary structure, i.e. a significant alpha helix character.[79]

Linear peptide epitope mimics containing carbohydrate residues have also been prepared. Of course, the presence of these carbohydrate residues increased the complexity of the synthesis of these mimics.[80,94]

Despite the abundance of examples of single peptides that were capable of eliciting an immunogenic response in animals, none have successfully survived clinical trials.[3,77]

10.5.2 Constraining Peptides for Optimization of Structural Mimicry

In several of the previous examples illustrating the use of single peptides as epitope mimics, the importance of a secondary structure already became apparent. In the influenza example, the peptide was of sufficient length to have a considerable percentage of alpha-helical character. Shorter peptides, which are more reluctant towards forming a helical structure, can be induced to do so by synthesizing peptide-based alpha-helical mimetics in which the side chains of two properly spaced amino acids are linked to stabilize this secondary structure. This can be done by a variety of chemical modifications, such as ring closing metathesis (often referred to as peptide stapling (Figure 10.4.)),[83,86,95] salt bridges, disulfide formation, and amide bond formation.[96,97] Another possibility is the use of small molecule alpha-helical mimetics.[26,97–100]

Loops are a common structural motif in epitopes, which is not surprising because these are common secondary structure elements at the surface of proteins amenable for post-translational modification but also for recognition by antibodies and other proteins. Clearly, the most prominent loops are the CDR-loops of the heavy and light chains of an antibody. It is increasingly realized that a cyclic peptide is a better mimic of a loop like structure. An additional advantage of cyclic peptide containing molecular constructs may be the reduced sensitivity to proteolytic degradation, facilitating the use in a biological context by prolonging the half-life.[44] There are a number of approaches to obtaining cyclic peptides. In most of these approaches, the linear peptide is synthesized first, followed by cyclization. The most popular ones are macrolactamization, disulfide formation, and thioether formation.[44,101] A successful example of this technique, especially with respect to synthetic vaccine applications, is the use of the CLIPS technology to constrain a peptide corresponding to the epitope of the follicle stimulating hormone (Figure 10.5). The constrained peptide was able to elicit antibodies, whereas the linear peptide was not.[102]

Other examples of the use of cyclic peptides are the phage-displayed cyclic hexapeptide library, of which some were able to bind to anti-rabies

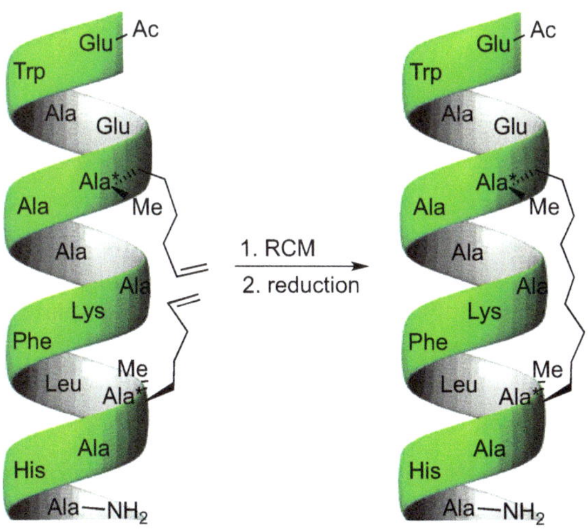

Figure 10.4 Strategy for stabilizing alpha-helices by an all-hydrocarbon crosslinker by Ring Closing Metathesis (RCM), nowadays often referred to as "stapling".[96]

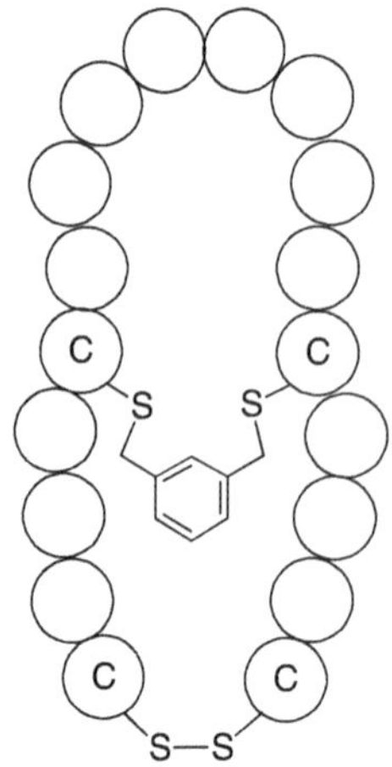

Figure 10.5 CLIPS technology, that is alkylating and connecting two (shown) or three cysteine residues by an aromatic (bis)electrophile as well disulfide formation, leading to a constrained peptide with perfect antigen properties.[102]

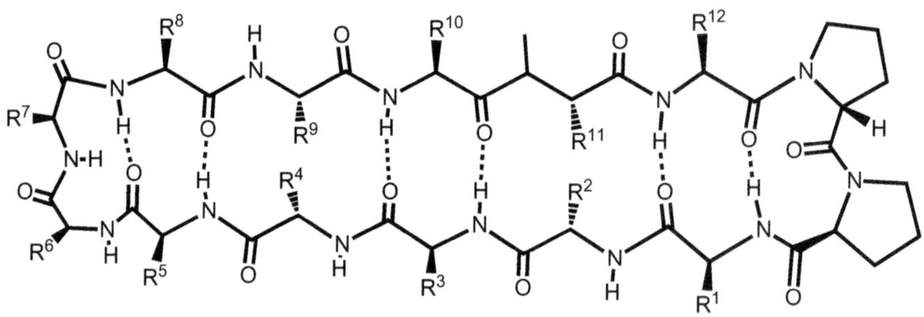

Figure 10.6 General structure of β-hairpin mimics developed and applied by Robinson *et al.*[31,104,105,120]

glycoprotein antibodies,[67] and the cyclic peptides designed by Villén *et al.* mimicking the epitope on the capsid of the foot-and-mouth disease virus.[103]

A specific loop-type is the β-hairpin, which can be mimicked by cyclization of the peptide with, for example, a D-proline-L-proline template (Figure 10.6). Examples are the preparation of epitope mimics of a malaria parasite protein and epitope mimics of the V3 loop of HIV gp120 protein.[31,104,105]

10.5.3 Assembly of Peptides for Mimicry of Discontinuous Epitopes

In attempts to closely mimic the tertiary structure of a discontinuous epitope, the individual peptides mimicking the corresponding segments of the discontinuous epitope have to be assembled on a single molecular entity. This can be a spacer or linker molecule, a molecular scaffold, a surface, or a dendrimer/polymer providing the confined presentation of the peptide mimicking epitopes. The molecular scaffold can also be composed of the peptide molecules themselves and obtained by linking epitope segments together by dimerization or via a linker of varying rigidity. There are large differences between all these approaches and a priori, it is hard to predict what will be the most successful approach.

10.5.3.1 Assembly of Peptides by the Preparation of Dimers or Multimers

In principle, the simplest approach for mimicking a discontinuous epitope, especially consisting of random coil peptide segment, is to judiciously combine the different peptide segments into a single construct by using spacers of appropriate length and rigidity. This approach has been followed in a number of cases. One example is the preparation of an immunogenic construct for foot-and-mouth-disease virus.[18,37,41,80] Connecting several identical peptide segments may lead to a multivalency effect, thereby increasing the affinity of the mimicked epitope.[106]

When connecting multiple peptides to a single spacer or linker molecule, it is important to be able to introduce the peptide sequentially. For this purpose, there are a large variety of orthogonal conjugation methods available. Examples are copper-catalyzed azide-alkyne click (CuAAC), strain-promoted azide-alkyne click (SPAAC), thiol-maleimide, thiol-ene click, oxime ligation, and native chemical ligation (NCL).[107,108] An example of the sequential introduction of three different moieties using orthogonal conjugation methods is shown in Figure 10.7.

10.5.3.2 Assembly of Peptides by Scaffolding

A scaffold, because of its more pre-organized character may represent the best approach to mimic the tertiary structure of the discontinuous epitope of a

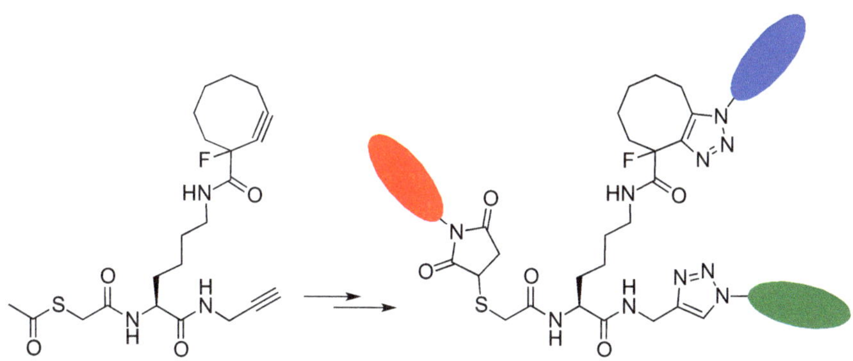

Figure 10.7 Orthogonally functionalized lysine-based trivalent molecule, suitable for the introduction of three different molecular entities.[107]

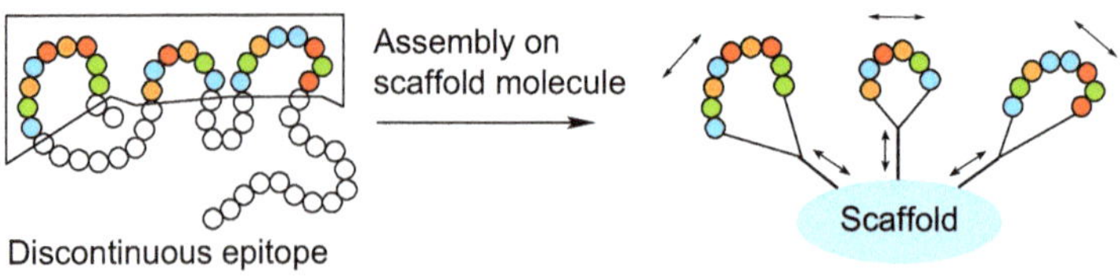

Figure 10.8 Discontinuous epitope mimicry by scaffolding of the corresponding cyclic peptides. Peptide fragments corresponding to the discontinuous epitope (colored) are synthesized as cyclic peptides. These peptides are then incorporated onto a scaffold molecule. The character and size of the cyclic peptides, the choice of the linkers as well as the scaffold determine the flexibility (indicated by the arrows) and therefore pre-organization of the entire molecular construct.

protein. In a way, the scaffold will act as a replacement of most of the framework of the native protein and the molecular construct of scaffold and peptides is in fact a *pars pro toto* of the entire protein.[78,109] The concept of discontinuous epitope mimics by scaffolding peptides is illustrated in Figure 10.8.

Assembly of cyclic peptides onto a scaffold clearly has advantages over assembly of linear peptides, such as a better mimicry of loops and increased proteolytic stabilization. Nevertheless, the number of examples of combinations of cyclic peptides and scaffold molecules is restricted, which points to the limited synthetic accessibility of cyclic peptides.[45,110]

10.5.3.3 Scaffolds

The most seminal contribution in this area was probably the cyclic peptide template TASP (Template Assembled Synthetic Protein), originally developed by Mutter and Vuilleumier[111] and further developed by, among others, Dumy *et al.* to the so-called RAFT (Regioselectively Addresssable Functionalized Template) (Figure 10.9).[112,113] Although it is, in principle, possible to attach four different groups to this cyclic decapeptide system, this template is a relatively large flexible system lacking sufficient pre-organization. On the other hand, the highly pre-organized calixarene (Figure 10.9) system, which is used by Hamilton *et al.* to which four, albeit identical, cyclic peptides can be introduced, is available.[114] Therefore, we have developed a relatively small molecular scaffold with sufficient pre-organization, which could be conveniently synthesized on a large scale for solid phase synthesis and library preparation purposes, but moreover would allow the introduction of at least *three different peptide segments*. This triazacylcophane-scaffold (TAC) has been applied successfully to incorporate both linear and cyclic peptides.[45,78,115]

Although the number of molecular scaffolds for multiple orthogonal conjugations is gradually increasing, it is still a challenge to incorporate *different* molecular entities onto a molecular scaffold. The choice of the scaffold

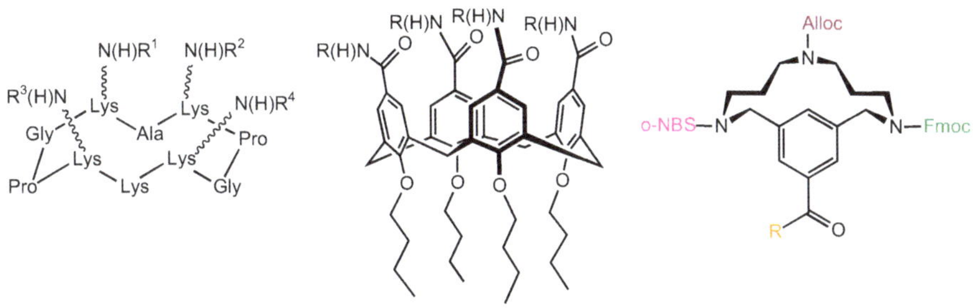

Figure 10.9 TASP/RAFT scaffold (left), calixarene scaffold (middle) and TAC-scaffold (right). The positions now occupied by the protecting groups in the TAC-scaffold allow introduction of *different* molecular moieties. The fourth position (orange R-group) is, in principle, also available.

seems an important aspect in the design of epitope mimics because each scaffold presents the (cyclic) peptides in a different way. It is hard to predict the required amount of rigidity and pre-organization for creation of the best mimic of the epitope, as was recently shown by Mulder *et al.* in a study comparing different scaffolds.[116] Adequate spacing and positioning of cyclic peptides are challenges that have hardly been touched, but will be extremely important for proper structural mimicry of discontinuous epitopes.

In addition to the synthetic organic scaffolds, there is an increasing use of a variety of biomolecules as scaffolds. One example is the family of cyclotides, a family of polycyclic peptides containing a cystine knot. The desired peptide sequences can be incorporated into the cyclotide via either chemical synthesis or recombinant expression.[117,118] Other examples are nucleotide (G-quadruplex),[119] engineered virus particles[120] and (parts) of proteins.[121,122] Several of these topics have been reviewed extensively.

10.5.4 Requirements for Discontinuous Epitope Containing Protein Mimics

The size and complexity of proteins show that, if one wishes to develop adequate protein mimics, the following requirements are essential. Firstly, a scaffold must provide a good platform with respect to flexibility, rigidity, positioning, and relative involvement of the different discontinuous epitope representing peptides. Secondly, it must be possible to introduce peptides of any size or character at desired positions on the scaffold. Thirdly, it has to be possible to construct these epitope loop mimics, which comprise cyclic peptides of any desired size and character. Fourthly, innovative screening approaches have to be developed to find the best protein mimic. Finally, the resulting protein mimics should be amenable to further development as (for example synthetic vaccine) other biologics or even synthetic antibodies, thus demanding reliable and reproducible state-of-the-art chemo- and biosynthesis.

10.6 Conclusions

Although the term "epitope" originates from the vast research area of immunology where the importance of highly selective and high affinity protein–protein interactions have always been evident, the same terms and principles operate in all other protein-directed biological processes and pathways.

To interfere with certain undesired protein–protein interactions or to mimic proteins in order to modulate protein–protein interactions, a comprehensive understanding of the interaction sites is crucial. A variety of complementary methods for localization and characterization of these interaction sites is available nowadays, however, none of these individual techniques can offer a *complete* understanding of the protein–protein interaction of interest. As a consequence, the detailed information on these interactions has to be obtained from the structure and activity of protein modulators and mimics. The possibilities for design, synthesis, and evaluation of these molecular constructs of "intermediate size" are gaining momentum, thereby taking up the great challenges of development of synthetic vaccines, synthetic antibodies, and other protein mimics.

References

1. T. J. Kindt, R. A. Goldsby and B. A. Osborne, *Kuby Immunology*, W. H. Freeman & Company, New York, NY, 6th edn, 2007, p. 84.
2. I. Roitt, J. Brostoff and D. Male, *Immunology*, Mosby, Edinburgh, UK, 6th edn, 2001, p. 73.
3. M. H. V. Van Regenmortel, *The Open Vaccine J*, 2009, **2008**, 33–44.
4. M. H. V. Van Regenmortel, in *Epitope Mapping Protocols*, ed. M. Schutkowski and U. Reineke, Humana Press, Totowa, NJ, 2nd edn, 2009, vol. 524, pp. 3–20.
5. M. P. H. Stumpf, T. Thorne, E. de Silva, R. Stewart, H. J. An, M. Lappe and C. Wiuf, *Proc Natl Acad Sci USA*, 2008, **105**, 6959–6964.
6. J. Eichler, *Curr Opin Chem Biol*, 2008, **12**, 707–713.
7. L. Lo Conte, C. Chothia and J. Janin, *J Mol Biol*, 1999, **285**, 2177–2198.
8. W. L. Delano, *Curr Opin Struct Biol*, 2002, **12**, 14–20.
9. Discontinuous epitopes are occasionally referred to as conformational epitopes (van Regenmortel, reference 4), which is probably an incorrect designation, since in a linear epitope, the conformation can be just as important.
10. H. M. Geysen, S. J. Rodda and T. J. Mason, *Mol immunol*, 1986, **23**, 709–715.
11. P. Delmastro, A. Meola, P. Monaci, R. Cortese and G. Galfrè, *Vaccine*, 1997, **15**, 1276–1285.
12. P. R. Dormitzer, G. Grandi and R. Rappuoli, *Nat Rev Microbiol*, 2012, **10**, 807–813.
13. P. M. Moyle and I. Toth, *ChemMedChem*, 2013, **8**, 360–376.
14. E. J. Sundberg, in *Epitope Mapping Protocols*, M. Schutkowski and U. Reineke, Humana Press, Totowa, NJ, 2009, vol. 524, pp. 23–36.

15. W. Kusharyoto, J. Pleiss, T. T. Bachmann and R. D. Schmid, *Protein Eng*, 2002, **15**, 233–241.
16. The term multivalency is usually reserved for the interaction of several identical sites or ligands. However, in essence many interactions in protein–protein interactions are of a multivalent nature, albeit of non-identical peptide sequences. See M. Mammen, S.-K. Choi and G. M. Whitesides, *Angew Chem Int Ed*, 1998, **37**, 2754–2794.
17. R. Pejchal, K. J. Doores, L. M. Walker, R. Khayat, P.-S. Huang, S.-K. Wang, R. L. Stanfield, J.-P. Julien, A. Ramos, M. Crispin, R. Depetris, U. Katpally, A. Marozsan, A. Cupo, S. Maloveste, Y. Liu, R. McBride, Y. Ito, R. W. Sanders, C. Ogohara, J. C. Paulson, T. Feizi, C. N. Scanlan, C.-H. Wong, J. P. Moore, W. C. Olson, A. B. Ward, P. Poignard, W. R. Schief, D. R. Burton and I. A. Wilson, *Science*, 2011, **334**, 1097–1103.
18. J. Villén, E. Borràs, W. M. M. Schaaper, R. H. Meloen, M. Dávila, E. Domingo, E. Giralt and D. Andreu, *ChemBioChem*, 2002, **3**, 175–182.
19. T. G. Johns, T. E. Adams, J. R. Cochran, N. E. Hall, P. A. Hoyne, M. J. Olsen, Y.-S. Kim, J. Rothacker, E. C. Nice, F. Walker, G. Ritter, A. A. Jungbluth, L. J. Old, C. W. Ward, A. W. Burgess, K. D. Wittrup and A. M. Scott, *J Biol Chem*, 2004, **279**, 30375–30384.
20. V. Chitarra, P. M. Alzari, G. A. Bentley, T. N. Bhat, J.-L. Eiselé, A. Houdusse, J. Lescar, H. Souchon and R. J. Poljak, *Proc Natl Acad Sci USA*, 1993, **90**, 7711–7715.
21. P. D. Kwong, R. Wyatt, J. Robinson, R. W. Sweet, J. Sodroski and W. A. Hendrickson, *Nature*, 1998, **393**, 648–59.
22. H. Hori, D. R. Keene, L. Y. Sakai, M. K. Wirtz, H. P. Bächinger, M. Godfrey and D. W. Hollister, *Mol Immunol*, 1992, **29**, 759–770.
23. F. A. Robey, T. Kelson-Harris, P. P. Roller and M. Robert-Guroff, *J Biol Chem*, 1995, **270**, 23918–23921.
24. C. Dreyfus, N. S. Laursen, T. Kwaks, D. Zuijdgeest, R. Khayat, D. C. Ekiert, J. H. Lee, Z. Metlagel, M. V Bujny, M. Jongeneelen, R. van der Vlugt, M. Lamrani, H. J. W. M. Korse, E. Geelen, Ö. Sahin, M. Sieuwerts, J. P. J. Brakenhoff, R. Vogels, O. T. W. Li, L. L. M. Poon, M. Peiris, W. Koudstaal, A. B. Ward, I. A. Wilson, J. Goudsmit and R. H. E. Friesen, *Science*, 2012, **337**, 1343–1348.
25. D. Corti, J. Voss, S. J. Gamblin, G. Codoni, A. Macagno, D. Jarrossay, S. G. Vachieri, D. Pinna, A. Minola, F. Vanzetta, C. Silacci, B. M. Fernandez-Rodriguez, G. Agatic, S. Bianchi, I. Giacchetto-Sasselli, L. Calder, F. Sallusto, P. Collins, L. F. Haire, N. Temperton, J. P. M. Langedijk, J. J. Skehel and A. Lanzavecchia, *Science*, 2011, **333**, 850–856.
26. J. Becerril and A. D. Hamilton, *Angew Chem Int Ed*, 2007, **46**, 4471–4473.
27. J. D. Sadowsky, W. D. Fairlie, E. B. Hadley, H.-S. Lee, N. Umezawa, Z. Nikolovska-Coleska, S. Wang, D. C. S. Huang, Y. Tomita and S. H. Gellman, *J Am Chem Soc*, 2007, **129**, 139–154.

28. S. Padavattan, S. Flicker, T. Schirmer, C. Madritsch, S. Randow, G. Reese, S. Vieths, C. Lupinek, C. Ebner, R. Valenta and Z. Markovic-Housley, *J Immunol*, 2009, **182**, 2141–2151.
29. V. Novitskaya, N. Makarava, A. Bellon, O. V. Bocharova, I. B. Bronstein, R. A. Williamson and I. V. Baskakov, *J Biol Chem*, 2006, **281**, 15536–15545.
30. R. Kayed, E. Head, F. Sarsoza, T. Saing, C. W. Cotman, M. Necula, L. Margol, J. Wu, L. Breydo, J. L. Thompson, S. Rasool, T. Gurlo, P. Butler and C. G. Glabe, *Mol Neurodegener*, 2007, **2**, 18.
31. J. A. Robinson, *J Pept Sci*, 2013, **19**, 127–140.
32. V. Tugarinov, A. Zvi, R. Levy and J. Anglister, *Nat Struct Biol*, 1999, **6**, 331–335.
33. A. Jackson, T. Klonisch, A. Lapthorn, P. Berger, N. Isaacs, P. Delves, T. Lund and I. Roitt, *J Reprod Immunol*, 1996, **31**, 21–36.
34. A. A. Bogan and K. S. Thorn, *J Mol Biol*, 1998, **280**, 1–9.
35. B. Ma, T. Elkayam, H. Wolfson and R. Nussinov, *Proc Natl Acad Sci USA*, 2003, **100**, 5772–5777.
36. Y. Li, H. Li, F. Yang, S. J. Smith-Gill and R. A. Mariuzza, *Nat Struct Biol*, 2003, **10**, 482–488.
37. C. Peri, P. Gagni, F. Combi, A. Gori, M. Chiari, R. Longhi, M. Cretich and G. Colombo, *ACS Chem Biol*, 2013, **8**, 397–404.
38. M. B. Irving, L. Craig, A. Menendez, B. P. Gangadhar, M. Montero, N. E. Van Houten and J. K. Scott, *Mol Immunol*, 2010, **47**, 1137–1148.
39. M. Xue, X. Shi, J. Zhang, Y. Zhao, H. Cui, S. Hu, H. Gao, X. Cui and Y.-F. Wang, *PloS One*, 2012, **7**, e49842.
40. Y. Cao, Z. Lu, P. Li, P. Sun, Y. Fu, X. Bai, H. Bao, Y. Chen, D. Li and Z. Liu, *J Virol Methods*, 2012, **185**, 124–128.
41. R. Franke, T. Hirsch, H. Overwin and J. Eichler, *Angew Chem Int Ed*, 2007, **46**, 1253–1255.
42. R. M. J. Liskamp, D. T. S. Rijkers and S. E. Bakker, in *Modern Supramolecular Chemistry: Strategies for Macrocycle Synthesis*, F. Diederich, P. J. Stang and R. R. Tykwinski, Wiley-VCH, Weinheim, Germany, 2008, pp. 1–27.
43. N. Sewald and H.-D. Jakubke, in *Peptides, Chemistry and Biology*, Wiley-VCH, Weinheim, Germany, 2002, p. 311.
44. C. J. White and A. K. Yudin, *Nat Chem*, 2011, **3**, 509–524.
45. G. E. Mulder, J. A. W. Kruijtzer and R. M. J. Liskamp, *Chem Comm*, 2012, **48**, 10007–10009.
46. Y. Li, H. Li, S. J. Smith-Gill and R. A. Mariuzza, *Biochemistry*, 2000, **39**, 6296–6309.
47. D. C. Ekiert, R. H. Friesen, G. Bhabha, J. P. J. Brakenhoff, M. Jongeneelen, W. Yu, C. Ophorst, F. Cox, H. J. W. M. Korse, B. Brandenburg, R. Vogels, T. Kwaks, R. Kompier, M. H. Koldijk, L. A. H. M. Cornelissen, L. L. M. Poon, M. Peiris, W. Koudstaal, I. A. Wilson and J. Goudsmit, *Science*, 2011, **333**, 843–850.

48. O. Rosen and J. Anglister, in *Epitope Mapping Protocols*, M. Schutkowski and U. Reineke, Humana Press, Totowa, NJ, 2nd edn, 2009, vol. 524, pp. 37–57.
49. B. Zilber, T. Scherf, M. Levitt and J. Anglister, *Biochemistry*, 1990, **29**, 10032–10041.
50. A. Zvi, I. Kustanovich, D. Feigelson, R. Levy, M. Eisenstein, S. Matsushita, P. Richalet-Sécordel, M. H. V. van Regenmortel and J. Anglister, *Eur J Biochem*, 1995, **229**, 178–187.
51. O. Rosen, M. Sharon, S. R. Quadt-Akabayov and J. Anglister, *Proc Natl Acad Sci USA*, 2006, **103**, 13950–13955.
52. A. Zvi, V. Tugarinov, G. A. Faiman, A. Horovitz and J. Anglister, *Eur J Biochem*, 2000, **267**, 767–779.
53. E. O. Hochleitner, C. Borchers, C. Parker, R. J. Bienstock and K. B. Tomer, *Protein Sci*, 2000, **9**, 487–496.
54. E. O. Hochleitner, M. K. Gorny, S. Zolla-Pazner and K. B. Tomer, *J Immunol*, 2000, **164**, 4156–4161.
55. C. E. Parker, D. Papac, S. K. Trojak and K. B. Tomer, *J Immunol*, 1996, **157**, 198–206.
56. C. Hager-Braun and K. B. Tomer, *Expert Rev Proteomics*, 2005, **2**, 745–756.
57. M. J. Chalmers, S. A. Busby, B. D. Pascal, G. M. West and P. R. Griffin, *Expert Rev Proteomics*, 2011, **8**, 43–59.
58. K. M. Downard, B. Morrissey and A. B. Schwahn, *Mass Spectrom Rev*, 2009, **28**, 35–49.
59. M. Dragusanu, B.-A. Petre, S. Slamnoiu, C. Vlad, T. Tu and M. Przybylski, *J Am Soc Mass Spectrom*, 2010, **21**, 1643–1648.
60. X. Lu, M. R. DeFelippis and L. Huang, *Anal Biochem*, 2009, **395**, 100–107.
61. N. Olsson, S. Wallin, P. James, C. A. K. Borrebaeck and C. Wingren, *Protein Sci*, 2012, **21**, 1897–1910.
62. R. H. Maier, C. J. Maier, R. Rid, H. Hintner, J. W. Bauer and K. Önder, *J Biomol Screen*, 2010, **15**, 418–426.
63. R. Volkmer, *ChemBioChem*, 2009, **10**, 1431–1442.
64. M. Hijnen, F. R. Mooi, P. G. M. Van Gageldonk, P. Hoogerhout, A. J. King and G. A. M. Berbers, *Infect Immun*, 2004, **72**, 3716–3723.
65. C. Katz, L. Levy-Beladev, S. Rotem-Bamberger, T. Rito, S. G. D. Rüdiger and A. Friedler, *Chem Soc Rev*, 2011, **40**, 2131–2145.
66. M. Paschke, *Appl Microbiol Biotechnol*, 2006, **70**, 2–11.
67. M. Houimel and K. Dellagi, *Vaccine*, 2009, **27**, 4648–4655.
68. C. Heinis, T. Rutherford, S. Freund and G. Winter, *Nat Chem Biol*, 2009, **5**, 502–507.
69. K. Dorgham, I. Dogan, N. Bitton, C. Parizot, V. Cardona, P. Debré, O. Hartley and G. Gorochov, *Aids Res Hum Retrov*, 2005, **21**, 82–92.
70. R. Liu and J. Hu, *J Proteomics Bioinform*, 2011, **04**, 10–15.
71. I. Mayrose, O. Penn, E. Erez, N. D. Rubinstein, T. Shlomi, N. T. Freund, E. M. Bublil, E. Ruppin, R. Sharan, J. M. Gershoni, E. Martz and T. Pupko, *Bioinformatics*, 2007, **23**, 3244–3246.

72. V. Moreau, C. Granier, S. Villard, D. Laune and F. Molina, *Bioinformatics*, 2006, **22**, 1088–1095.
73. S. A. Plotkin, *Clin Vaccine Immunol*, 2009, **16**, 1709–1719.
74. R. Rappuoli, C. W. Mandl, S. Black and E. De Gregorio, *Nat Rev Immunol*, 2011, **11**, 865–872.
75. R. Knittelfelder, A. B. Riemer and E. Jensen-Jarolim, *Expert Opin Biol Ther*, 2009, **9**, 493–506.
76. A. B. Riemer and E. Jensen-Jarolim, *Immunol Lett*, 2007, **113**, 1–5.
77. D. Hans, P. R. Young and D. P. Fairlie, *Med Chem*, 2006, **2**, 627–646.
78. M. Hijnen, D. J. Van Zoelen, C. Chamorro, P. Van Gageldonk, F. R. Mooi, G. Berbers and R. M. J. Liskamp, *Vaccine*, 2007, **25**, 6807–6817.
79. T. T. Wang, G. S. Tan, R. Hai, N. Pica, L. Ngai, D. C. Ekiert, I. A. Wilson, A. García-Sastre, T. M. Moran and P. Palese, *Proc Natl Acad Sci USA*, 2010, **107**, 18979–18984.
80. N. Gaidzik, U. Westerlind and H. Kunz, *Chem. Soc. Rev.*, 2013, **42**, 4421–4442.
81. D. J. Craik, D. P. Fairlie, S. Liras and D. Price, *Chem Biol Drug Des*, 2013, **81**, 136–147.
82. At the moment, most of the drugs on the market are either small molecules (<500 Da) or biomolecules (>5000 Da e.g. antibodies). Interest in the intermediate size molecules, specifically peptides, has grown immensely in the last few years.
83. C. J. Brown, S. T. Quah, J. Jong, A. M. Goh, P. C. Chiam, K. H. Khoo, M. L. Choong, M. A. Lee, L. Yurlova, K. Zolghadr, T. L. Joseph, C. S. Verma and D. P. Lane, *ACS Chem. Biol.*, 2013, **8**(3), 506–512.
84. C. E. Schafmeister, J. Po and G. L. Verdine, *J Am Chem Soc*, 2000, **122**, 5891–5892.
85. P. S. Kutchukian, J. S. Yang, G. L. Verdine and E. I. Shakhnovich, *J Am Chem Soc*, 2009, **131**, 4622–4627.
86. L. D. Walensky, A. L. Kung, I. Escher, T. J. Malia, S. Barbuto, R. D. Wright, G. Wagner, G. L. Verdine and S. J. Korsmeyer, *Science*, 2004, **305**, 1466–1470.
87. H. Yin and A. D. Hamilton, *Angew Chem Int Ed*, 2005, **44**, 4130–4163.
88. D.-J. Van Zoelen, M. R. Egmond, R. J. Pieters and R. M. J. Liskamp, *ChemBioChem*, 2007, **8**, 1950–1956.
89. C. Hunke, T. Hirsch and J. Eichler, *ChemBioChem*, 2006, **7**, 1258–1264.
90. P. Timmerman, S. G. Shochat, J. Desmet, R. Barderas, J. I. Casal, R. H. Meloen and D. Altschuh, *J Mol Recognit*, 2010, **23**, 559–568.
91. P. Timmerman, R. Barderas, J. Desmet, D. Altschuh, S. Shochat, M. J. Hollestelle, J. W. M. Höppener, A. Monasterio, J. I. Casal and R. H. Meloen, *J Biol Chem*, 2009, **284**, 34126–34134.
92. J. P. M. Langeveld, J. I. Casal, A. D. M. E. Osterhaus, E. Cortés, R. de Swart, C. Vela, K. Dalsgaard, W. C. Puijk, W. M. M. Schaaper and R. H. Meloen, *J Virol*, 1994, **68**, 4506–4513.

93. R. Cardoso, M. I. Homsi-Brandeburgo, V. M. Rodrigues, W. B. Santos, G. L. R. Souza, C. R. Prudencio, A. C. S. Siquieroli and L. R. Goulart, *Toxicon*, 2009, **53**, 254–261.
94. A.-B. M. Abdel-Aal, D. El-Naggar, M. Zaman, M. Batzloff and I. Toth, *J Med Chem*, 2012, **55**, 6968–6974.
95. H. Jo, N. Meinhardt, Y. Wu, S. Kulkarni, X. Hu, K. E. Low, P. L. Davies, W. F. Degrado and D. C. Greenbaum, *J Am Chem Soc*, 2012, **134**, 17704–17713.
96. R. M. J. Liskamp, D. T. S. Rijkers, J. A. W. Kruijtzer and J. Kemmink, *ChemBioChem*, 2011, **12**, 1626–1653.
97. V. Haridas, *Eur J Org Chem*, 2009, **2009**, 5112–5128.
98. L. R. Whitby, K. E. Boyle, L. Cai, X. Yu, M. Gochin and D. L. Boger, *Bioorg Med Chem Lett*, 2012, **22**, 2861–2865.
99. F. Campbell, J. P. Plante, T. A. Edwards, S. L. Warriner and A. J. Wilson, *Org Biomol Chem*, 2010, **8**, 2344–2351.
100. K. Long, T. A. Edwards and A. J. Wilson, *Bioorganic & Medicinal Chemistry*, 2013, **21**, 4034–4040.
101. P. Timmerman, J. Beld, W. C. Puijk and R. H. Meloen, *ChemBioChem*, 2005, **6**, 821–824.
102. P. Timmerman, W. C. Puijk and R. H. Meloen, *J Mol Recognit*, 2007, **20**, 283–299.
103. J. Villén, R. A. Rodríguez-Mias, J. I. Núñez, E. Giralt, F. Sobrino and D. Andreu, *Chem Biol*, 2006, **13**, 815–823.
104. J. A. Robinson, S. Demarco, F. Gombert, K. Moehle and D. Obrecht, *Drug Discov Today*, 2008, **13**, 944–951.
105. J. A. Robinson, *Curr Opin Chem Biol*, 2011, **15**, 379–386.
106. P. Van de Vijver, M. Schmitt, D. Suylen, L. Scheer, M. C. L. G. D. Thomassen, L. J. Schurgers, J. H. Griffin, R. R. Koenen and T. M. Hackeng, *J Am Chem Soc*, 2012, **134**, 19318–19321.
107. D. M. Beal, V. E. Albrow, G. Burslem, L. Hitchen, C. Fernandes, C. Lapthorn, L. R. Roberts, M. D. Selby and L. H. Jones, *Org Biomol Chem*, 2012, **10**, 548–554.
108. D. M. Beal and L. H. Jones, *Angew Chem Int Ed*, 2012, **51**, 6320–6326.
109. R. Franke, T. Hirsch and J. Eichler, *J Recept Signal Transduction Res*, 2006, **26**, 453–460.
110. R. Franke, C. Doll, V. Wray and J. Eichler, *Org Biomol Chem*, 2004, **2**, 2847–2851.
111. M. Mutter and S. Vuilleumier, *Angew Chem Int Ed*, 1989, **28**, 535–554.
112. D. Boturyn, J.-L. Coll, E. Garanger, M.-C. Favrot and P. Dumy, *J Am Chem Soc*, 2004, **126**, 5730–5739.
113. Y. Singh, G. T. Dolphin, J. Razkin and P. Dumy, *ChemBioChem*, 2006, **7**, 1298–1314.
114. H. S. Park, Q. Lin and A. D. Hamilton, *J Am Chem Soc*, 1999, **121**, 8–13.
115. C. Chamorro, J. A. W. Kruijtzer, M. Farsaraki, J. Balzarini and R. M. J. Liskamp, *Chem Commun*, 2009, 821–823.

116. G. E. Mulder, L. van Ufford, J. van Ameijde, A. J. Brouwer, J. A. W. Kruijtzer and R. M. J. Liskamp, *Org. Biomol. Chem.*, 2013, **11**, 2676–2684.
117. V. Baeriswyl and C. Heinis, *ChemMedChem*, 2013, **8**, 377–384.
118. T. L. Aboye, H. Ha, S. Majumder, F. Christ, Z. Debyser, A. Shekhtman, N. Neamati and J. A. Camarero, *J Med Chem*, 2012, **55**, 10729–10734.
119. P. S. Ghosh and A. D. Hamilton, *J Am Chem Soc*, 2012, **134**, 13208–13211.
120. A. Ghasparian, T. Riedel, J. Koomullil, K. Moehle, C. Gorba, D. I. Svergun, A. W. Perriman, S. Mann, M. Tamborrini, G. Pluschke and J. A. Robinson, *ChemBioChem*, 2011, **12**, 100–109.
121. A. Angelini, P. Diderich, J. Morales-Sanfrutos, S. Thurnheer, D. Hacker, L. Menin and C. Heinis, *Bioconjugate Chem*, 2012, **23**, 1856–1863.
122. A. Nuccitelli, R. Cozzi, L. J. Gourlay, D. Donnarumma, F. Necchi, N. Norais, J. L. Telford, R. Rappuoli, M. Bolognesi, D. Maione, G. Grandi and C. D. Rinaudo, *Proc Natl Acad Sci USA*, 2011, **108**, 10278–10283.

CHAPTER 11

Cystine-Knot Micro-Proteins

BILL ELDRIDGE,* SIMON ROBINS AND DUNCAN MCGREGOR

Cyclogenix Ltd, Rowett Institute, Greenburn Road, Aberdeen, UK
*Email: bill.eldridge@cyclogenix.com

11.1 Cystine-Knot Micro-Proteins

The Cystine-Knot (CK) micro-protein family, incorporating knottins and cyclotides, have the potential to display a plethora of therapeutically useful biological activities. They are typically around 30–40 amino acid residues in length and have a common stable tertiary fold that is formed and stabilized by a characteristic pattern of disulphide bonds (I–IV, II–V, III–VI for triple-bonded molecules), and a compact knotted triple-stranded β-sheet structure. Tethered peptide loops emerge from this structure that frequently carry a natural pharmacophoric sequence, such as a protease inhibitor etc.[1] The essential difference between a "knottin" and a "cyclotide" is that the cyclotide is naturally cyclic (N–C-terminally joined), while a knottin is a linear form. Within certain families, both cyclic and linear forms have been found, with very little sequence variation e.g. cyclic MCOTI-I, MCOTI-II and linear MCOTI-III.[2] The CK structure is found in over 30 different protein families, among which conotoxins, spider toxins, squash inhibitors, agouti-related proteins (AGRP) and plant cyclotides are the most populated families (Figure 11.1).[3]

The distinguishing structural characteristic of members of the CK micro-protein family is the length of the peptide loops between adjacent cysteine residues. Despite the rigid nature of the overall structure, a high degree of variability exists within the loops with respect to length and amino acid sequence. Thousands of different CK micro-proteins exist in plants and animals

RSC Drug Discovery Series No. 36
Biotherapeutics: Recent Developments using Chemical and Molecular Biology
Edited by Lyn H. Jones and Andrew J. McKnight

Published by the Royal Society of Chemistry, www.rsc.org

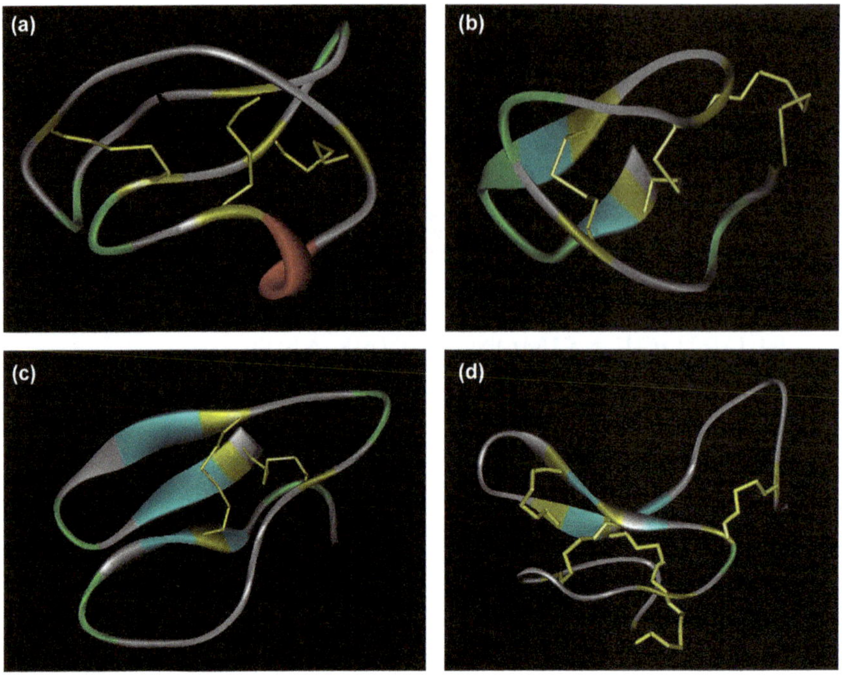

Figure 11.1 Ribbon representations of (a) MCOTI II (PDB code: 1HA9[14]), (b) ω-conotoxin (PDB code: 1OMG[15]), (c) cellobiohydrolase (PDB code: 1CBH[16]), and (d) agouti-related protein (PDB code: 1HYK[17]).

that define a natural combinatorial library of cystine-intervening sequences that are built around the central CK motif, conferring particular biological activities. Both open-chain and cyclic peptides containing the CK motif are remarkably stable towards extreme pH, chemical and thermal denaturation, and proteolytic attack.[4,5] This is probably a result of the enormous conformational rigidity that is introduced by covalent disulphide linkage of the knotted core.[6,7]

11.1.1 Conotoxins and Venom-Derived CK Micro-Proteins

Naturally occurring CK micro-proteins come from phylogenetically diverse sources, including spiders, cone shells, plants and fungi, and display a plethora of pharmacologic activities.[8–13]

Many CK toxin micro-proteins target voltage-gated ion channels,[18] including the conotoxins from marine cone snails (e.g. ω-conotoxins,[19] κ-conotoxin PVIIA,[20] δ-conotoxin TXVIA[15] and conotoxins GS[21]), as well as x-agatoxins, robustoxin, versutoxin, and protein 5 from spiders.[23] Perhaps the most interesting example is the marketed micro-protein Ziconotide. This is a ω-conotoxin MVII calcium-channel blocker derived from the mollusc *Conus magus*. It is an intrathecally administered conotoxin that has already been approved for the treatment of severe chronic pain.[23,24] This peptide inhibits presynaptic voltage-gated calcium channels and thus neurotransmission across nerve synapses. By contrast with these ion channel inhibitors, the venom from *Psalmopoeus cambridgei*, a tarantula that is native to the West Indies, has been reported to contain CK micro-proteins that target the capsaicin receptor TRPV1, an excitatory channel expressed by sensory neurons of the pain pathway, and function as TRP agonists. These vanillotoxins agonise excitatory channels on somatosensory neurons to elicit pain and inflammation in mammals.[25]

11.1.2 Human CK Micro-Proteins

A CK micro-protein of human origin is the C-terminal domain of human AGRP.[17] AGRP is an endogenous antagonist of the melanocortin-3 and melanocortin-4 G-protein-coupled receptors.[26] The C-terminal domain of AGRP (AGRP 87–132) possesses five disulphide bonds and a well-defined 3D structure that displays full activity like the full-length protein.[27] A minimized 34-residue domain has been shown to fold autonomously into the cysteine-knot structure.[28]

11.1.3 Plant-Derived CK Micro-Proteins

Cyclotides are perhaps the most prevalent and well-known CK micro-proteins. They are a family of plant micro-proteins characterised by the cyclic CK structure (CCK).[29,30] The CCK motif comprises a head-to-tail cyclised peptide backbone that is cross-braced by a knotted arrangement of disulphide bonds, with cysteines paired in the same pattern as found in linear CK micro-proteins. The prototypic member, Kalata B1, was originally discovered because of its oral availability and bioactivity. Kalata B1 is derived from the plant *Oldenlandia affinis*, the leaves of which are boiled to make a tea that contains Kalata B1 as the bioactive ingredient. The tea is taken by women in West Africa to accelerate childbirth.[31] Cyclotides occur in a large number of plants from the Violaceae (violet) and Rubiaceae (coffee) families and one member of the Cucurbitaceae (squash) family, and recent studies suggest that they might also occur in other plant families. They have a diverse range of biological activities including uterotonic activity, anti-HIV, neurotensin inhibition, antimicrobial and insecticidal activity.[32,33] The unusual cyclic nature of the cyclotides has resulted in this CK family being well studied, and several attempts have been made to use members of this family as scaffolds for drug discovery.[34,35]

Cyclotides of the squash family of protease inhibitors have been reported to be extraordinarily stable. They have been shown to be refractory to the action of chaotropic agents 6M guanidine hydrochloride and 8M urea, to temperatures approaching boiling, to acid, as well as to a range of proteases. Head-to-tail cyclisation was shown to have minimal effect on stability in these studies. Removal of the disulphide bond by reduction resulted in a significant decrease in resistance to chemical or enzymatic breakdown compared to the oxidized species.[14] Studies of the related squash inhibitor cyclotides MCOTI-I, MCOTI-II and EETI-II (linear or head-to-tail cyclised CK micro-proteins) again demonstrated remarkable stability, indicating that the main determinant for stability and for protease resistance is the CK knotted structure.[36]

11.2 Use as Scaffolds

11.2.1 Loop Grafting

Cyclotides have proven to be highly amenable to modification of loops via insertion of bioactive peptides, replacing the wild type sequences. The inhibitor loop of EETI-II was replaced with significantly longer antibody epitope sequences up to 17 amino acids (replacing the native five amino acids) while retaining the micro-protein structure.[37]

Bioactive peptides derived from the myelin oligodendrocyte glycoprotein and a VEGF A competitor peptide have also been used to replace one of the loops of Kalata B1 and were shown to maintain bioactivity whilst demonstrating significant improvements in stability *in vivo*.[38] Recently, the Kalata B1 structure has been modified to generate a cyclic orally available bradykinin inhibitor, which has been shown to be effective in mouse models of inflammatory pain.[39] Truncations of the native loop sequences are equally tolerated, for example, Thongyoo *et al.* derived potent β-tryptase and human leukocyte elastase inhibitors via truncations of one of the loops flanking the cognate trypsin inhibitor loop.[40]

11.2.2 Display Libraries Using Loop Replacement

The CK structure is amenable to modifications of the wild type loop sequences and has been used to generate libraries containing random peptide loops.[41–44] Model "spiked" selections using an antibody epitope in loop 1 of EETI-II, and bacterial surface display was used to enrich cells displaying specific epitopes from a background of cells displaying an irrelevant epitope.[37] EETI-II has also been used as a scaffold for display of randomised peptide loop libraries using mRNA display.[41] Here the inhibitor loop was replaced with a randomised library of six amino acids and used to select novel binding molecules to the cognate target of EETI (trypsin). Similarly Souriau *et al.* replaced a loop within a CK micro-protein with a randomised ten amino acid sequence and used phage display to demonstrate the selection of novel binding moieties from this library.[42] Kalata B1 has also been used as a scaffold for loop replacement using bacterial surface display.[45] Here, seven amino acids were replaced with a randomised sequence and used to select thrombin binders and inhibitors.

The human CK AGRP micro-protein has also been used as a scaffold for the display of randomised peptidic loops using yeast display to generate novel integrin-binding moieties. A natural six amino acid loop was replaced with a nine amino acid loop containing the cognate "RGD" integrin-binding motif, flanked by randomised peptide sequences. Screening of the resulting libraries produced AGRP-based micro-proteins binding $\alpha v\beta 3$ integrins with affinities ranging from 15 nM to 780 pM.[44]

In the earliest example, a related CK micro-protein, the C-terminal knottin-like domain of cellobiohydrolase I that contains only the two core disulphide bonds, was used as a scaffold to generate a phage display library of variants of one of the faces of the molecule. Novel peptides binding either the cognate target "cellulose" or alkaline phosphatase, an unrelated target, were selected from this library.[46]

11.2.3 MCOTI-II as a Loop Replacement Scaffold

Like other CK micro-proteins previously used as display scaffolds, such as the closely related EETI-II, MCOTI II forms a stable and versatile scaffold that can be used to provide a structured framework for bioactive peptides to be grafted into. The peptides are then afforded increased protection from environmental conditions and as such may be able to survive longer in the blood or other tissues, therefore improving their efficacy. At Cyclogenix, we have created large ($\sim 10^{11}$ individual clones) loop replacement libraries of randomised sequence encoding peptides of varying length, replacing loops 1, 2 and 5 within the CK scaffold of MCOTI II (Figure 11.2).

The DNA sequence for MCOTI II scaffold was modified via PCR using degenerate oligonucleotides. Modified sequences were cloned into M13 phagemid vector such that the first five amino acid residues of loop 6 served as a linker region, fusing the scaffold to pIII. The last three amino acid residues were expressed at the N-terminus of the peptide (Figure 11.3).

We have demonstrated that these loop replacements are well tolerated and libraries are capable of producing novel specific binding molecules to selected target proteins. Moreover, there is strong evidence that library members are folded correctly into the cognate CK structure whilst fused to phage particles, other proteins and as discrete molecules following selection and chemical synthesis. For example, model selections were carried out against streptavidin, using each loop replacement library. After two rounds, enriched clones were

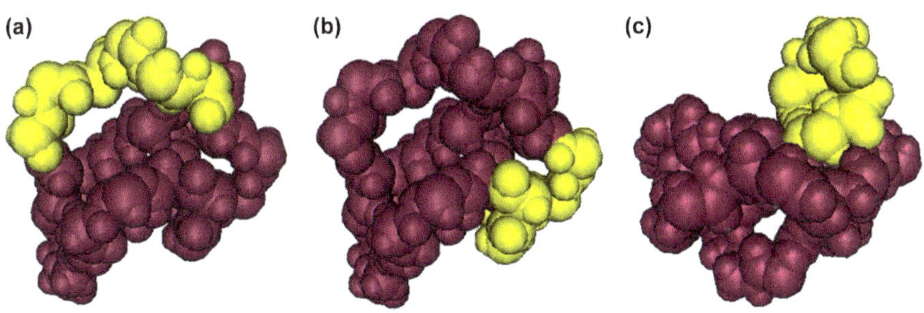

Figure 11.2 Space-filling model representations of MCOTI II showing selected loops (yellow) for library design: (a) loop 1, (b) loop 2 and (c) loop 5. Each loop was replaced with randomised peptide sequences of 10, 12 and 14 amino acids.

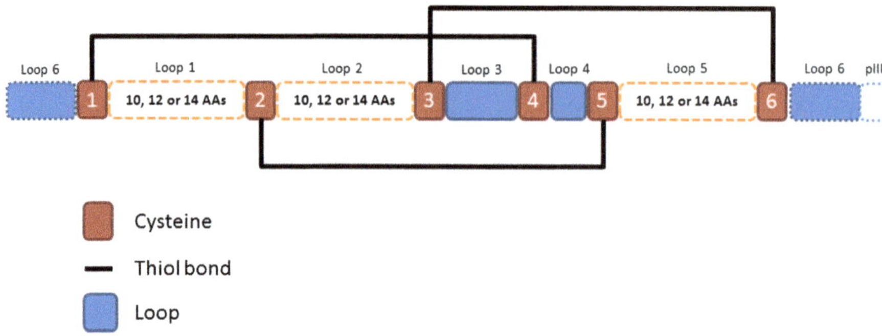

Figure 11.3 Schematic diagram of randomised loop replacement libraries within the MCOTI II scaffold. Randomised peptides of either 10, 12 or 14 amino acids were inserted between cysteines 1 and 2, 2 and 3 or 5 and 6. The first five residues of loop 6 were used to provide a linker between the cyclotide and pIII of M13 phage to facilitate phage display.

analysed for binding via ELISA and then genetic material from individual binding clones was sequenced. Figure 11.4 provides a schematic diagram of the sequences recovered from each library. Sequencing showed that the position of the cognate "HPQ" streptavidin-binding motif was strictly contained within the central 3–4 residues of the loop, suggesting that peptides displayed on the phage particles were highly structured and that randomised regions within the scaffold were being forced into a uniform domain.

This finding confirmed previous reports that loop replacements within the CK structure were well tolerated and that it is possible to derive novel biological activities from a scaffold based on this structure.[37,41–44,46] They also highlight the great flexibility the scaffold has with regard to choice of loops for modification. Each loop library tested provided multiple binding clones with the same "HPQ" motif, but flanked by differing amino acid sequences, demonstrating that multiple clones had been enriched from within each library.

Cyclogenix have used the randomised libraries to develop panels of CK micro-proteins that specifically bind to over 20 different therapeutically relevant targets (e.g. TrkA, Figure 11.5).

11.3 Stability of Selected CK Micro-Proteins

The CK structure is reported to be highly stable to thermal, chemical and enzymatic degradation due to the presence of the CK.[3,29] In order to evaluate the stability of novel cyclotides derived from selections, we subjected them to a range of denaturing conditions; linear cyclotides derived from selections were synthesised by FMOC chemistry and refolded under oxidising conditions. Selected CK micro-proteins demonstrated remarkable stability, for example over 80% of full-length peptide was recovered after incubating cyclotides in 0.1M HCl for 24 h at room temperature (Figure 11.6). Similarly, incubating the peptides at 107 °C yielded >60% full length peptide after 6 h (Figure 11.7).

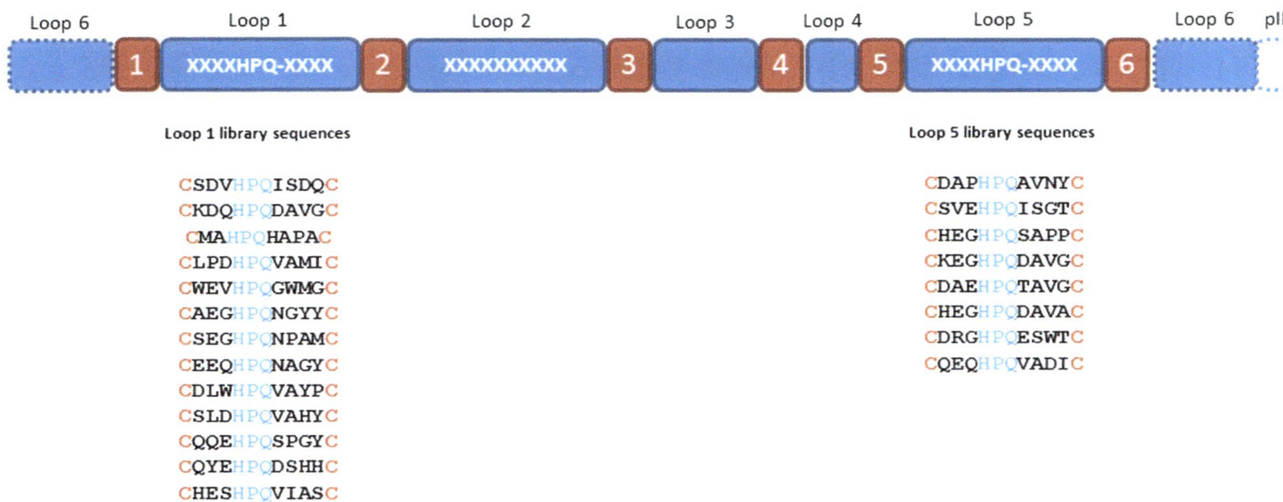

Figure 11.4 Schematic diagram demonstrating the position of streptavidin-binding motifs within modified MCOTI II loops after two rounds of selection using ten amino acid loop replacement libraries. "HPQ" motif in all binding clones was central in the loop, suggesting that when displayed on the surface of phage particles these peptides are highly structured, dictating the position of the motif to provide high affinity binding. No "Loop 2" binding clones were selected against streptavidin, suggesting defined structural requirements for binding of this target.

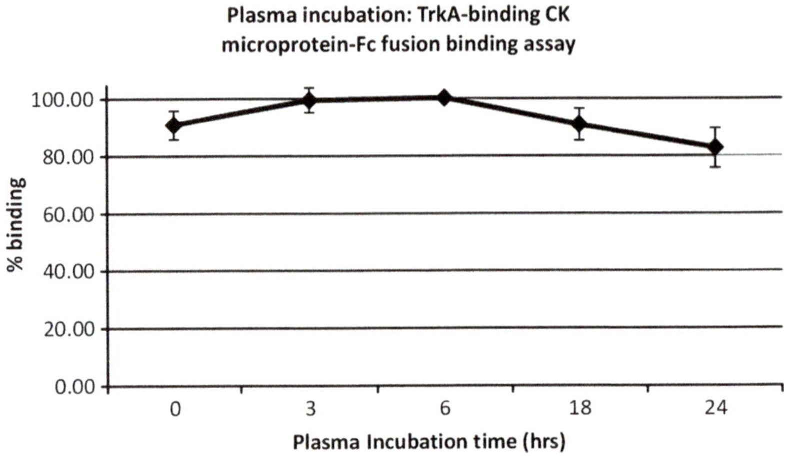

Figure 11.5 ELISA binding assay of a TrkA-binding CK micro-protein fused to human IgG1 Fc. The % binding was measured after incubation in human plasma at 37 °C. Little or no reduction in signal is observed throughout the time course up to 24 h.

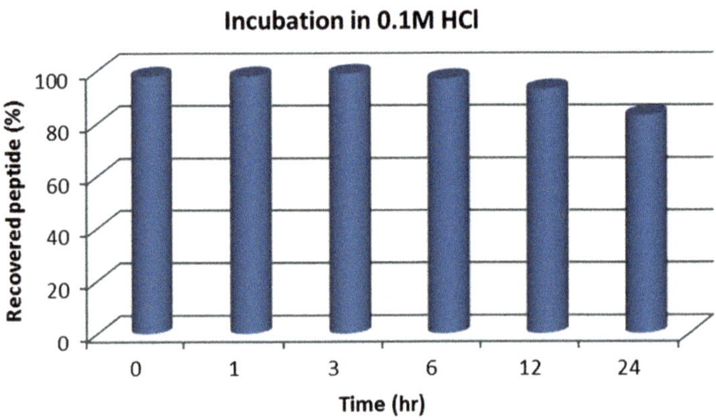

Figure 11.6 Selected CK micro-protein incubated in 0.1M HCl at room temperature for up to 24 h the % soluble full-length peptides was then determined via LC/MS.

Incubation in human plasma for 24 h resulted in limited degradation compared with a linear peptide control (insulin) of the same size (Figure 11.8).

Due to their unique properties of stability and relatively small size, CK micro-proteins have potential applications in hitherto unapproachable areas for biotherapeutics. Their ability to withstand environmental extremes has encouraged researchers to examine their development as orally available biotherapeutics with some promising results.[5,47] It is possible therefore to envisage the use of these micro-proteins as either orally available biotherapeutics in themselves, or as delivery agents for other therapeutics where oral delivery is not normally possible, e.g. cytokines or antibodies.

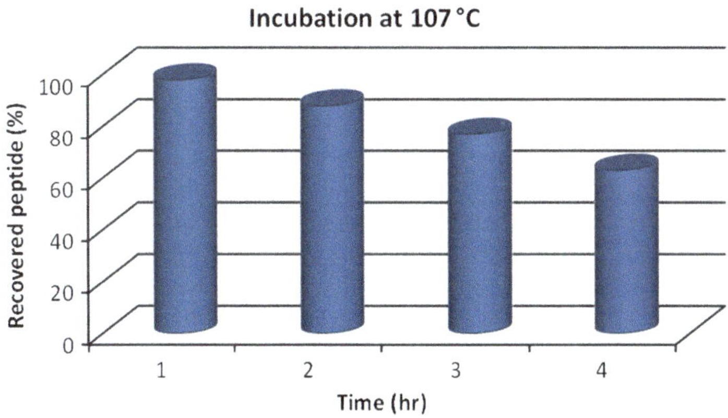

Figure 11.7 Selected CK micro-protein incubated at 107 °C for up to 6 h. The % full-length soluble peptides was then determined via LC\MS.

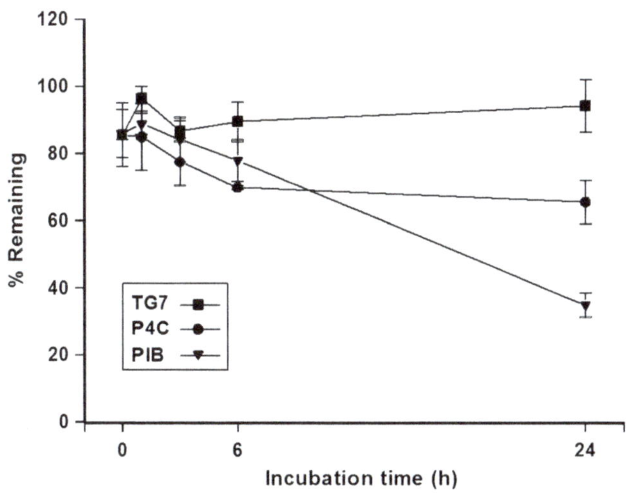

Figure 11.8 Selected CK micro-proteins (TG7 and P4C) were synthesised via FMOC, refolded under oxidising conditions and incubated in human plasma for up to 24 h at 37 °C. After 24 h, close to 100% of CK micro-protein remained intact, compared to the linear control, Porcine Insulin B-chain (PIB), which showed significant reduction to 35%.

11.4 Potential Therapeutic Applications of CK Micro-Proteins

Specific examples of CK micro-proteins as marketed biotherapeutic agents exist[24,25] and some early examples of potential diagnostic agents have been reported.[48] However, the CK structure has yet to be fully exploited for medicinal and diagnostic use.

Once in the bloodstream, CK micro-proteins have relatively high clearance rates, which may be advantageous for use as imaging agents, but may limit their use as biotherapeutics. A widely used solution to high clearance rates is to fuse these moieties to proteins with long plasma half-lives, such as immunoglobulin Fc domains, albumin or albumin-binding proteins,[49] thereby coupling the extended residence of these plasma proteins with the novel bioactivity conferred by the micro-protein.[50–52] Cyclogenix have already fused a CK micro-protein to human IgG Fc and demonstrated that not only is such a molecule bioactive, but it also remains stable in plasma for extended periods (Figure 11.5). Fusing CK micro-proteins to, for instance, Fc domains, also opens the possibility of generating multi-specific moieties. Here, multiple micro-proteins may be fused in tandem to an Fc domain and may be particularly effective in treatment of certain cancers for example, where metastases overcome the action of mono-specific treatments, without creating manufacturing problems that make bispecific antibodies even more expensive than monoclonal counterparts.[53,54] Given their inherent resistance to proteolysis, it is unlikely that CK micro-proteins will be immunogenic. Moreover, if delivered orally, it may even be possible to induce tolerance.[55–59]

The cell-penetrating capabilities of these molecules may also be exploited for delivery of bioactive compounds to specific organs including the brain. With developments in display technologies, it may be possible to select CK micro-proteins that target and transmigrate from the blood into specific organs.[60–62]

These interesting properties may allow CK micro-proteins to address the shortcomings of current biotherapeutic agents, particularly with regard to delivery, tissue and cell penetration and multi-specific targeting.

References

1. L. Chiche, A. Heitz, J. C. Gelly, J. Gracy, P. T. Chau, P. T. Ha, J. F. Hernandez and D. Le-Nguyen, *Curr Protein Pept Sci*, 2004, **5**, 341–349.
2. J. F. Hernandez, J. Gagnon, L. Chiche, T. M. Nguyen, J. P. Andrieu, A. Heitz, T. Trinh Hong, T. T. Pham and D. Le Nguyen, *Biochemistry*, 2000, **39**, 5722–5730.
3. J. Gracy, D. Le-Nguyen, J. C. Gelly, Q. Kaas, A. Heitz and L. Chiche, *Nucleic Acids Res*, 2008, **36**, D314–319.
4. D. C. Ireland, M. L. Colgrave and D. J. Craik, *Biochem J*, 2006, **400**, 1–12.
5. M. Werle, K. Kafedjiiski, H. Kolmar and A. Bernkop-Schnurch, *Int J Pharm*, 2007, **332**, 72–79.
6. M. L. Colgrave and D. J. Craik, *Biochemistry*, 2004, **43**, 5965–5975.
7. R. Kratzner, J. E. Debreczeni, T. Pape, T. R. Schneider, A. Wentzel, H. Kolmar, G. M. Sheldrick and I. Uson, *Acta Crystallogr Sect D Biol Crystallogr*, 2005, **61**, 1255–1262.
8. Y. Q. Tang, J. Yuan, G. Osapay, K. Osapay, D. Tran, C. J. Miller, A. J. Ouellette and M. E. Selsted, *Sciences (NY)*, 1999, **286**, 498–502.
9. M. Trabi and D. J. Craik, *Trends Biochem Sci*, 2002, **27**, 132–138.

10. M. Maqueda, A. Galvez, M. M. Bueno, M. J. Sanchez-Barrena, C. Gonzalez, A. Albert, M. Rico and E. Valdivia, *Curr Protein Pept Sci*, 2004, **5**, 399–416.
11. D. J. Craik, *Sciences (NY)*, 2006, **311**, 1563–1564.
12. H. E. Hallen, H. Luo, J. S. Scott-Craig and J. D. Walton, *Proc Natl Acad Sci USA*, 2007, **104**, 19097–19101.
13. M. S. Donia, J. Ravel and E. W. Schmidt, *Nat Chem Biol*, 2008, **4**, 341–343.
14. A. Heitz, J. F. Hernandez, J. Gagnon, T. T. Hong, T. T. Pham, T. M. Nguyen, D. Le-Nguyen and L. Chiche, *Biochemistry*, 2001, **40**, 7973–7983.
15. T. Kohno, J. I. Kim, K. Kobayashi, Y. Kodera, T. Maeda and K. Sato, *Biochemistry*, 1995, **34**, 10256–10265.
16. J. Kraulis, G. M. Clore, M. Nilges, T. A. Jones, G. Pettersson, J. Knowles and A. M. Gronenborn, *Biochemistry*, 1989, **28**, 7241–7257.
17. K. A. Bolin, D. J. Anderson, J. A. Trulson, D. A. Thompson, J. Wilken, S. B. Kent, I. Gantz and G. L. Millhauser, *FEBS Letters*, 1999, **451**, 125–131.
18. R. S. Norton and P. K. Pallaghy, *Toxicon*, 1998, **36**, 1573–1583.
19. P. K. Pallaghy and R. S. Norton, *J Pept Res*, 1999, **53**, 343–351.
20. M. J. Scanlon, D. Naranjo, L. Thomas, P. F. Alewood, R. J. Lewis and D. J. Craik, *Structure (London, UK)*, 1997, **5**, 1585–1597.
21. J. M. Hill, P. F. Alewood and D. J. Craik, *Structure (London, UK)*, 1997, **5**, 571–583.
22. P. K. Pallaghy, K. J. Nielsen, D. J. Craik and R. S. Norton, *Protein Sci*, 1994, **3**, 1833–1839.
23. C. I. Schroeder, C. J. Doering, G. W. Zamponi and R. J. Lewis, *Med Chem (Shariqah, UAE)*, 2006, **2**, 535–543.
24. G. P. Miljanich, *Curr Med Chem*, 2004, **11**, 3029–3040.
25. J. Siemens, S. Zhou, R. Piskorowski, T. Nikai, E. A. Lumpkin, A. I. Basbaum, D. King and D. Julius, *Nature*, 2006, **444**, 208–212.
26. A. M. Wilczynski, C. G. Joseph and C. Haskell-Luevano, *Med Res Rev*, 2005, **25**, 545–556.
27. J. C. McNulty, D. A. Thompson, K. A. Bolin, J. Wilken, G. S. Barsh and G. L. Millhauser, *Biochemistry*, 2001, **40**, 15520–15527.
28. P. J. Jackson, J. C. McNulty, Y. K. Yang, D. A. Thompson, B. Chai, I. Gantz, G. S. Barsh and G. L. Millhauser, *Biochemistry*, 2002, **41**, 7565–7572.
29. D. J. Craik, N. L. Daly, T. Bond and C. Waine, *J Mol Biol*, 1999, **294**, 1327–1336.
30. M. E. Felizmenio-Quimio, N. L. Daly and D. J. Craik, *J Biol Chem*, 2001, **276**, 22875–22882.
31. L. Gran, K. Sletten and L. Skjeldal, *Chem Biodivers*, 2008, **5**, 2014–2022.
32. D. J. Craik, J. S. Mylne and N. L. Daly, *Cell Mol Life Sci*, 2010, **67**, 9–16.
33. J. P. Tam, Y. A. Lu, J. L. Yang and K. W. Chiu, *Proc Natl Acad Sci USA*, 1999, **96**, 8913–8918.

34. T. L. Aboye, H. Ha, S. Majumder, F. Christ, Z. Debyser, A. Shekhtman, N. Neamati and J. A. Camarero, *J Med Chem*, 2012, **55**, 10729–10734.
35. D. J. Craik, J. E. Swedberg, J. S. Mylne and M. Cemazar, *Expert Opin Drug Discovery*, 2012, **7**, 179–194.
36. A. Heitz, O. Avrutina, D. Le-Nguyen, U. Diederichsen, J. F. Hernandez, J. Gracy, H. Kolmar and L. Chiche, *BMC Struct Biol*, 2008, **8**, 54.
37. A. Christmann, K. Walter, A. Wentzel, R. Kratzner and H. Kolmar, *Protein Eng*, 1999, **12**, 797–806.
38. S. Gunasekera, F. M. Foley, R. J. Clark, L. Sando, L. J. Fabri, D. J. Craik and N. L. Daly, *J Med Chem*, 2008, **51**, 7697–7704.
39. C. T. Wong, D. K. Rowlands, C. H. Wong, T. W. Lo, G. K. Nguyen, H. Y. Li and J. P. Tam, *Angew Chemie Int Ed Engl*, 2012, **51**, 5620–5624.
40. P. Thongyoo, C. Bonomelli, R. J. Leatherbarrow and E. W. Tate, *J Med Chem*, 2009, **52**, 6197–6200.
41. R. Baggio, P. Burgstaller, S. P. Hale, A. R. Putney, M. Lane, D. Lipovsek, M. C. Wright, R. W. Roberts, R. Liu, J. W. Szostak and R. W. Wagner, *J Mol Recog*, 2002, **15**, 126–134.
42. C. Souriau, L. Chiche, R. Irving and P. Hudson, *Biochemistry*, 2005, **44**, 7143–7155.
43. A. Wentzel, A. Christmann, R. Kratzner and H. Kolmar, *J Biol Chem*, 1999, **274**, 21037–21043.
44. A. P. Silverman, A. M. Levin, J. L. Lahti and J. R. Cochran, *J Mol Biol*, 2009, **385**, 1064–1075.
45. J. A. Getz, J. J. Rice and P. S. Daugherty, *ACS Chem Biol*, 2011, **6**, 837–844.
46. G. P. Smith, S. U. Patel, J. D. Windass, J. M. Thornton, G. Winter and A. D. Griffiths, *J Mol Biol*, 1998, **277**, 317–332.
47. M. Werle, T. Schmitz, H. L. Huang, A. Wentzel, H. Kolmar and A. Bernkop-Schnurch, *J Drug Target*, 2006, **14**, 137–146.
48. H. Kolmar, *Curr Pharm Des*, 2011, **17**, 4329–4336.
49. J. Feldwisch and V. Tolmachev, *Methods Mol Biol*, 2012, **899**, 103–126.
50. H. Huang, J. Y. Lai, J. Do, D. Liu, L. Li, J. Del Rosario, V. R. Doppalapudi, S. Pirie-Shepherd, N. Levin, C. Bradshaw, G. Woodnutt, R. Lappe and A. Bhat, *Clinical Cancer Res*, 2011, **17**, 1001–1011.
51. M. Peeters, A. H. Strickland, M. Lichinitser, A. V. Suresh, G. Manikhas, J. Shapiro, W. Rogowski, X. Huang, B. Wu, D. Warner, R. Jain and N. C. Tebbutt, *Br J Cancer*, 2013.
52. R. Hecht, Y. S. Li, J. Sun, E. Belouski, M. Hall, T. Hager, J. Yie, W. Wang, D. Winters, S. Smith, C. Spahr, L. T. Tam, Z. Shen, S. Stanislaus, N. Chinookoswong, Y. Lau, A. Sickmier, M. L. Michaels, T. Boone, M. M. Veniant and J. Xu, *PloS One*, 2012, **7**, e49345.
53. B. N. Rexer and C. L. Arteaga, *Crit Rev Oncog*, 2012, **17**, 1–16.
54. L. Moserle and O. Casanovas, *J Intern Med*, 2013, **273**, 128–137.
55. A. Mitchison and J. Sieper, *Zeitschrift fur Rheumatologie*, 1995, **54**, 141–144.

56. K. Kobayashi, T. Yoshida, K. Takahashi and M. Hattori, *Bioconjugate Chem*, 2003, **14**, 168–176.
57. J. Contreras, A. Y. Elnagar, S. F. Hamm-Alvarez and J. A. Camarero, *J Controlled Release*, 2011, **155**, 134–143.
58. K. P. Greenwood, N. L. Daly, D. L. Brown, J. L. Stow and D. J. Craik, *Int J Biochem Cell Biol*, 2007, **39**, 2252–2264.
59. L. Cascales, S. T. Henriques, M. C. Kerr, Y. H. Huang, M. J. Sweet, N. L. Daly and D. J. Craik, *J Biol Chem*, 2011, **286**, 36932–36943.
60. Y. H. Chen, M. Chang and B. L. Davidson, *Nat Med*, 2009, **15**, 1215–1218.
61. A. N. Veleva, D. B. Nepal, C. B. Frederick, J. Schwab, P. Lockyer, H. Yuan, D. S. Lalush and C. Patterson, *Molecules (Basel, Switzerland)*, 2011, **16**, 900–914.
62. B. Du, H. Han, Z. Wang, L. Kuang, L. Wang, L. Yu, M. Wu, Z. Zhou and M. Qian, *Mol Cancer Res*, 2010, **8**, 135–144.

Subject Index

References to tables and charts are in **bold** type